Limitations and Potentials for Biological Nitrogen Fixation in the Tropics

BASIC LIFE SCIENCES
Alexander Hollaender, General Editor
Associated Universities, Inc.
Washington, D.C.

1973: Volume 1 • GENE EXPRESSION AND ITS REGULATION
Edited by F. T. Kenney, B. A. Hamkalo, G. Favelukes, and J. T. August

Volume 2 • GENES, ENZYMES, AND POPULATIONS
Edited by A. M. Srb

1974: Volume 3 • CONTROL OF TRANSCRIPTION
Edited by B. B. Biswas, R. K. Mandal, A. Stevens, and W. E. Cohn

Volume 4 • PHYSIOLOGY AND GENETICS OF REPRODUCTION (Parts A and B)
Edited by E. M. Coutinho and F. Fuchs

1975: Volume 5 • MOLECULAR MECHANISMS FOR REPAIR OF DNA (Parts A and B)
Edited by P. C. Hanawalt and R. B. Setlow

Volume 6 • ENZYME INDUCTION
Edited by D. V. Parke

1976: Volume 7 • NUTRITION AND AGRICULTURAL DEVELOPMENT
Edited by N. Scrimshaw and M. Béhar

1977: Volume 8 • GENETIC DIVERSITY IN PLANTS
Edited by Amir Muhammed, Rustem Aksel, and R. C. von Borstel

Volume 9 • GENETIC ENGINEERING FOR NITROGEN FIXATION
Edited by Alexander Hollaender, R. H. Burris, P. R. Day, R. W. F. Hardy, D. R. Helinski, M. R. Lamborg, L. Owens, and R. C. Valentine

1978: Volume 10 • LIMITATIONS AND POTENTIALS FOR BIOLOGICAL NITROGEN FIXATION IN THE TROPICS
Edited by Johanna Döbereiner, Robert H. Burris, Alexander Hollaender, Avilio A. Franco, Carlos A. Neyra, and David Barry Scott

---

A Continuation Order Plan is available for this series. A continuation order will bring delivery of each new volume immediately upon publication. Volumes are billed only upon actual shipment. For further information please contact the publisher.

Limitations and Potentials for Biological Nitrogen Fixation in the Tropics

Edited by

Johanna Döbereiner
EMBRAPA
Rio de Janeiro, Brazil

Robert H. Burris
University of Wisconsin
Madison, Wisconsin

Alexander Hollaender
Associated Universities, Inc.
Washington, D.C.

and

Avilio A. Franco
Carlos A. Neyra
David Barry Scott

PLENUM PRESS · NEW YORK AND LONDON

Library of Congress Cataloging in Publication Data

International Latin American Symposium, 15th Brasília, Brazil, 1977.
Limitations and potentials for biological nitrogen fixation in the tropics.
(Basic life sciences; v. 10)
Includes index.
1. Nitrogen–Fixation–Congresses. 2. Microorganisms, Nitrogen-fixing–Congresses.
3. Agriculture–Tropics–Congresses. I. Döbereiner, Johanna. II. Title.
QR89.7.I57 1977 589'.7'04133 77-28218
ISBN-13: 978-1-4615-8959-4 e-ISBN-13: 978-1-4615-8957-0
DOI: 10.1007/978-1-4615-8957-0

Proceedings of a Conference on Limitations and Potentials for Biological Nitrogen Fixation in the Tropics, held in Brasilia, Brazil, July 18–22, 1977

Softcover reprint of the hardcover 1st edition 1978
A Division of Plenum Publishing Corporation
227 West 17th Street, New York, N.Y. 10011

Organizing Committee

GENERAL CO-ORDINATION:
JOHANNA DÖBEREINER, EMBRAPA, Rio de Janeiro, Brazil
MILTON THIAGO DE MELLO, Universidade de Brasília, Brazil
ROBERT H. BURRIS, University of Wisconsin, Madison, U.S.A.
ALEXANDER HOLLAENDER, Associated Universities, Washington, D.C., U.S.A.

SCIENTIFIC CO-ORDINATION:
CARLOS A. NEYRA, EMBRAPA, Rio de Janeiro, Brazil
DAVID BARRY SCOTT, EMBRAPA, Rio de Janeiro, Brazil
ROBERTO MEIRELLES DE MIRANDA, Universidade de Brasília, Brazil

ADMINISTRATION CO-ORDINATION:
AVILIO A. FRANCO, EMBRAPA, Rio de Janeiro, Brazil
JOSÉ CARMINE DIANESE, Universidade de Brasília, Brazil

RECEPTION AND INFORMATION:
HELVÉCIO DE-POLLI, EMBRAPA, Rio de Janeiro, Brazil
CHRISTINE SCOTT, EMBRAPA, Rio de Janeiro, Brazil
FRANCISCO CUPERTINO PEREIRA, Universidade de Brasília, Brazil

TREASURER:
PAULO AUGUSTO DA EIRA, EMBRAPA, Rio de Janeiro, Brazil

EDITORS OF SYMPOSIUM PROCEEDINGS:
JOHANNA DÖBEREINER, EMBRAPA, Rio de Janeiro, Brazil
ROBERT H. BURRIS, University of Wisconsin, Madison, U.S.A.
DAVID BARRY SCOTT, EMBRAPA, Rio de Janeiro, Brazil
ALEXANDER HOLLAENDER, Associated Universities, Washington, D.C., U.S.A.

Supporting Institutions

Conselho Nacional de Desenvolvimento Científico e Tecnológico, (CNPq), Brazil
Empresa Brasileira de Pesquisa Agropecuária (EMBRAPA), Brazil
Universidade de Brasília (UnB), Brazil
Universidade Federal Rural do Rio de Janeiro (UFRRJ), Brazil
Energy Research and Development Administration (ERDA), U.S.A.
National Academy of Sciences (NAS—USA) with grant funds provided by the United States Agency for International Development
National Science Foundation (NSF), U.S.A.
Organization of American States (OAS)
United Nations Development Program (PNUD)
United Nations Environmental Programme (UNEP)

Organized by the Program for International Cooperation in Training and Research on Nitrogen Fixation in the Tropics

Acknowledgment for the Brasilia Symposium

Special thanks to Dr. M. T. de Mello from the University of Brasilia for the excellent organization of the Symposium, the staff of EMBRAPA at Km 47 Rio de Janeiro for their help and cooperation, and Mary Jo Marcouiller and Joann Trodahl from the University of Wisconsin for the transcribing of the proceedings.

Preface

The 15th Latin American Symposium was held in Brasilia (FD) on July 18-22, 1977, on a topic of great interest for agriculture, especially in the tropics. Many new developments have taken place in the field of research in N_2 fixation during the last few years. They were made possible by the improved methods of measuring of nitrogenase activity, progress in genetic engineering fields and the increased interest in taking advantage of natural sources for biological nitrogen fixation. The approach used in this Symposium together with the one held four months earlier in Brookhaven on "Genetic Engineering for Nitrogen Fixation" gives an interesting picture of the present status of nitrogen fixation from two diverse approaches.

This is my 20th year visiting Latin America. I am most impressed with the tremendous development which has taken place during these years in Latin American science. I want to congratulate our Brazilian colleagues for arranging this excellent and timely symposium and its excellent organization.

These symposia are a cooperative effort between our Latin colleagues and scientists all over the world. They are made possible by excellent local support and support by a number of international agencies and several groups in the United States.

Earlier symposia in this series are listed on pp. viii-x.

Alexander Hollaender

1961

I. International Symposium on Tissue Transplantation--Santiago, Viña del Mar, and Valparaiso, Chile. Published in 1962 by the University of Chile Press, Santiago; edited by A. P. Cristoffanini and Gustavo Hoecker; 269 pp.

1962

II. International Symposium on Mammalian Cytogenetics and Related Problems in Radiobiology--São Paulo and Rio de Janeiro, Brazil. Published in 1964 by The Macmillian Company, New York, under arrangement with Pergamon Press, Ltd., Oxford; edited by C. Pavan, C. Chagas, O. Frota-Pessoa, and L. R. Caldas; 427 pp.

1963

III. International Symposium on Control of Cell Division and the Induction of Cancer--Lima, Peru, and Calí, Colombia. Published in 1964 by the U.S. Department of Health, Education, and Welfare as National Cancer Institute Monograph 14; edited by C. C. Congdon and Pablo Mori-Chavez; 403 pp.

1964

IV. International Symposium on Genes and Chromosomes, Structure and Function--Buenos Aires, Argentina. Published in 1965 by the U.S. Department of Health, Education, and Welfare as National Cancer Institute Monograph 18; edited by J. I. Valencia and Rhoda F. Grell, with the cooperation of Ruby Marie Valencia; 345 pp.

1965

V. International Symposium on the Nucleolus--Its Structure and Function--Montevideo, Uruguay. Published in 1966 by the U.S. Department of Health, Education, and Welfare as National Cancer Institute Monograph 23; edited by W. S. Vincent and O. L. Miller, Jr.; 610 pp.

1966

VI. International Symposium on Enzymatic Aspects of Metabolic Regulation--Mexico City, Mexico. Published in 1967 by the U.S. Department of Health, Education, and Welfare as National Cancer Institute Monograph 27; edited by M. P. Stulberg; 343 pp.

1967

VII. International Symposium on Basic Mechanisms in Photochemistry and Photobiology--Caracas, Venezuela. Published in 1968 by Pergamon Press as Volume 7, No. 6, Photochemistry and Photobiology; edited by J. W. Longworth; 326 pp.

1968

VIII. International Symposium on Nuclear Physiology and Differentiation--Belo Horizonte, Minas Gerais, Brazil. Published in 1969 by the Genetics Society of America as a supplement to Genetics, Volume 61, No. 1; edited by R. P. Wagner; 469 pp.

1969

IX. International Symposium on Fertility of the Sea--São Paulo, Brazil. Published in 1971 by Gordon and Breach Science Publishers, New York; edited by J. D. Costlow; 2 volumes, 622 pp.

1970

X. International Symposium on Visual Processes in Vertebrates--Santiago, Chile. Published in 1971 by Pergamon Press as Volume 11, Supplement No. 3, Vision Research; edited by Thorne Shipley and J. E. Dowling; 477 pp.

1971

XI. International Symposium on Gene Expression and Its Regulation--La Plata, Argentina. Published in 1973 by Plenum Publishing Corporation, New York, as Vol. 1 Basic Life Sciences Series (Alexander Hollaender, General Editor); edited by Francis T. Kenney, Barbara A. Hamkalo, Gabriel Favelukes, and J. Thomas August.

1972

XII. International Symposium on Genes, Enzymes, and Populations--Calí, Columbia. Published in 1973 by Plenum Publishing Corporation, New York, as Vol. 2 Basic Life Sciences Series (Alexander Hollaender, General Editor), edited by Adrian M. Srb; 359 pp.

1973

XIII. International Symposium on Physiologic and Genetic Aspects of Reproduction--Salvador da Bahia, Brazil. Published in 1974 by Plenum Publishing Corporation, New York, as Vol. 5

(Parts A and B) Basic Life Sciences Series (Alexander Hollaender, General Editor); edited by Elsimar M. Coutinho and Fritz Fuchs; 871 pp.

1974

XIV. International Symposium on Nutrition and Agricultural and Economic Development in the Tropics--Guatemala. Published in 1976 by Plenum Publishing Corporation, New York, as Vol. 7 Basic Life Science Series (Alexander Hollaender, General Editor); edited by N. Scrimshaw and M. Behar; 500 pp.

Contents

Opening Address 1
José Carlos de Almeida Azevedo

"Cerrado": A Region of High Agricultural Potential that Requires Nitrogen 5
Ady Raul da Silva

Potential for Nitrogen Fixation in Tropical Legumes and Grasses 13
Johanna Döbereiner

Free-Living Bacteria 25
Roger Knowles

Nitrogen Fixation by Soil Algae of Temperate and Tropical Soils 41
W. D. P. Stewart, M. J. Sampaio, A. O. Isichei, and R. Sylvester-Bradley

Contribution of the Legume-*Rhizobium* Symbiosis to the Ecosystem and Food Production 65
Avilio A. Franco

Plant Influence in Symbiotic Nitrogen Fixation 75
Joachim F. W. von Bülow

Plant Photosynthesis 95
C. C. Black, R. H. Brown, and R. C. Moore

Interactions of Plant Photosynthesis with Dinitrogen Fixation and Nitrate Assimilation 111
Carlos A. Neyra

Some Aspects of the *Alnus*-Type Root Nodule Symbiosis 121
C. Miguel, A. Cañizo, A. Costa, and C. Rodriguez-Barrueco

Legumes and Acid Soils 135
C. S. Andrew

Micronutrient Requirements of Legume-*Rhizobium* Symbiosis in the Tropics 161
Avilio A. Franco

Ecology of Legume-*Rhizobium* Symbiosis 173
Eli Sidney Lopes

Nitrogenase Systems 191
R. H. Burris, T. Ljones, and D. W. Emerich

Relationship between Hydrogen Metabolism and Nitrogen Fixation in Legumes 209
Harold J. Evans, Tomás Ruiz-Argüeso, and Sterling A. Russell

Ammonia Assimilation in N_2-Fixing Systems 223
D. Barry Scott

Genetics and Regulation of Nitrogen Fixation 237
Winston J. Brill

Leghaemoglobin, Oxygen Supply and Nitrogen Fixation: Studies with Soybean Nodules 247
F. J. Bergersen

Nitrogen Fixation by *Rhizobium* spp. in Laboratory Culture Media 263
F. J. Bergersen and A. H. Gibson

Limiting Factors in Grass Nitrogen Fixation 275
J. Balandreau, P. Ducerf, Ibtissam Hamad-Fares, Pierrette Weinhard, G. Rinaudo, C. Millier, and Y. Dommergues

Physiology and Biochemistry of *Spirillum lipoferum* 303
R. H. Burris, Stephen L. Albrecht, and Yaacov Okon

Taxonomy of the Root-Associated Nitrogen Fixing Bacterium *Spirillum lipoferum* 317
Noel R. Krieg and Jeffrey J. Tarrand

Abstracts of Original Papers 335

Abstracts of Posters 355

Symposium Participants 377

Index 389

OPENING ADDRESS

J.C. de A. Azevedo

Rector of the University of Brasília

Ladies and Gentlemen:

Day after day Nature reveals challenges to the perception of man. It is very curious to notice that the scales of biological and cultural evolution differ so much: at least two million years passed from *Australopithecus* to *Homo sapiens*; yet in much less than fifteen thousand years *Homo sapiens* changed into today's man, a member of a civilized community that understands that the search for truth and the dignity of mankind are the highest ideals of his life.

Ten thousand years ago man learned the rudiments of agriculture; yet, despite the advances in genetics and in the technologies of planting and harvesting, during those millenia, agriculture has not developed much. More serious than this, it does not yet allow us to face the increasing demand for food that is getting more severe day after day. With a two percent a year increase, human population will double in a few decades and therefore the production of food must grow by a higher factor, otherwise the hunger that now exists in the world will increase in a frightening way.

My duty in this opening session is limited to welcoming our guest scientists to Brasilia in the name of his excellency the Vice President of the Republic, General Adalberto Pereira dos Santos; our Minister of Agriculture, Dr. Allyson Paulinelli, and the head of The National Council for the Scientific and Technological Development - CNPq, Dr. José Dion de Mello Telles. However, I beg your permission to take a few minutes of the time devoted to scientific activities in order to draw your kind attention to a matter that is peculiar to Brazil.

Everyone knows the manifold difficulties we face in our determination to expand the "green frontier" - that is to increase the area of cultivated land. In terms of today's technology and economy, there are very few new areas available. There are also indications that cultivated lands are decreasing in developed countries such as the United States that has lost hundreds of square kilometers of cultivable soils in the last few years.

In Brazil, where there is immense insolation, and where the seasons are not so sharply defined, there is plenty of water and the climate is mild; the inhospitable and wild "cerrado" alone covers about two million square kilometers of our country and has large areas that can be easily cultivated mechanically. Large tracts of these lands surround our Capital.

Therefore the solution of the problems that motivated our honored guest scientists to come to our Capital is of great importance if Brazil is to produce adequate food for its people. It is also important for the whole world to incorporate those vastnesses of the "cerrado" region into its "green frontiers."

Unfortunately our country faces many difficulties, and despite the important and irrefutable progress we have achieved since 1964, our technology lags far behind our needs. We have an inadequate number of competent professionals in our universities and research centers, and it unfortunately appears as if our country was not endowed with all the generosities of Nature: we do not have enough fossil fuels and therefore we lack fertilizers.

Despite the many successes of our agriculture - under President Geisel's Government we attained the second position among crop product exporters - we must take care that we do not lose this privileged position. Mexico and the Phillipines now face many difficulities, which are aggravated by large population growth, and they have changed from grain exporters to importers in the last few years.

The shortage of fertilizers is so huge in our country that it clearly hampers our development and substantially affects our trade balance.

All advances that changed the course of mankind were the results of the competence and dedication of a few individuals such as Newton, Einstein, Mendel, Enders, Priestley, Pasteur and others of the same stature, who were concerned as are the scientists who now kindly listen to me, with the destinies of mankind.

Our country - so well represented in this ceremony by our illustrious Vice President of the Republic - trusts faithfully in your work in order to solve the two gravest problems of today; to assure daily food and a dignified life for all. These are the intrinsic human rights that precede all others; no human right is more important than to feed the children, the sufferers and the forlorn by fortune. No human right is more important than health, that cannot exist without enough food, and without which life loses its meaning and dignity.

Here are congregated many specialized scientists who are studying the mechanism of biological fixation of nitrogen, one of the more fascinating possibilities that science has opened up to us. Though ignorant of these matters, I am convinced that agriculture - that began not more than ten thousand years ago in the harsh and depleted lands of the Middle East - is passing through one of its most exciting phases. Unveiling the potentialities of biological nitrogen fixation, the scientists will not only be increasing food production for mankind, but they will be preserving our soils and landscapes and keeping unaltered our climate. There can be no more important contribution of Science to mankind than this.

In the name of the Vice President General Adalberto Pereira dos Santos, our Minister of Agriculture Dr. Allysson Paulinelli, the President of CNPq, José Dion de Mello Telles, the organizers of this meeting, Johanna Döbereiner, Milton Thiago de Mello, Robert H. Burris and Alexander Hollaender, and in the name of the University of Brasilia and my own, I welcome you all to Brasilia and also thank the following organizations that made this meeting possible:

Universidade de Brasilia
Universidade Federal Rural do Rio de Janeiro
Empresa Brasileira de Pesquisa Agropecuária
Conselho Nacional de Desenvolvimento Científico e Technológico - CNPq
National Science Foundation
Energy Research and Development Administration
Organization of American States
United Nations Environmental Program

"CERRADO": A REGION OF HIGH AGRICULTURAL POTENTIAL THAT REQUIRES NITROGEN

Ady Raul de Silva

Centro de Pesquisa Agropecuária do Cerrado

Empresa Brasileira de Pesquisa Agropecuária (CPAC/EMBRAPA)

Planaltina, D.F., Brasil

There are around 183 million hectares under the vegetation called "cerrado," in Brasil, from the Northern Territory of Amapá to as far south as the State of São Paulo. Its main part is in the central part of the country in the States of Minas Gerais, Goiás, Mato Grosso and the Federal District, where Brasilia is located (Ferri, 76).

This huge region is being utilized in an extensive system of cattle production, with the exception of relatively small areas where an intensive agriculture is practiced that has productivity equivalent to the agriculture in our traditional areas of original fertile soils.

Among the difficulties for efficient utilization of the "cerrados," are their low soil fertility and small holding capacity for potash and nitrogen. This probably will require that potash and nitrogen applications be rather frequent.

Under these circumstances, atmospheric nitrogen fixation will be very important. The abundance of legumes among the "cerrado" native vegetation supports the hypothesis that *Rhizobium* is well adapted to "cerrado" soil conditions in spite of its low pH and aluminum toxicity.

Nitrogen fixation through the *Spirillum* spp. has been verified in wheat and corn in "cerrado" soils (EMBRAPA, 76), and there is hope that it may play an important role in the nitrogen supply.

Since nitrogen is the most expensive nutrient and has to be brought to the "cerrado" region, atmospheric nitrogen fixation would be an important factor in reducing costs of agricultural production.

The "cerrado" designates four different types of vegetation: (1) grassland, (2) grassland with small trees, (3) the typical "cerrado" vegetation with many small trees and (4) a vegetation with large trees but smaller and different from forest trees. Under the "cerrado" vegetation the soils are of different types, but they have common characteristics: they are very low in plant nutrients, mainly in phosphorus, nitrogen, potash, calcium, magnesium and in several cases in micronutrients such as zinc; besides, they are acid and in many cases show aluminum toxicity, not because of the presence of large amounts of aluminum but because of the low amount of bases, resulting in a high degree of aluminum saturation.

The low utilization of the "cerrado" region for farming is due to the need of fertilizer application, soil acidity correction and its location in the central part of the country.

Brazil, because of its large area, traditionally developed agriculture production in natural fertile soils because of the high cost of fertilizers, in relation to agriculture production, due to: a) fertilizer importation; b) high cost of transportation, distribution and commercialization. Besides, the low value of the land, in relation to the cost of fertilizers, made it a better option to use more land than to add fertilizers, which results in larger profits in the long run, because of the increasing value of the land.

Today, the overall situation is different. Most of the fertile soils are already in agriculture, many for so long, that they are in need of increasing amounts of fertilizers and soil acidity correction. Consequently, the price of land is very high.

The very rapid agricultural expansion, caused by an economic policy of developing agricultural production, increased the prices of land and provided facilities for the occupation of new land under advantageous conditions.

Also, the migration of the rural population to the cities to find a better standard of living and improved social securities, made labour expensive. In the past it was abundant and cheap.

Mechanization in the traditional agricultural areas is not easy in many regions, because the majority of the fertile soils are in rough lands. Traditional agriculture also developed a farming system with limited mechanization, due to the fact that the soils were originally covered by forest; clearing it completely was very expensive. The agricultural production of coffee, cotton and many other crops was done by hand or with little mechanization.

The above explains why the "cerrado" region was not utilized before.

For "cerrado" production one factor is very important: the ratio of the price of fertilizer and limestone versus the price of agricultural products at the farm.

The "cerrado" soils are acid and very poor in nutrients. So, they need a large amount of fertilizers and limestone when they are first used; this is the opposite to what happens in fertile soils, which do not need any fertilizer or limestone initially but require them later, after several years of intensive use.

The unknown possibilities of the "cerrado" region and the high initial investment in fertilizers and limestone, made many people in the past (and even today) doubt the economic wisdom of its utilization.

The Federal Government has underway a program of agricultural research to evaluate the potential of the "cerrado," to establish the systems of production adequate for the region and to study the basic factors of agricultural production. Besides, for specific and limited areas it is giving temporary subsidies, and good financing conditions for fertilizers, limestone, clearing of the land, soil bed preparation, and construction of facilities and homes on the farms.

An important point has convinced thousands of farmers from other areas to move to the "cerrado:" the cost of the land is still so cheap (2 to 20 times less than the fertile and cultivated soils of their own farms), that it is convenient to buy the "cerrado" land, because the difference allows one to fertilize and add the limestone needed and still to have a larger and better property than the original farm. Besides, the farmer knows that it will not be very long until the "cerrado" land will have a value much higher than the price of acquisition.

Summarizing, the "cerrados" were not being used because of (1) and abundance of fertile soils elsewhere, (2) the high price of fertilizers at farm in relation to the low price of the products from the farm, (3) the abundance and low cost of manpower, (4) the high price of agricultural farm machinery and its operation in relation to the price of agricultural products at the farms, and (5) last but not less important, the lack of knowledge about the potentials of the "cerrado" territory.

The "cerrados" are being occupied now because an economic policy to support agriculture has improved the value of agricultural products of the farms. This gave origin to an agricultural expansion, increased the demands for new lands, increased the prices of the cultivated lands, improved the price ratio of agricultural products over input prices, increased the price of manpower and made farm mechanization a more convenient solution to this. The financial policy, with abundant credit, made the expansion possible.

The above explanation, although logical, may raise doubts as to what will happen now. To see what will happen in the future, it is best to look in the past. The grassland territory of the State of Rio Grande do Sul was very low in plant nutrients and acidic in the 1950's; today, more than 4 million hectares are cropped with soybeans and wheat, every year. The same happened with the grassland territory of the State of Paraná in the sixties. In the last 5 years, more than 2 million hectares of grassland of the southern part of the State of Mato Grosso was occupied by farmers, and each year more are moving into regions of poor soils that are suitable for mechanized farming.

In the future, the still practically untouched "cerrados" areas will be occupied in response to the market, prices and technological advances.

An estimate of the total potential will be made to give an idea about what it will mean to Brazil and to the world economy.

Considering the "cerrado" area of 183 million hectares (Ferri, 76) and the one-third of it (55 million hectares) now being used for agricultural production (equivalent to the area now used for crop producton) it will mean an increase of 162% over 1970 when 34 million hectares were under cultivation (Anuário Estatístico do Brásil, 1975).

The potential for production on those 55 million hectares, if a medium technology and a large area for corn production is considered (a crop well fitted to the overall region), may be estimated at 165 million tons or 10 times today's Brazilian production. Measured as soybean production, whose viability probably will be feasible in the whole area but has been proved only in a part of the region, the total production would be 82.5 million tons or 7 times larger than the largest Brazilian soybean crop obtained until now. Considering world production, these figures for corn will mean 51.2% of the 1975 production and for soybean 120.7%.

Such percentages never will be accomplished on the 55 million hectares considered in the hypothesis, because they will not be cultivated with just one crop. However, these figures give an idea on the potential for grain production on a world basis.

For the world supply it is important to compare the potential and its influence on the international market, the difference between local production and local consumption, i.e., the surplus to be exported. Assuming that Brazil is already exporting corn and soybeans, and that the traditional areas of agriculture still have available land and that the yields still are low, it is possible to establish that the traditional area will be able to supply Brazil and that the "cerrado" production will be exported. The 165 million tons mentioned above is from an area 3.2 times larger than the corn acreage in the world in 1975 (FAO). The 82.5 million tons of soybeans is 4 times larger than the total soybeans produced commercially in the world in 1975 (IFAO).

For the above figures, the hypothesis was based on the utilization of only one-third of the "cerrado" for crop production. Let us assume that 15% of the remaining two-thirds will not be used. Still, 100.5 million hectares will be available for pasture and forestry. Now, assuming that half of that area will be used for pastures, they will be able to support with proper intermediate technology (1 head for 3 hectares), more than 25 million heads of cattle. The other half (50 million hectares of forest), also with medium to low level technology, will yield 750 million cubic meters of wood annually, on the basis of a relative low yield of 15 m^3/ha a year.

POSSIBILITIES AND DIFFICULTIES FOR THE INTENSIVE USE OF THE "CERRADOS"

Because of the heterogeneity of the "cerrado" there are many alternative uses, so that it is not an easy task to evaluate the possibilities and difficulties, especially if one tries to evaluate uses in terms of money.

A valid approach was made by the author (Silva, 76) by comparing the factors for production in the "cerrado" with those in areas in Brazil where crops are already well established and in economical production.

One of the regions of "cerrado" that has been used for crop production for a long time is that of the State of São Paulo, called "mogiana" near the region of Triângulo in the State of Minas Gerais. The yield of corn, soybeans, cotton, and pastures in the "cerrado" has been the same as that in fertile soils which occur side by side in the region and have been subject to mechanized farming.

A common practice for using "cerrado" is by clearing, planting rice for two years and then transforming into pasture with the grasses (Hyparrhenia rufa (Ness) Staff) or Brachiaria sp (mainly decumbens and brizanta) and a rather low amount of fertilizer (around 30 kg/ha of P_2O_5) and without application of any limestone. This very simple technology allows an increase of cattle production from 5 ha/1 head to 2 ha/1 head of cattle. This type of utilization of "cerrado" has been made on thousands of hectares in the States of Minas Gerais, Mato Grosso and Goiás.

In the "cerrados" of the Federal District with the soil conditions described (Brasil, 70), which are considered typical and representative of a very large area (Cline and Buol, 73), with a so-called corrective fertilization of 200 kg/ha of P_2O_5, besides a partial correction of the acidity (just enough to neutralize the alumnium toxicity, usually 2 to 4 t lime/ha), the soil fertility reaches a level equivalent to soils in use for many years in the State of Rio Grande do Sul for soybean and wheat production. Yields of 4 tons of corn and 2.2 tons of soybeans have been obtained in the rainy season without irrigation. In the dry season, with irrigation, 2 to 2.5 tons of wheat are obtained isolated or in rotation with corn and soybeans (Silva, 76, EMBRAPA 76).

The main factors to improve the soil fertility of the "cerrados" are phosphorus and limestone. There is a strong interaction between these two factors.

In large areas of the "cerrados" in the States of Minas Gerais, Goiás and Mato Grosso there are many deposits of limestone spread in the region. So, there is no need for long distance transportation of limestone, and this is very important in reducing costs for the farmers. More important, mines of phosphates in Brazil are located in the "cerrado" region: Araxá, Patos de Minas and Tapira, in the state of Minas Gerais, and Catalão, in the State of Goiás.

The clearing of "cerrado," the seed bed preparation and the mechanization of farming is easy and cheap in the "cerrado" region, because the land is rather flat in large areas and the physical properties of the soil are favorable.

The local market for the products is growing fast since the founding of Brasilia, where there is already a population of 1 million persons. There are many other cities in the "cerrado" region: Belo Horizonte, Goiânia, Uberaba, Uberlândia, Sete Lagoas, Campo Grande, Patos de Minas, Anápolis, Rondonópolis and others. The population of Goiás and Mato Grosso is growing very quickly due to the migration from other states.

The highway system is reasonable and there is a trend to improve highways to connect Brasilia with all regions of the country.

The farm roads are easy to build and preserve because of the nature of the soil, its good drainage and abundant material for road building and preservation. Naturally, the exising roads will not be adequate for the production of 50 million hectares, but they will have to be built at the same time that the production is increasing.

Regarding the environmental conditions, there are difficulties in the climate and in the soils. The climate is the same as that of many important agricultural areas used for crop production for half a century or more in Brazil. The main difficulty is the dry periods during the rainy season ("veranicos"). These vary in intensity and condition the potential crop production directly or by influencing insects and diseases.

The small water holding capacity of the soils of the "cerrado" is characteristic and it increases the effect of the dry periods on yields. It has been estimated that 8 consecutive days without rain exhausts the water capacity available for plants to a 50 cm depth of the soil. A study of 42 years of climatological data in the region of Brasilia, has shown the frequency of dry spells presented in Table 1.

Although there is an urgent need for simple but comprehensive agricultural research, research underway allows us to establish systems of production for immediate use as was done for wheat (Silva, 76).

There is already growing knowledge for successfully growing many crops. There is research and farm production of coffee, corn, soybeans, rice, beans, cassava, sorghum and several forages, besides large plantations of *Eucalyptus* and *Pinus* for wood and cellulose production.

TABLE 1

Frequency of dry periods ("veranicos") in rainy seasons during 42 years in the Brasilia region (Wolf)

Consecutive days without rain	Frequency
8 or more	3 per year
10 or more	2 per year
13 or more	1 per year
18 or more	2 in 7 years
22 or more	1 in 7 years

There is a limitation of specially trained manpower personnel; the traditional farmer of the region has not the knowledge and the experience for intensive agricultural production. However, the official extension services and several companies are hiring agricultural university graduates to provide technical assistance. In addition, the migration of farmers from areas of mechanized agriculture and intensive production provides farmers with experience for the region.

The geographic position of the most important areas of the "cerrados" in the central part of the country is a factor that makes it difficult to export its production. Recent progress in the transportation of iron ore (a product of low value per ton) for export from distances of over 500 km gives hope that it will be economical to export grains, considering also that important grain export countries also have their crop production in central areas (such as the United States and Canada).

REFERENCES

1. BRASIL. Ministério da Agricultura. Equipe de Pedologia e Fertili dade do Solo. Levantamento semidetalhado dos solos de áreas do Ministério da Agricultura no Distrito Federal. Rio de Janeiro, Boletim Técnico, 8, 1970, 135 p.

2. CLINE, M.G. & BUOL, S.W. Soils of the Central Plateau of Brazil and extension of results of field research conducted near Planaltina, Federal District, to them. Ithaca, N.Y. EE.UU., Cornell University, Agronomy Memo 73-13, 1973.

3. EMBRAPA. Centro de Pesquisa Agropecuária dos Cerrados. Relatório técnico anual. Brasília, CPAC, 1976 - 150 p.

4. FERRI, M.G. Ecologia dos cerrados. In: Simposio Sobre O Cerrado, 4, Brasília, 1976. Anais. Belo Horizonte, Itatiaia, São Paulo, Ed. da Universidade de São Paulo, 1977 - 15-36 pp.

5. FUNDAÇÃO IBGE. Anuário estatístico do Brasil - 1975. Rio de Janeiro, Directoria de Divulgação, 1976 - 1017 p.

6. SILVA, A.R. da, LEITE, J.C., MAGALHAES, J.C.A. & NEUMAIR, N. A Cultura do trigo irrigada nos cerrados do Brasil Central Brasília, CPAC, Circular Técnica, 01, 1976 - 70 p.

7. WOLF, J.M. Probabilidades de ocorrência de períodos secos na estação chuvosa para Brasília, DF., in press.

POTENTIAL FOR NITROGEN FIXATION IN TROPICAL LEGUMES AND GRASSES

Johanna Döbereiner

Programa Fixação Biológica de Nitrogênio

Convenio CNPq-EMBRAPA-UFRRJ, Km 47
Seriopédica 23460, Rio de Janeiro - Brazil

INTRODUCTION

In nature, biological fixation of atmospheric nitrogen is widespread. The process accomplishes the reduction of N_2 to NH_3, using energy materials provided directly or indirectly by plant photosynthesis. Industrially the manufacture of one ton of nitrogen fertilizer requires an energy equivalent of seven barrels of oil.

Equilibrium ecosystems and the many lands cropped by traditional agriculture are based largely on nitrogen inputs from biological fixation. The last decade revealed many new opportunities to explore biological nitrogen fixation for intensive agriculture. Fortunately it is in the tropics, where it is most needed, that new ways for substantial protein yield increases can be expected through intelligent manipulation of the various biological N_2-fixing systems.

The South-American highland savannas called CERRADOS or LLANOS represent the largest available area in the world where modern agriculture could be extended without destroying forests (28). Almost two million Km^2 of cerrados with mean population densities of less than two persons per Km^2 and agricultural use of less than 2% have been shown to be suitable for mechanized agriculture, once economically feasible practices of correcting the serious major and minor element deficiencies are established. There are few alternatives for lime, phosphate and trace elements, but nitrogen requiremenets will have to be provided mostly by biological fixation if economically viable systems are to be established. Intensive agriculture in only half of this area would need 10 million tons of nitrogenous fertilizer (100 kg N/ha) per year, one fifth of the world production, if there were no biological fixation.

Present research data indicate that at least two thirds of this amount can be supplied by intelligent use of legumes and crop rotations. Exploitation of nitrogen fixation in grasses and cereals represents one of today's major challenges.

THE LEGUME SYMBIOSIS

The legume-*Rhizobium* symbiosis is the most elaborate and most efficient association between plants and bacteria and for this reason has been most studied. There is little doubt that most legumes can obtain the nitrogen they need for growth and grain yield, from biological fixation in their root nodules. Nitrogen from the soil or fertilizer competes with biological fixation and is used by the plant in preference to free nitrogen from the air. Twelve million tons of soybeans are now produced annually in Brazil without any mineral nitrogen fertilizer. Without biological fixation, this would require 5 million tons of nitrogen fertilizer.

In most cases, even in legumes, the potential for N_2 fixation is substantially higher than the amount actually fixed. The inoculation of legumes, in contrast to general thought, is usually not enough to bring about striking yield increases, especially in the tropics where *Rhizobium* strains and legumes species with little symbiotic specificity are much more common than in temperate regions. In such circumstances Date (12) concluded that not more than 5% of the nodules are formed by the inoculated bacteria. Therefore it does not seem justified to spend disproportionate efforts to inoculate such species as cowpea (*Vigna sinensis*), siratro (*Phaseolus atropurpureus*), kudzu (*Pueraria javanica*) or perennial soybean (*Glycine wightii*), with selected *Rhizobium* strains. In contrast all efforts are justified to guarantee efficient nodulation of more specific species such as stylo (*Stylosanthes guyanensis*), centro (*Centrosema pubescens*) or soybeans (*Glycine max*). In establishing soybeans in a new area, such as, for example, the cerrado areas, the selection of appropriate strains can decide between failure and success of many succeeding soybean crops. Once certain strains are established in a soil, they are difficult to replace. Characteristics to be observed here are plant genotype interactions, response to extreme temperatures, tolerance of acid soil (28, 58) and good nodulation. In addition, more efficient N_2 fixation can be due in part to recycling more hydrogen from nodules which otherwise represents an energy-wasting process (55).

Apart from a few cases such as those mentioned above, the most spectacular increases of nitrogen input from biological fixation are obtained by manipulating the whole soil-plant-bacteria system. The first steps in this direction are experiments to identify the major limiting factors. Table 1 gives an example how elimination of Mn toxicity by liming and correcting phosphate deficiency, both of which affect N_2 fixation much before plant growth, can affect N_2 fixation. Trace element deficiencies are wide-spread in the leached oxysols of the tropics. An example showing that input of N through

TABLE 1

Effect of liming and phosphate fertilizer on N_2 fixation of *Glycine weightii* in soil with Mn toxicity (58).

Plant cultivar	$CaCO_3$ (t/ha)	Super phosphate (t/ha)	Mn in plants (ppm)	Nodules (mg/pot)	Plant-N (mg/pot)
Tinareo	0	0	970	4	4
	0	1.3	786	4	7
	4	0	613	13	15
	4	1.3	260	113	69
Sp-1	0	0	1605	3	6
	0	1.3	940	5	6
	4	0	76	5	8
	4	1.3	233	185	65

biological fixation may be increased by as much as 6-fold by improved trace element supply is given in Table 2. Besides treatment effects, the interaction of species x treatments was highly significant.

A survey in Rio de Janeiro State revealed that 88% of the soils were Mo-deficient (50). Problems of Mo assimilation seem to be a major difficulty in dry bean (*Phaseolus vulgaris*) N_2 fixation, especially in acid soils. When supplied with Mo and phosphate, beans are able to obtain all the nitrogen necessary to produce 3000 kg seed/ha which is 5 times the Brazilian average yield. The nitrogen incorporated into seed of these plants was up to 100 kg/ha (Table 3). Even so, under unfavorable conditions, complementation with low mineral nitrogen might be justified for this crop. Studies of the interactions of the activities of the key enzymes in nitrogen metabolism, nitrogenase and nitrate reductase, indicate that application of 40 kg N/ha during pod filling should be much more effective than at sowing. This was in fact confirmed in a field experiment and is a good example how apparently basic enzyme studies can have immediate application in the field (25).

TABLE 2

Limiting factors for legume growth and nitrogen fixation in a cerrado soil classified as Red Latosol (means of 3 pots) (24).

	Glycine weightii (mg/pot)		Macroptilium atropurpureum (mg/pot)	
	Nodules	Plant N	Nodules	Plant N
complete	232.2	127	223.9	128
complete + N	1.7	238	15.9	189
minus P	1.5	43	4.6	34
minus K	85.8	134	264.3	157
minus S	183.2	136	229.6	126
minus Zn, Cu, B, Mo, Fe	0.7	20	94.1	104
minus lime (Ca, Mg)	179.5	43	236.6	69
Blank	0.4	16	10.9	25

TABLE 3

Effects of Mo and lime on bean yield in N deficient red yellow podzolic soil without N fertilizer. Beans were grown in large pots with 9 kg soil; yields/ha were extrapolated by area. Values are means of 4 pots (26).

Lime t/ha	Dry weight of beans/pot		N removed in seeds (kg/ha)	
	Blank	250 g Mo/ha	Blank	250 g Mo/ha
0	4.1	7.0	10	18
1	7.6	22.5	19	57
2	14.2	31.1	36	79
4	37.3	39.6	95	101

The possibility of reforestation with legume trees has been neglected. A number of species are available for tropical regions, for example the well known acacias. These and other Mimosoideae require specific Rhizobium strains and therefore inoculation of seed beds can be of advantage (18).

The use of legumes in crop rotation or as green manure is a traditional practice which is coming back in many places because of rising fertilizer prices. Table 4 gives an example of the superiority of Crotalaria juncea over other legumes and sorghum as a green manure for field beans (Phaseolus vulgaris). In eight additional field experiments in São Paulo State, green manuring with C. juncea increased the bean yield by 43% on the average (36). In cerrado soils, this practice seems vital for obtaining reasonable bean yields (29). Few people know that Crotalaria spp. also furnishes excellent cellulose for making fine paper.

TABLE 4

Effect of green manure incorporated immediately before planting on the yield of field beans (Phaseolus vulgaris) in São Paulo, Brazil (38).[a]

Green manure	Fresh weight incorporated (t/ha)	Bean seed yield (kg/ha)
Crotalaria juncea	58	800
Cajanus cajan	25	637
Lablab purpureus	32	591
Tephrosia candida	14	609
Sorghum vulgare	56	416
None	--	418
L.S.D. (Ducan)	--	165

[a]Means of 3 field experiments.

THE GRASS ASSOCIATIONS

The ultimate limit to legume N_2 fixation and growth is photosynthesis (31). Legumes possess the less efficient C_3 photosynthetic pathway and attempts to overcome this limitation by photorespiration inhibitors or increased CO_2 concentrations have so far not yielded promising results for practical agriculture. The potential seed or dry matter yield of legumes is therefore much below that of grain crops, especially the tropical species like maize, sorghum, sugar cane or millets which have the more efficient C_4 dicarboxylic acid photosynthetic pathway. These plants should be expected to have a surplus of photosynthates to use for biological N_2 fixation. Earlier papers where nitrogen balance studies indicated substantial nitrogen gains under tropical grasses (39, 43) and reports on selective stimulation of N_2 fixing bacteria in the rhizosphere of sugar cane and Paspalum notatum (17, 19) were widely neglected until we reported (21) the observation of an intracellular association of Azospirillum (at the time called Spirillum lipoferum), with Digitaria decumbens and with maize (9). These reports received great prominence and there have been inevitable reactions. The time seems ripe now for realistic assessments of possibilities for utilization in agriculture of whatever contribution can be expected from biological N_2 fixation in associations with cereals and forage grasses. In order to make use of the already-existing associations and to enhance their N_2 fixation, it is necessary to understand their nature. Much has been done in this direction in the last three years, in numerous laboratories all over the world and only a few highlights can be mentioned here. Many other aspects will be discussed in detail during this symposium.

Table 5 summarizes results obtained with the acetylene reduction assay in intact soil-plant cores and corresponding results obtained with excised, preincubated roots. The latter assay is much easier to perform and many more samples can be analyzed, although it has been criticized due to the lag in C_2H_2 reduction which occurs after disturbing the system and the necessarily prolonged preincubation period which is required to eliminate it. Considerable rates of N_2 fixation seem to occur in many forage grasses and in wheat (40) and significant correlations of intact core assays with the excised root assay were observed in three sets of data. Large overestimates with excised root assays seem to occur in maize (although the mean value calculated from the data presented by Tjepkema and Van Berkum (61) in Table 5 are equivalent) and sorghum, although in these two crops the possibility of cores giving underestimates still has to be considered. Several papers in this symposium will discuss experiments carried out to elucidate the lag phase in C_2H_2 reduction, but a definite explanation seems further away than before. The mean seasonal N_2ase activities, if extrapolated from core assays, are low, although during certain stages of plant growth (mainly the reproductive periods of grains) much higher values can be observed, thus indicating the potential. Nitrogen fixation and incorporation into plant tissue within hours has now been confirmed with $^{15}N_2$ experiments with <u>Digitaria decumbens</u> and <u>Paspalum notatum</u> (14).

In many systems, with <u>Paspalum notatum</u> (19) and perhaps sugar cane (5) as exceptions, <u>Azospirillum spp</u>. seem to be the major N_2-fixing bacteria present (9, 21, 30, 50, 56). Mature roots, surface sterilized with chloramine T or other disinfectants, were shown to contain viable <u>Azospirillum</u> even after 1 h (30, 53, 57) exposure, and tetrazolium-reducing bacteria very similar to this organism were found within the stele and inner cortex cells (48).

<u>Azospirillum spp</u>. (syn <u>Spirillum lipoferum</u> Beijerinck) is a remarkable organism. When it was rediscovered (21), apart from the original description by Beijerinck in 1925 (5) and a brief note by Becking (4), little was known about these N_2-fixing bacteria. Since then quite a few laboratories in the world have taken up work with this organism, judging from the numbers of strain requests we have been receiving. Hundreds of strains have been isolated from many countries, and the intrinsic classification problems have been solved very elegantly by the proposal of a new genus containing two species <u>Azospirillum brasilense</u> and <u>Azospirillum lipoferum</u> (32, 33). A wealth of information is available on the physiology and biochemistry

TABLE 5

Nitrogen (C_2H_2) fixation in intact soil-plant systems and excised roots preincubated under microaerophilic conditions for 8 to 16 hours.

Plant	No. of cores	nmol C_2H_2/(g roots x h) Excised roots	Intact system	Estimated N_2 fixation g/ha/per day*	Reference
Paspalum notatum	2	20	31	78	(22)
	14		836±239	239±68	(15)
Digitaria spp.	13	--	1512±234	432±67	(20)
Digitaria decumbens cv slenderstem	7	--	738±281	210±80	(15)
Digitaria decumbens cv transvala	13	--	848±79	242±43	
Digitaria decumbens	10	66	283	340†	(1)
Triticum estivum	10	193±72	660±140	229†	(20)
Triticum estivum	40	631	942±132	400‡	(40)
Zea mays	18	201±104	193±135	4	(64)
Sorghum vulgare	16	781±533	68±3	1	(64)

*Calculated from results obtained during 24 h assays with intact systems for C_2H_2/N_2=3/1 (core area basis).

†Correlation between the two types of assay r = 0.87

‡Correlation between the two types of assay r = 0.83

of this organism (10, 41, 42, 43, 44, 45, 57), much of which will be reviewed in this symposium (11). One of the most interesting novel features is its capacity to bring about both fixation of N_2 and dissimilation of NO_3^- either to NO_2^- or to gaseous products (N_2O and N_2). Thus it participates in all but one (nitrification) of the processes of nitrogen transformation in nature. With its wide distribution in soils, and especially on grass roots, *Azospirillum spp.* seem to play a role in many other steps of nitrogen nutrition in plants besides N_2 fixation. There are strains of these bacteria which reduce NO_3^- to NO_2^- only and others which are able under certain conditions to dissimilate soil and fertilizer NO_3^- into gaseous products which are returned to the atmosphere and lost from the agricultural systems (41). Anaerobic NO_3^--dependent N_2 fixation was demonstrated (42, 57). The isolation of *Azospirillum* mutants which have lost both or only one of these steps will be reported in this symposium (35). Most of them are able to fix nitrogen in the presence of high levels of soil nitrate and possibly could be explored for simultaneous use of fertilizer and molecular nitrogen by plants. Plant genotype effects on N_2 fixation in *Brachiaria* cores indicate negative interaction with denitrification (51).

One of the intrinsic problems of all nitrogen fixing bacteria is the tremendous sensitivity to O_2 of the nitrogen fixing enzyme, nitrogenase. Efficient metabolism in microorganisms, as in all living beings, on the other hand is dependent on respiration which requires oxygen. Some nitrogen fixing bacteria have evolved elaborate mechanisms which protect their nitrogenase from oxygen without depriving the cell of it. Others are dependent on a symbiont or specific structures to afford protection from oxygen. This is the case for the legume symbiosis and also for several algal associations (61). These systems are much more efficient in terms of energy conversion as long as they are in a symbiotic state (62, 63). *Rhizobium* has recently been shown to fix nitrogen even in culture media, but conditions must be used which protect its nitrogenase from oxygen. *Azospirillum* grown in culture medium shows many similarities to *Rhizobium*. The most remarkable similarity is the inability to fix N_2 under oxygen concentrations which are necessary for optimal growth, indicating evolutionary advantages of environments which offer some kind of oxygen protection in exchange for products of nitrogen fixation. In the legume symbiosis, leghaemoglobin assists by maintaining optimum O_2 concentrations for production of ATP for N_2-fixation and facilitating the flux of O_2 at these low concentrations of free, dissolved O_2 (7). As yet unidentified O_2-carrying proteins may be involved in the non-leguminous angiosperm root nodules (8) and in *Trema* nodules formed by *Rhizobium* (2, 65). The recent finding that *Azospirillum* can use nitrate instead of oxygen for nitrogen fixation (42, 57) may indicate another alternative.

Several other grass-bacteria associations have been found recently. Larson and Neal (34) described a very specific association of one wheat line with a Bacillus sp.. This wheat line contained increased numbers of N_2-fixing Bacillus sp. and nitrate-reducing bacteria and a lower incidence of total bacteria in the rhizosphere. A number of water plants and weeds have also been shown to exhibit substantial nitrogenase activity. The tropical marine angiosperm Thalassia testudinum is an example (49). There N_2-fixing bacteria "other than Azotobacter or Clostridium" were indicated as being responsible for the activity. Spartina alterniflora a Canadian salt marsh C_4 grass was shown to have an association with organisms similar to Azospirillum (47).

Perhaps the most important N_2-fixing grass system is rice. For instance a total of 23 rice crops, in an eleven year experiment in the Philippines, were obtained from a non-fertilized field with no apparent decline in nitrogen fertility of the soil. About 45 to 60 kg N/ha/crop were removed in straw and grain (68). Blue-green algae and photosynthetic bacteria account for a large part of the N_2 fixation in paddy rice (23, 62, 68). Very promising results have been reported from experiments on the use in agriculture of the waterfern Azolla which harbors a N_2-fixing alga in its leaves (5, 62, 63). The physiology of these systems is similar to the other symbiotic associations (52). Results from intact soil-rice systems in the field from which algae have been removed, indicate about 50 to 200 g/ha fixed per day at the flowering stage (68). Balandreau (3) reported that 25 to 30 kg N/ha can be fixed per growing season by the system. Methane-oxidizing bacteria which are able to fix N_2 were also found in rice paddies. The large amount of CH_4 which can accumulate in these systems should not be overlooked as a potential carbon source for N_2 fixation (13). Unfortunately most research on N_2 fixation in rice has been confined to Asia, although this grain also is a major food all over Latin America.

In many of the above summarized grass associations, agronomic studies on the inferences from fertilizers or pesticides and especially from plant genotype treatments should furnish short term results which may lead to agricultural systems which maximize grass N_2 fixation before the exact nature of these associations is known. Molybdenum spraying of the leaves of sorghum and maize grown in the field increased nitrogenase activity in the roots (67). High nitrogen fertilizer levels inhibit nitrogenase activity on roots of sorghum, maize, rice and many forage grasses, but low levels may even increase it (67, 68). Certain herbicides increase nitrogenase activity on maize roots grown in the field (37). Plant genotype effects have now been shown in the major grains maize, wheat and rice (9, 40, 68) and forage grasses (20, 51). Their manipulation remains difficult as long as the physiology of the associations is not fully understood.

It is the aim of this paper to call attention to some of the many ways that can be explored to contribute nitrogen via biological fixation to agricultural systems. The rapidly growing interest in this field will no doubt help to answer economically important problems in tropical agriculture in the near future.

REFERENCES

1. Abrantes, G. T. V., Day, J. M., Döbereiner, J. (1975) Bull. Int. Inf. Biol. Sol, Lyon No. 21,1.
2. Appleby, C. (1974) I Intern. Symp. N_2-Fixation, Pullman, U.S.A.
3. Balandreau, J. (1975) Thesis D.Sc. Univ. Nancy, France.
4. Becking, J. H. (1963) Antonie van Leeuwenhoek 29, 326.
5. Becking, J. H. (1974) I Intern. Symp. N_2 fixation (Newton, W. E. & Nyman, C. J. eds.) Washington State Univ. Press. p. 581.
6. Beijerinck, M. W. (1925) Centralbl. Bact. Parasitenkde Abt. II 63, 353.
7. Bergersen, F. J. (1977) see this Symposium.
8. Bond, G. (1974) Root nodule symbiosis with actinomycete-like organisms, p. 342. In. The Biology of Nitrogen Fixation (Quispel, A., ed). North Holland, Amsterdam.
9. Bülow, J. F. W. and Döbereiner, J. (1975) Proc. Natl. Acad. Sci. (USA) 72, 2389.
10. Burris, R. H. (1976) II Intern. Symp. N_2-Fixation, Salamanca, Spain.
11. Burris, R. H. (1977) see this Symposium,
12. Date, R. (1971) XII Pacific Science Congress, Canberra, Australia.
13. De-Bont, J. A. and Mulder, E. G. (1976) Appl. Environm. 231, 640.
14. De-Polli, H., Matsui, E., Döbereiner, J. and Salati, E. (1976) Soil Biol. Biochem. 9, 119.
15. De-Polli, H. (1976) M.Sci. thesis, Escola Superior de Agricultura Luiz de Queiroz, Piracicaba, SP, Brazil.
16. Döbereiner, J. (1961) Plant and Soil 14, 211.
17. Döbereiner, J. (1966) Pesq. Agropec. Brasil 1, 357.
18. Döbereiner, J. (1967) Pesq. Agropec. Brasil 2, 301.
19. Döbereiner, J. (1970) Centralbl. Bact. Parasitenkde. Abt. II, 124.
20. Döbereiner, J. (1976) in "Genetic Control of Diversity of Plants," Int. Sym. Lahore, Pakistan.
21. Döbereiner, J. and Day, J. M. (1976) Proc. I Intern. Symp. N_2 Fixation (Newton, W. E. & Nyman, C. J., eds.), p. 518.
22. Döbereiner, J., Day, J. M. and Dart, P. J. (1972). J. Gen. Microbiol. 71, 103.
23. Elnawany, A. S. (1976) In Environmental Role of Nitrogen-Fixing Blue-Green Algae and Asymbiotic Bacteria, Intern. Symp. Uppsala, Sweden.
24. Franca, G. E., Carvalho, M. M. de (1970) Pesq. Agropec. Brasil, 5, 147.

25. Franco, A. A., Pereira, J. C. and Neyra, C. A. (1977) see this Symposium.
26. Franco, A. A. and Day, J. M. (1975) V American Rhizobium Conf. Raleigh, USA.
27. Galetti, P., Franco, A. A., Azevedo, H. and Döbereiner, J. (1971) Pesq. Agropec. Brasil. Sér Agron. 6, 1.
28. Goodland, R. J. and Irwin, H. S. (1977) In "Extinction is Forever" N.Y. Botanical Garden, p. 214.
29. Guazzelli, R. J. and Vieira, I. F. (1970) Ann. Rep. Inst. Pesq. Agropec. Oeste, Sete Logoas, Brazil.
30. Hamad-Fares. I., Diem. H. G., Rougier, M., Balandreau, J. P. and Dommergues, Y. (1976) In Environmental Role of Nitrogen-Fixing Blue-Green Algae and Asymbiotic Bacteria, Intern. Symp. Uppsala, Sweden.
31. Havelka, U. D. and Hardy, R. W. F. (1974) I Intern. Symp. N_2-Fixation (Newton, W. E. & Nyman, C. J., eds.) Washington State Univ. Press. p. 518.
32. Krieg, N. R. (1977) Conf. Gen. Engin. N_2 Fixation, Brookhaven Nat. Lab., N.Y.
33. Krieg, N. R. (1977) see this Symposium.
34. Larson, R. I. and Neal, J. L. (1976) In Environmental Role of Nitrogen-Fixing Blue-Green Algae and Asymbiotic Bacteria, Intern, Symp. Uppsala, Sweden.
35. Magalhães, L. M. S., Neyra, C. A. and Döbereiner, J. (1977) see this Symposium.
36. Mascarenhas, A. A., Miyasaka, S., Lovadini, L. A. C., Freire, E. S., Sobrinho, J. T., Cruz, L. P., Nery, C. and Andtade, F. G. (1967) Bragantia 26, 219.
37. Marriel, I. E. and Cruz, J. C. (1977) see this Symposium.
38. Miyasaka, S., Freire, E. S., Mascarenhas, H. A. A., Nery, C., Sordi, G. (1966) Bragantia 25, 277.
39. Moore, A. V. (1966) Soils and Fertilizers 29, 113.
40. Nery, M., Abrantes, G. T. V., Santos, D. dos, and Döbereiner, J. (1977) Rev. Brasil Cien. Solo 1, 15 (see also this Symposium).
41. Neyra, C. A., Döbereiner, J., Laande, R. and Knowles, R. (1977) Can. J. Microbiol. 23, 300.
42. Neyra, C. A. and Van Berkum, P. (1977) Can J. Microbiol. 23, 306.
43. Okon, Y., Albrecht, S. L. and Burris, R. H. (1976) J. Bacteriol. 127, 1248.
44. Okon, Y., Albrecht, S. L. and Burris, R. H. (1976) J. Bacteriol. 128, 592.
45. Okon, Y., Houchins, J. P., Albrecht, S. L. and Burris, R. H. (1976) J. Gen. Microbiol. 98, 87.
46. Parker, C. A. (1957) J. Soil Sci. 8, 48.
47. Patriquin, D. G. (1976) In Environmental Role of Nitrogen-Fixing Blue-Green Algae and Asymbiotic Bacteria, Intern. Symp. Uppsala, Sweden.
48. Patriquin, D. G. and Döbereiner, J. (1977) see this Symposium.
49. Patriquin, D. G. and Knowles, R. (1972) Marine Biol. 16, 49.

50. Peres, J. R. R., Nery, M., Franco, A. A. (1975) Ann. XV Congr. Brasil. Cien. Solo, Campinas, Brazil. p. 163.
51. Pereira, P. A. A., Neyra, C. A. and Döbereiner, J. (1977) see this Symposium.
52. Peters, G. A. (1974) I Intern. Symp. N_2 Fixation (Newton, W. E. and Nyman, C. J., eds.) Washington State Univ. Press, p. 592.
53. Reynders, L. and Vlassak, K. (1976) Agriculture (Netherlands) 24, 5.
54. Ruschel, A. P., Victoria, R. L., Salati, E., and Henis, Y. (1976) In Environmental Role of Nitrogen-Fixing Blue-Green Algae and Asymbiotic Bacteria, Intern. Symp. Uppsala, Sweden.
55. Schubert, K. and Evans, H. (1976) Proc. Nat. Acad. Sci. USA 73, 1207.
56. Scott, C. A., Magalhães, F. M. M., Divan, D. L. S., Scott, D. B. (1977) see this Symposium.
57. Scott, D. B. and Scott, C. A. (1977) see this Symposium.
58. Souto, S. M. and Döbereiner, J. (1969) Pesq. Agropec. Brasil 4, 129.
59. Subba-Rao, N. S. (1977) Intern. Symp. Improving Crop Productivity and Animal Productivity by Molecular and Allied Techniques, New Delhi.
60. Silva, M. F. S. and Döbereiner, J. (1977) see this Symposium.
61. Stewart, W. D. P. and Rewell, P. (1977) Nature 265, 371.
62. Stewart, W. D. P. (1977) Ambio 6, 166.
63. Stewart, W. D. P. (1977) see this Symposium.
64. Tjepkema, J. D. and Van Berkum, P. (1977) Appl. Environ. Microbiol. 33, 626.
65. Trinick, M. J. (1974) I Intern. Symp. N_2-Fixation (Newton, W. E. and Nyman, C. J., eds.) Washington State Univ. Press, p. 507.
66. Vlassak, personal communication.
67. Van Berkum, P., Neyra, C. A. and Bülow, J. F. (1976) XI Reun. Brasil. Milho e Sorgo, Piracicaba, Brazil.
68. Watanabe, Y. and Kuk-Ki-Lee (1975) In Biological Nitrogen Fixation in Farming Systems in the Tropics. IITA, Ibadan, Nigeria.

FREE-LIVING BACTERIA

Roger Knowles

Department of Microbiology, Macdonald Campus of McGill University, Ste Anne de Bellevue, Qué. H0A 1C0, Canada

INTRODUCTION

Free-living or asymbiotic N_2-fixing bacteria are very widely distributed taxonomically, geographically and ecologically. Before considering a number of N_2 fixation systems we will review the current list of N_2-fixing bacteria and the habitats in which they are observed to occur. Since N_2 fixation is most frequently limited by the availability of energy, usually in the form of photosynthetic products, the degree of coupling between the N_2-fixing system and the source of photosynthate is clearly critical. We will discuss systems in which this degree of coupling is relatively loose. Other papers in this symposium deal with photosynthetic bacteria and with some root-associated bacteria, systems in which coupling to photosynthate is relatively close. This paper will therefore make merely brief references to such systems.

TAXONOMIC DISTRIBUTION AND ECOLOGICAL OCCURRENCE

Some earlier reports of N_2 fixation by certain microorganisms consistently resisted confirmation and such organisms are therefore not included in the annotated list which follows. Some more recent, though unconfirmed, reports however are included. It must be remembered then that the list is somewhat fluid, particularly in view of the known genetic properties of the Enterobacteriaceae and related organisms. In these bacteria the possibility for plasmid-mediated transfer of genetic information suggests that the list as reported here will remain far from fixed. In what follows, all N_2-fixing free-living procaryotes other than the cyanobacteria are included and the terminology follows that of Buchanan and Gibbons (15) so far as possible. No attempt is made to document with all relevant references.

Aerobes

Azotobacteraceae. A wide range of carbon compounds is utilized. Are fully aerobic to microaerophilic when fixing N_2. Some show conformational and respiratory protection of nitrogenase which enable them to grow at ambient O_2 concentration. Many produce slime which may act as an O_2 diffusion barrier (15).
Azotobacter chroococcum, beijerinckii, vinelandii, paspali
Azomonas agilis, insignis, macrocytogenes
Beijerinckia indica, mobilis, fluminensis, derxii
Derxia gummosa

Coryneform
Corynebacterium autotrophicum. A H_2-utilizing autotroph which is microaerophilic (85)
Mycobacterium flavum and other spp. (33) Taxonomy uncertain (10) Microaerophilic (9)

Methylobacteria. Use only C-1 compounds and are microaerophilic when fixing N_2 (84)
Methylococcus capsulatus (21)
Methylomonas methanitrificans (synonymous with *Methylosinus trichosporium* and *Pseudomonas methanitrificans* (15, 20)

Spirillaceae. Taxonomy uncertain. Microaerophilic (50)
Spirillum (Azospirillum?) lipoferum (30, 79, 80)
Aquaspirillum fasciculus, peregrinum (50)

Chemolithotrophs
Thiobacillus ferrooxidans (54)

Facultative Anaerobes

Fix N_2 only under anaerobic conditions. Not all species or strains fix N_2.

Bacillaceae
Bacillus polymyxa, macerans, circulans (?) (41, 53, 60)

Enterobacteriaceae
Citrobacter freundii, intermedius (53, 61)
Enterobacter aerogenes, cloacae (2, 46, 53, 61)

Erwinia herbicola (Enterobacter agglomerans) (2, 61, 70)
Escherichia coli (8, 61) and by genetic manipulation (28)
Klebsiella pneumoniae, rubiacearum (?) (48, 56)
Salmonella typhimurium by genetic manipulation (69)

Phototrophic non-sulfur bacteria (51, 67)
Rhodopseudomonas capsulata, gelatinosa, palustris, sphaeroides
Rhodospirillum rubrum
Rhodomicrobium vannielii

Obligate Anaerobes

Bacillaceae
Clostridium butyricum, butylicum, pasteurianum, acetobutylicum, etc. (57, 74)
Desulfotomaculum orientis, ruminis (16, 43)

Uncertain affiliation
Desulfovibrio desulfuricans, vulgaris, gigas (73, 82). Was a component of *Methanobacterium omelianskii* (13, 14)

Phototrophic sulfur bacteria
Chlorobium limicola, thiosulfatophilum (?). (C. limicola was a component of *Chloropseudomonas ethylicum* (37))
Chromatium minutissimum, vinosum (52)
Ectothiorhodospira shaposhnikovii (88)

In view of the very great taxonomic and metabolic diversity amongst N_2-fixing bacteria it is not surprising that they have been isolated from many different habitats. Some indication of this ecological diversity is given by Table I which is modified from Knowles (47). More complete details of the literature on which the table is based can be found in the above publication. Table I shows that groups such as *Azotobacter, Klebsiella, Clostridium* and *Desulfovibrio* occur in a wide variety of habitats, where they are found sometimes in relatively high numbers. Other groups are found mainly in soil, whereas the phototrophic bacteria occur predominantly in aquatic environments. The Table further shows that many genera are associated with the root region of terrestrial plants (47).

SOME N_2 FIXATION SYSTEMS IN RELATION TO THEIR ENERGY SOURCES

Heterotrophic N_2-fixing bacteria depend directly or indirectly on plant photosynthesis for their supply of energy and carbon. The degradation of plant residues thus supports N_2 fixers in many

Table I. Occurrence of N_2-fixing bacteria

Group	Soil		Freshwater		Marine		Phyllo-plane	Animal	Miscel-laneous
	Root free	Root zone	Water	Sedt.	Water	Sedt.			
AEROBES									
Azotobacter	+	+	+	+	+	+	+		
Azomonas			+				+		+
Beijerinckia	+	+					+		
Derxia	+	+							
Corynebacterium	+			+					
Mycobacterium	+								
Methylobacteria	+		+						
Spirillum	+	+							
FACULTATIVE ANAEROBES									
Bacillus	+	+						+	
Citrobacter								+	+
Enterobacter	+	+				+		+	+
Escherichia								+	+
Klebsiella	+	+	+		+		+	+	+
Rhodopseudomonas	+	+	+	+					
Rhodospirillum			+	+	+				
Rhodomicrobium			+	+					
ANAEROBES									
Clostridium	+	+	+	+		+	+	+	
Desulfovibrio	+		+	+	+	+			
Desulfotomaculum	+		+	+	+			+	
Chlorobium			+	+		+			
Chromatium			+			+			

habitats in which the ratio of carbon to nitrogen is high: environments receiving CH_4 from anaerobic metabolism; cellulose rich habitats in soil, mud, decaying wood and the guts of certain animals such as termites; and process waters and effluents in the pulp and paper industry.

Autotrophic or chemolithotrophic bacteria such as *Corynebacterium* apparently use hydrogen, another and very important intermediate or product of anaerobic decomposition processes.

Heterotrophic N_2 fixers present inside or near plant roots are considerably closer to the ultimate source of photosynthate than are the organisms referred to above and might therefore be more effective and efficient in their ability to fix N_2.

Bacteria which are themselves photosynthetic are perhaps less likely to be limited by available energy and carbon and may be important in various aquatic and semi-aquatic systems. However, since they will be the subject of a later paper in the symposium they will not be again referred to here.

Some of the N_2 fixation systems referred to briefly above will next be discussed in more detail.

Systems Supported by Hydrogen or Methane

The important role played by H_2 as a product of the anaerobic metabolism of organic compounds and as an electron donor for several groups of aerobic and anaerobic bacteria is summarized in Table II which is modified from Schlegel (76). Of all of the groups of H_2 utilizers, only the denitrifiers and the true methanogenic bacteria are not reported to include N_2-fixing strains (some strains of N_2-fixing *Spirillum lipoferum* also denitrify but are not reported to utilize H_2). Unfortunately, it is not known to what extent N_2 fixation by H_2 utilizers occurs in natural systems. Certainly, at least some of the N_2 fixation in fresh water and marine sediment systems of up to 3 g N m^{-2} yr^{-1} (reviewed in 47 and 49) could be due to such bacteria. Hydrogen was reported to support C_2H_2-reducing activity in a marine carbonate sand containing sulfide-rich microenvironments (65).

As indicated in Table II one of the products of anaerobic H_2 utilization is CH_4. In aerobic environments this CH_4 supports the growth of the methylotrophs some of which also fix N_2. Microbial activities are stimulated in the presence of natural gas (1) and reported increases in total soil N due to exposure to CH_4 (20) may extrapolate to give a N_2 fixation of about 100 g N m^{-2} yr^{-1}.

Table II. Production of H_2 in anaerobic metabolism and its role as electron donor for groups of bacteria some of which include N_2 fixers (modified from reference 73).

				Ref.
		Aerobic:		
		$CO_2 \xrightarrow{(N_2)}$	Cell-C + H_2O	85
		$CH_4 \xrightarrow{(N_2)}$	Cell-C + H_2O	24
Cellulose		**Anaerobic:**		
Starch				
Organic acids etc.	— H_2 →	NO_3^-, NO_2^-, NO, N_2O →	N_2, NH_3	
		SO_4^{2-}, SO_3^{2-}, $S_2O_3^{2-} \xrightarrow{(N_2)}$	H_2S, S°	15
		CO_2, CO, HCOOH →	CH_4	
		CO_2, organic acids $\xrightarrow[h\nu]{(N_2)}$	Cell-C	15

In a fresh water lake in Canada it was shown that after overturn, when combined nitrogen was present throughout the water column, CH_4 oxidation occurred at all depths regardless of the state of O_2 saturation (75). However, during summer stratification, when combined nitrogen was depleted in the epilimnion and upper hypolimnion, CH_4 oxidation occurred within a very narrow depth range where the O_2 concentration was 0 to 2 mg per liter (Fig. 1). It appeared that the bacteria responsible, being obliged to fix N_2, were now strictly microaerophilic (75). There are unfortunately no systematic studies of N_2 fixation by methylotrophs in nature, whether terrestrial, sediment or water column.

The study of N_2 fixation by CH_4-utilizing bacteria is hampered by several problems involving the C_2H_2 reduction assay.

1. The inhibition of CH_4 oxidation by C_2H_2 (26, 27) deprives the bacteria of their source of C and energy and both growth and C_2H_2 reduction stops. Methanol utilization is not so inhibited by C_2H_2 and growth and C_2H_2 reduction can proceed (27). However, it is difficult to arrive at an added CH_3OH concentration which can be realistically equated to the *in situ* availability of CH_4. It was pointed out that N_2ase activity in pure cultures of CH_4 bacteria could be assayed using N_2O instead of C_2H_2 as substrate (21) but this method could not be employed in natural systems because of the possible reduction of N_2O to N_2 by denitrifying bacteria.

2. Ethylene is co-oxidized or metabolized by CH_4-utilizing and other bacteria (25, 26, 34, 80, 87) but not by CH_4 bacteria

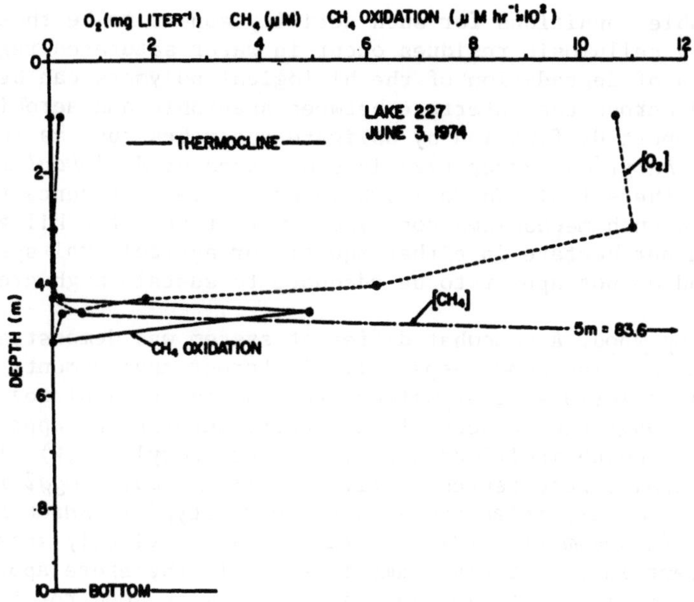

FIG. 1. Depth profile of CH_4 oxidation in a lake during summer stratification. Due to limitation of combined N the CH_4 oxidizers were N_2-fixing and were therefore microaerophilic (Reproduced with permission from ref. 75).

growing on CH_3OH (26). The cooxidation of C_2H_4 is also inhibited by C_2H_2 and thus, during a C_2H_2 reduction assay using CH_3OH as substrate, cooxidation of the C_2H_4 produced would not occur to invalidate the assay (25). However it is clear that it is not possible to correct C_2H_2 assay data using a control in which disappearance of added C_2H_4 is determined (80).

3. A third problem, which in view of the above discussion may not be relevant, is the observation that C_2H_2 inhibits the production of CH_4 in sediment (64) and in rice paddy soil (71). It is interesting that its production is also inhibited by all of the naturally occurring nitrogen oxides (NO_3^-, NO_2^-, NO and N_2O) (5) and by sulfate (86).

From the above it seems clear that routine C_2H_2 assays of sediment or other materials under aerobic conditions would not include activity due to the methylotrophs. Further study appears desirable.

Systems Supported by Plant Residues

Plant residues. Degradation of plant material supports N_2 fixation in ways other than through the agency of H_2 or CH_4. The most favorable conditions for such activity appear to be those in which highly cellulosic residues occur in water saturated regions. The products of degradation of the biological polymers can be transferred across the interface between anaerobic and aerobic zones to support N_2 fixation by obligate anaerobes such as the *Clostridium butylicum* group (72) in some cases or *Azotobacter* species in others (55). There is, however, no real evidence that in the field such mechanisms contribute more than a few kilograms of nitrogen per hectare in either aquatic or agricultural systems (47, 66) and do not appear to be adequate to sustain high crop yields.

Decaying wood. A somewhat different system was demonstrated to exist in decaying wood, especially in forest environments (77, 78). Associations occur between various normally nitrogen-limited rot fungi and members of the Enterobacteriaceae approximately 50% of which individuals could reduce acetylene (2). The N_2-fixing strains were representatives of *Enterobacter agglomerans (Erwinia herbicola), Enterobacter aerogenes,* atypical *Enterobacter,* and *Klebsiella pneumoniae* (2). The latter was previously shown to be abundant in forest environments (31). It therefore appeared that such N_2-fixing bacteria, in association with wood rot fungi, would promote decay of living trees in an otherwise rather nitrogen-limited environment and could therefore be of considerable commercial significance (2). However, it was not claimed that this represented a significant N input on an ecosystem level, a supposition supported by other work on decaying chestnut logs in which N_2 fixation in this habitat extrapolated to about 0.9 kg N $ha^{-1}yr^{-1}$ (17).

Paper industry process waters. Another system supported ultimately by cellulose and lignin rich materials from the forest environment is the various process waters and effluents of the pulp and paper industry. Nitrogen-fixing *Azotobacter (Azomonas)* and *Bacillus* (48), *Klebsiella pneumoniae* (12, 48, 61), *Enterobacter aerogenes, E. cloacae, Erwinia herbicola, Citrobacter freundii, C. intermedius,* and *Escherichia coli* (61) were all observed, frequently in high numbers, in paper mill process waters and effluents. The circumstantial evidence and assays carried out on effluents directly (48) suggest that such N_2 fixation could support the growth of these organisms as well as that of non-N_2-fixing bacteria such as *Acinetobacter* (61). This may be a factor contributing to the problem of slime production in the pulp and paper mill environment.

Gastro-intestinal tracts of animals. The gastro-intestinal tracts of animals feeding on high carbon diets might be expected to harbour N_2-fixing bacteria and to exhibit N_2 fixation activity. Indeed N_2-fixing *Klebsiella pneumoniae* is reported from faeces and intestine contents (8) and from guts of soil animals (19); *Escherichia coli* from faeces and intestine contents (8); *Citrobacter freundii* from termite guts (35); *Enterobacter cloacae* from guts of soil animals (19) and from faeces and intestine contents (8);

Table III. N_2 fixation activities associated with the gastrointestinal tracts of animals.

Animal	Activity	Ref.
	mg N animal^{-1}d^{-1}	
Steer rumen	10	40
Guinea pig	0.76	8
Sheep rumen	0.4	42
	1.0	32
	1.0	36
	32 - 750	44
Termites:	ng N animal^{-1}d^{-1}	
Coptotermes	4.5	11
	5.5	70
	(2.7 μg g^{-1}d^{-1})	
Mastotermes	Up to 1,230	35
	(Up to 24.6 μg g^{-1}d^{-1})	
Kalotermes	(0.8-18 μg g^{-1}d^{-1})	7
Soil animals	0.28-0.47 ng g^{-1}d^{-1}	17
Shipworms	1.2-36,000 μg g^{-1}d^{-1}	18
Sea urchins	55 μg animal $^{-1}$d^{-1}	38

Desulfotomaculum ruminis from rumen contents (68); *Bacillus macerans* and four *Clostridium* species from guts of soil animals (19); and a spirillum-like organism from ship worm gut (18). This latter report is interesting in that the bacterium was a Gram-negative facultative anaerobe, fixing N_2 only under anaerobic conditions, and having the ability to degrade cellulose. If confirmed, this would be the first report of the isolation of a cellulolytic N_2-fixing bacterium.

The N_2 fixation activities actually recorded for animal systems are in general very low (Table III). For example, whereas the daily intake of nitrogen by the sheep is of the order of 10 to 40 g N (36, 44), the estimated amounts of N fixed range from 0.4 to 750 mg N animal^{-1}d^{-1}, very small percentages of the daily requirements. Even the high value of 750 mg reported for a sheep fed a diet containing molasses along with N_2-fixing cells of *Bacillus macerans* (44) nevertheless represents a small part of the daily requirement. The possibility of such diet manipulation may, however, be worth investigating further, especially in view of the current interest in some tropical areas in feeding cattle with sugar cane residues. In the case of termites, most authorities agree that the observed fixation could supply but a small proportion of their daily required N, and the same would seem to apply to the soil animals.

The highest activities recorded for shipworms (Table III) were associated with animals extracted from a log floating in the nutrient-poor Sargasso Sea and would result in a doubling of cell-N in 1.4 days. For worms of coastal species near Woods Hole the doubling of cell-N would require 32 days (18). The activities associated with sea urchins (38) were inversely related to the N content of the macro-algae on which they were feeding (Patriquin, personal communication) and were estimated to represent about 8-15% of the daily N requirement.

The significance of the contribution of N_2 fixation to the N nutrition of animals seems to be greater in the marine than in the terrestrial species, if such a generalization can be made on the basis of the few studies made. Whether biological N_2 fixation in the rumen of important agricultural species can ever be enhanced may depend on possible problems such as the low availability of molecular N_2 in the rumen and the recycling of combined N to the rumen via the rumen wall.

Systems Associated with Living Plants

The N_2-fixing bacteria observed within the root region of plants (Table I) include aerobes such as *Azotobacter*, *Beijerinckia*, *Derxia* and *Spirillum*, facultative anaerobes such as *Bacillus*, *Enterobacter*, *Klebsiella* and *Rhodopseudomonas*, and obligate anaerobes such as clostridia and probably *Desulfovibrio*. These bacteria may be present in locations (Fig. 2) which differ in the availability of organic carbon and in the intensity of microbial competition for this carbon. Thus they may colonize the root tip mucigel where there is a continuous supply of polysaccharide and sloughed-off root cap cells (4), as well as the mucigel found further along the root (22). The rhizoplane or root surface is a location which, operationally, is frequently indistinguishable from the mucigel and adhering debris. Outside the root, N_2 fixers are reported from the rhizosphere region (Fig. 2) which is that part of the soil influenced by the exudations and metabolic activities of the root and in which the availability of carbon, oxygen and combined nitrogen

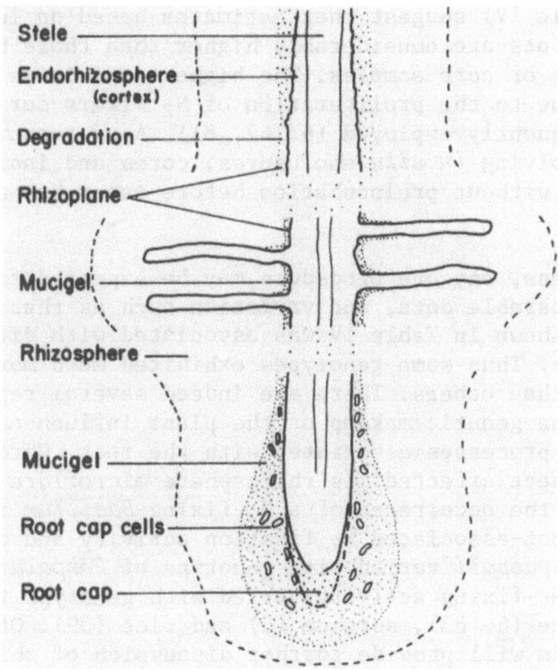

FIG. 2. Diagrammatic representation of the root showing the regions subject to colonization by N_2-fixing bacteria.

may be very different from that in the bulk soil. Thus the root acts as a net source of organic carbon, a net sink of inorganic nitrogen, and either a source or sink of oxygen depending on whether the system is more or less anaerobic and water-saturated, or aerobic and terrestrial (4).

Recently it has become more clear that bacteria can invade plant root cells by penetrating the cell walls (63), and the observations that *Spirillum lipoferum* frequently forms colonies inside cortex and endodermis cells (29) as well as within the stele (Patriquin - personal communication) of *Digitaria* and *Zea* lead to considerable interest in this so-called associative or endorhizosphere system (Fig. 2). The extent to which internal infections occur in other plants is not yet clear and certainly attempts to induce such associations by inoculation of plants and soils have not met with unqualified success (6, 62).

Optimum assay procedures for measuring N_2 fixation associated with plant root systems are not yet established and observed activities vary with the procedures employed. For example, recently reported activities of *Zea mays* in three different locations (Table IV) suggest that estimates based on isolated and preincubated roots are considerably higher than those based on *in situ* enclosures or core samples. The higher activities of isolated roots may be due to the proliferation of N_2 fixers during the preincubation frequently employed (6, 62, 81). A systematic comparison of methods involving *in situ* enclosures, cores and isolated excised roots with and without preincubation before assay has not been reported.

Nevertheless, any one procedure may be expected to yield reasonably comparable data, and variation such as that represented by the ranges shown in Table IV was associated with different plant genotypes. Thus some genotypes exhibited much more N_2 fixation activity than others. There are indeed several reports suggesting that the genetic makeup of the plant influences N_2 fixation and other processes associated with the root. Chromosome substitution in wheat affected its rhizosphere microflora (58, 59) and controlled the occurrence of a N_2-fixing *Bacillus* in the rhizosphere (60). Root-associated N_2 fixation activity and occurrence of *Azotobacter paspali* varied with genotype of *Paspalum notatum* (23) and root N_2-fixing activity varied with genotype from 3- to 20-fold in maize (6, 83), sorghum (6) and rice (39). Other papers in the symposium will provide further discussion of this and other points raised briefly here.

Table IV. Estimates of N_2 fixation (C_2H_2-C_2H_4 assay) by *Zea mays*.

Method	Location	g N $ha^{-1}d^{-1}$	Ref.
In situ	France	8	3
In situ and cores	Brazil	3	81
Cores	Oregon	0.9	6
Isolated roots*	Brazil	35	81
Isolated roots*	Oregon	5 - 135	6
Isolated roots*	Brazil	105 - 2300	83

*Root preincubated for 10-16 h before assay.

The foregoing suggests that it might be useful for plant breeders to attempt to select for increased root-associated N_2-fixation. Such selection would probably involve selection for increased root "leakiness" which would inevitably result in decreased overall yield unless photosynthetic capacity was at the same time increased. Nevertheless, in the near future the availability and cost of inorganic nitrogenous fertilizers might be such that reduced yields coupled with greater plant self-sufficiency for nitrogen might well be a most desirable attribute.

ACKNOWLEDGEMENT

I thank the Canada-Brazil Exchange Programme and CNPq - Brazil for their support of my participation in this Symposium.

REFERENCES

1. Adamse, A.D., Hoeks, J., de Bont, J.A.M., and van Kessel, J.F. (1972) *Arch. Mikrobiol.* 83, 32.
2. Aho, P.E., Seidler, R.J., Evans, H.J., and Raju, P.N. (1974) *Phytopathology* 64, 1413.
3. Balandreau, J. (1975) *Activité nitrogenasique dans la rhizosphère de quelques graminées.* Thèse de Docteur ès Sciences Naturelles, Nancy, France.
4. Balandreau, J., and Knowles, R. (1971) in *Interactions between plant roots and microorganisms* (Dommergues, Y., and Krupa, S., eds.) pp. 000-000, Elsevier Co., Holland. (in press)
5. Balderston, W.L., and Payne, W.J. (1976) *Appl. Environ. Microbiol.* 32, 264.
6. Barber, L.E., Tjepkema, J.D., Russell, S.A. and Evans, H.J. (1976) *Appl. Environ. Microbiol.* 32, 108.
7. Benemann, J.R. (1973) *Science* 181, 164.
8. Bergersen, F.J., and Hipsley, E.H. (1970) *J. Gen. Microbiol.* 60, 61.
9. Biggins, D.R., and Postgate, J.R. (1969) *J. Gen. Microbiol.* 56, 181.
10. Biggins, D.R., and Postgate, J.R. (1971) *J. Gen. Microbiol.* 65, 119.
11. Breznak, J.A., Brill, W.J., Mertins, J.W., and Coppel, H.C. (1973) *Nature* 244, 577.
12. Brown, C., and Seidler, R.J. (1973) *Appl. Microbiol.* 25, 900.
13. Bryant, M.P., Campbell, L.L., Reddy, C.A., and Crabill, M.R. (1977) *Appl. Environ. Microbiol.* 33, 1162.
14. Bryant, M.P., Wolin, E.A., Wolin, M.J., and Wolfe, R.S. (1967) *Arch. Mikrobiol.* 59, 20.
15. Buchanan, R.E., and Gibbons, N.E. (1974) *Bergey's Manual of Determinative Bacteriology*, 8th Ed., Williams and Wilkins Co., Baltimore.

16. Campbell, L.L., and Postgate, J.R. (1965) *Bacteriol. Rev.* 29, 359.
17. Carnaby, B.W., and Waide, J.B. (1973) *Plant Soil* 39, 445.
18. Carpenter, E.J., and Culliney, J.L. (1975) *Science* 187, 551.
19. Citernesi, U., Neglia, R., Seritti, A., Lepidi, A.A., Filippi, C., Bagnoli, G., Nuti, M.P., and Galluzzi, R. (1977) *Soil Biol. Biochem.* 9, 71.
20. Coty, V.F. (1967) *Biotech. Bioeng.* 9, 25.
21. Dalton, H., and Whittenbury, R. (1976) *Arch. Mikrobiol.* 109, 147.
22. Dart, P.J., and Mercer, F.V. (1964) *Arch. Mikrobiol.* 47, 344.
23. Day, J.M., Neves, M.C.P. and Döbereiner, J. (1975) *Soil Biol. Biochem.* 7, 107.
24. De Bont, J.A.M. (1976) *Antonie van Leeuwenhoek J. Microbiol. Serol.* 42, 255.
25. De Bont, J.A.M. (1976) *Can. J. Microbiol.* 22, 1060.
26. De Bont, J.A.M., and Mulder, E.G. (1974) *J. Gen. Microbiol.* 83, 113.
27. De Bont, J.A.M., and Mulder, E.G. (1976) *Appl. Environ. Microbiol.* 31, 640.
28. Dixon, R.A., and Postgate, J.R. (1972) *Nature* 237, 102.
29. Döbereiner, J., and Day, J.M. (1976) in *Proc. Internat. Symp. N_2 fixation, Pullman, U.S.A.* (Newton, W.E., and Nyman, C.J., eds.) pp. 518-538, Washington State Univ. Press.
30. Döbereiner, J., Marriel, I.E., and Nery, M. (1976) *Can. J. Microbiol.* 22, 1464.
31. Duncan, D.W., and Razzel, W.E. (1972) *Appl. Microbiol.* 24, 933.
32. Elleway, R.F., Sabine, J.R., and Nicholas, D.J.D. (1971) *Arch. Mikrobiol.* 76, 277.
33. Fedorov, M.V., and Kalininskaya, T.A. (1961) *Microbiology (Transl. of Mikrobiologiya)* 30, 7.
34. Flett, R.J., Rudd, J.W.M., and Hamilton, R.D. (1975) *Appl. Microbiol.* 29, 580.
35. French, J.R.J., Turner, G.L., and Bradbury, J.F. (1976) *J. Gen. Microbiol.* 95, 202.
36. Granhall, U., and Ciszuk, P. (1971) *J. Gen. Microbiol.* 65, 91.
37. Gray, B.H., Fowler, C.F., Nugent, N.A., and Fuller, R.C. (1972) *Biochem. Biophys. Res. Commun.* 47, 322.
38. Guerinot, M.L., Fong, W., and Patriquin, D.G. (1977) *J. Fisheries Res. Board Can.* 34, 416.
39. Hamad-Fares, I. (1976) *La fixation de l'azote dans la rhizosphère du riz.* Thèse de Docteur ès Sciences Naturelles, Nancy, France.
40. Hardy, R.W.F., Holsten, R.D., Jackson, E.K., and Burns, R.C. (1968) *Plant Physiol.* 43, 1185.
41. Hino, S., and Wilson, P.W. (1958) *J. Bacteriol.* 75, 403.
42. Hobson, P.N., Summers, R., Postgate, J.R., and Ware, D.A. (1973) *J. Gen. Microbiol.* 77, 225.
43. Iizuka, H., Okazaki, H., and Seto, N. (1969) *J. Gen. Appl. Microbiol.* 15, 11.

44. Jones, K., and Thomas, J.G. (1974) *J. Gen. Microbiol.* 85, 97.
45. Jurgensen, M.F., and Davey, C.B. (1971) *Plant Soil* 34, 341.
46. Koch, B.L., and Oya, J. (1974) *Soil Biol. Biochem.* 6, 363.
47. Knowles, R. (1977) in *A Treatise on Dinitrogen Fixation. Section IV: Agronomy and Ecology* (Hardy, R.W.F., and Gibson, A.H., eds.) pp. 33-83, Wiley-Interscience, New York.
48. Knowles, R., Neufeld, R., and Simpson, S. (1974) *Appl. Microbiol.* 28, 608.
49. Knowles, R., and Wishart, C. (1977) *Environ. Pollution* 13, 133.
50. Krieg, N.R. (1976) *Bacteriol. Rev.* 40, 55.
51. Lindstrom, E.S., Lewis, S.M., and Pinsky, M.J. (1951) *J. Bacteriol.* 61, 481.
52. Lindstrom, E.S., Tove, S.R., and Wilson, P.W. (1950) *Science* 112, 197.
53. Line, M.A., and Loutit, M.W. (1971) *J. Gen. Microbiol.* 66, 309.
54. Mackintosh, M.E. (1971) *J. Gen. Microbiol.* 66, i.
55. Magdoff, F.R., and Bouldin, D.R. (1970) *Plant Soil* 33, 49.
56. Mahl, M.C., Wilson, P.W., Fife, M.A., and Ewing, W.H. (1965) *J. Bacteriol.* 89, 1482.
57. Mishustin, E.N., and Shilnikova, V.K. (1969) in *Soil Biology, Reviews of Research,* pp. 65-124, U.N.E.S.C.O., Paris.
58. Neal, J.L Jr., Atkinson, T.G., and Larson, R.I. (1970) *Can. J. Microbiol.* 16, 153.
59. Neal, J.L., Larson, R.I., and Atkinson, T.G. (1973) *Plant Soil* 39, 209.
60. Neal, J.L., and Larson, R.I. (1976) *Soil Biol. Biochem.* 8, 151.
61. Neilson, A.H., and Sparell, L. (1976) *Appl. Environ. Microbiol.* 32, 197.
62. Okon, Y., Albrecht, S.L., and Burris, R.H. (1977) *Appl. Environ. Microbiol.* 33, 85.
63. Old, K.M., and Nicholson, T.H. (1975) *New Phytologist* 74, 51.
64. Oremland, R.S., and Taylor, B.F. (1975) *Appl. Microbiol.* 30, 707.
65. Patriquin, D.G., and Knowles, R. (1975) *Marine Biol. (Berlin)* 32, 49.
66. Paul, E.A. (1976) in *Environmental Biogeochemistry, Vol. 1. Carbon Nitrogen, Phosphorus, Sulfur and Selenium Cycles* (Nriagu, J.O., ed.) p. 225 Ann Arbor Science, Michigan.
67. Pfennig, N. (1967) *Ann. Rev. Microbiol.* 21, 285.
68. Postgate, J.R. (1970) *J. Gen. Microbiol.* 65, 137.
69. Postgate, J.R., and Krishnapillai, V. (1977) *J. Gen. Microbiol.* 98, 379.
70. Potrikus, C.J., and Breznak, J.A. (1977) *Appl. Environ. Microbiol.* 33, 392.
71. Raimbault, M. (1975) *Ann. Microbiol. (Inst. Pasteur),* 126A, 247.
72. Rice, W.A., and Paul, E.A. (1972) *Can. J. Microbiol.* 18, 715
73. Riederer-Henderson, M-A., and Wilson, P.W. (1970) *J. Gen. Microbiol.* 61, 27.

74. Rosenblum, E.D., and Wilson, P.W. (1949) *J. Bacteriol.* 57, 413.
75. Rudd, J.W.M., Furutani, A., Flett, R.J., and Hamilton, R.D. (1976) *Limnol. Oceanogr.* 21, 357.
76. Schlegel, H.G. (1974) *Tellus* 26, 11.
77. Seidler, R.J., Aho, P.E., Raju, P.N., and Evans, H.J. (1972) *J. Gen. Microbiol.* 73, 413.
78. Sharp, R.F. (1975) *Soil Biol. Biochem.* 7, 9.
79. Sylvester-Bradley, R. (1976) *J. Gen. Microbiol.* 97, 129.
80. Sylvester-Bradley, R. (1976) *Investigations into the measurement of nitrogen fixation in freshwater lake sediments.* Ph.D. thesis, University of Edinburgh.
81. Tjepkema, J., and Van Berkum, P. (1977) *Appl. Environ. Microbiol.* 33, 626.
82. Trüper, H.G., Kelleher, J.J., and Jannasch, H.W. (1969) *Arch. Mikrobiol.* 65, 208.
83. Von Bülow, J.F.W., and Döbereiner, J. (1975) *Proc. Nat. Acad. Sci. U.S.* 72, 2389.
84. Whittenbury, R., Phillips, K.C., and Wilkinson, J.T. (1970) *J. Gen. Microbiol.* 61, 205.
85. Wiegel, J., and Schlegel, H.G. (1976) *Arch. Microbiol.* 107, 139.
86. Winfrey, M.R., and Zeikus, J.G. (1977) *Appl. Environ. Microbiol.* 33, 275.
87. Yoshida, T., and Suzuki, T. (1975) *Soil Sci. Plant Nutr. (Tokyo)* 21, 129.
88. Zakhvateva, B.V., Malafeeva, I.V., and Kondrat'eva, E.N. (1970) *Microbiology (Transl. of Mikrobiologiya)* 39, 661.

NITROGEN FIXATION BY SOIL ALGAE OF TEMPERATE AND TROPICAL SOILS

W.D.P. Stewart, M.J. Sampaio, A.O. Isichei
and R. Sylvester-Bradley*

Department of Biological Sciences
University of Dundee, Scotland
*Instituto Nacional de Pesquisas da Amazonia
Manaus, Amazonas, Brazil

INTRODUCTION

Blue-green algae are common components of the microbial flora of the soil in many parts of the world (7,8,18,30). In the tropics most attention has been paid to their role in rice paddy soils where, free-living (26,27) and in symbiotic association with the water-fern *Azolla* (3,19,21,36) they contribute substantial amounts of nitrogen to the ecosystem. In this paper we present information on the occurrence, activity, and factors affecting soil algae from tropical savanna regions of Nigeria and from the Amazon region of Brazil. The findings are compared with observations made on algae from temperate soils in Scotland. These studies complement ones from tropical (e.g. 22,26,27,37) and temperate (e.g. 6,9,12,28) regions.

THE STUDY AREAS AND THE OCCURRENCE OF POTENTIAL NITROGEN-FIXING ALGAE

Heterocystous, non-heterocystous and unicellular blue-green algae are now known to fix N_2 (see 35). Of these the heterocystous forms which invariably fix N_2 under aerobic and anaerobic conditions are ecologically the most important (see 18, 26,30) although the non-heterocystous strains which fix N_2 under anaerobic conditions only (23,24,29,33,34) may also be important in environments such as marine salt marshes (31).

The Scottish samples provide good examples from a temperate maritime climate and the sites chosen for study are shown in Fig. 1. At each sampling site of 10 km^2, the following soil classes were selected, if present: arable soils, bogland, coniferous woodland, deciduous woodland, fresh-water marshes, grassland, heathland, river banks, rock outcrops, marine rocky shores, marine sand-dunes and salt marshes. Samples were collected using a sterile cork-borer, or sterile scalpel, transferred to sterile containers and returned to the laboratory. They were then moistened with nitrogen-free medium (2) and incubated for 2 - 4 weeks in petri dishes at 3000 lux and 25°C. Table 1 shows the algae which predominated in the various samples at the end of this period. *Nostoc commune* was the most common heterocystous alga present, with *Calothrix* dominating in the rocky shore samples. *Oscillatoria tenuis* was the most common non-heterocystous form and in maritime marshy areas *Nodularia spumigena* and *Gloeocapsa* were particularly common.

Fig. 1. Map of Scotland showing the major sampling areas (●).

TABLE 1

The dominant algae detected in algal samples from the various habitats in Scotland after incubation of samples in nitrogen-free medium, at 25°C and 3000 lux for 2 - 4 weeks.

Habitat	Dominant algal genera
Arable land	*Nostoc, Anabaena, Cylindrospermum, Oscillatoria*
Acid bogland	*Oscillatoria, Anabaena, Lyngbya, Nostoc*
Coniferous woodland	*Oscillatoria*
Deciduous woodland	*Nostoc*
Freshwater marsh	*Oscillatoria, Phormidium, Nostoc, Anabaena*
Permanent grassland	*Nostoc, Oscillatoria, Lyngbya, Aulosira, Anabaena, Phormidium*
Heathland	*Nostoc*
Riverbanks	*Oscillatoria, Nostoc, Phormidium*
Rock outcrops	*Anabaena, Anabaenopsis, Nostoc, Phormidium, Oscillatoria, Lyngbya*
Marine rocky shores	*Calothrix*
Sand dunes	*Nostoc, Anabaena, Lyngbya, Oscillatoria*
Salt marshes	*Phormidium, Oscillatoria, Lyngbya, Anabaena, Nostoc, Nodularia, Gloeocapsa*

Studies on Nigerian algae were concerned with those which occurred as soil crusts (Plates 1 and 2). Fifty sampling sites were chosen and the major ones are shown in Fig. 2. These cover the 5 main types of savanna found in the country (derived savanna, Southern Guinea savanna, Northern Guinea savanna, Sudan savanna which covers about two-thirds of all savanna land in Nigeria, and Sahel savanna) as well as the predominantly forested areas of the south-west of the country. There is a decrease in rainfall from the south-west where the rainy season lasts for about 11 months, to the north-east where the rainy season lasts only for about three months and where the soils are extremely dry for most of the

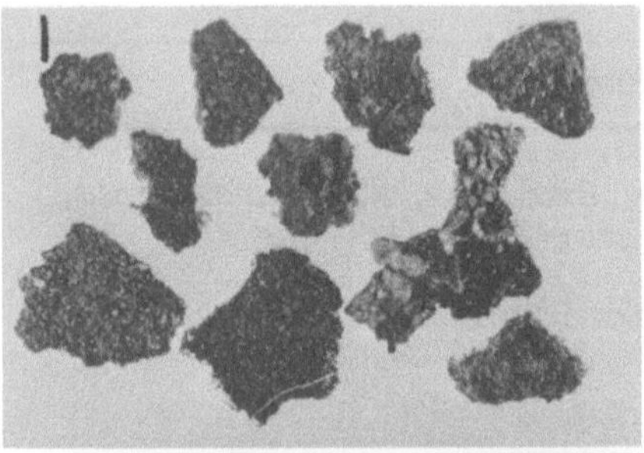

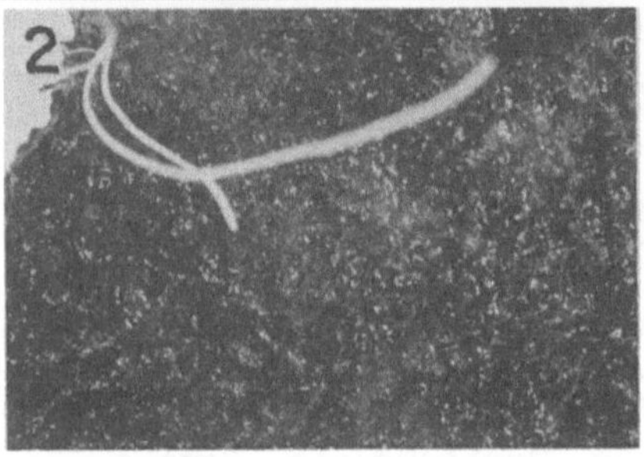

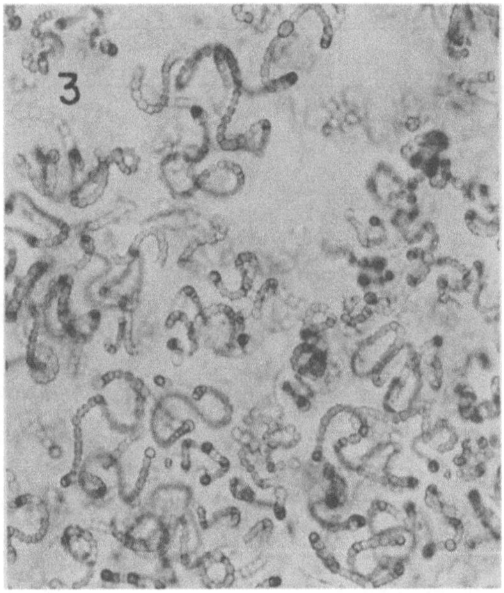

Plate 1. Soil crusts from Nigeria (x 0.18).

Plate 2. Soil crust from Nigeria showing a gelatinous blue-green algal mat (x 0.9).

Plate 3. Light micrograph of a *Nostoc* sp from a Nigerian soil crust (x 270).

year (see 15). The diversity of soil types in Nigeria is great but most are ferruginous ferratites and, apart from areas where there are iron pans, the soil can usually support good algal crusts when moisture is available. Large areas of savanna are burnt annually and blue-green algae develop prominently in open areas after the start of the rainy season (see 15). They are least abundant in the dry Sahel.

On direct microscopical examination of the soil crusts, it was found that all were dominated by one algal genus *Scytonema*, together with small quantities of non-heterocystous Oscillatoriaceae. and occasionally with species of *Tolypothrix* and/or *Nostoc* (Plate 3). The *Scytonema* (Plates 4-6) shows typical false branching, numerous heterocysts and frequently brown, blackish, or ochre, thick mucilaginous sheaths (see e.g. Plate 6). The presence of one dominant alga only in Nigerian soil crusts contrasts with our findings for Scottish soils where species diversity is greater and where *Nostoc* is usually the dominant heterocystous alga.

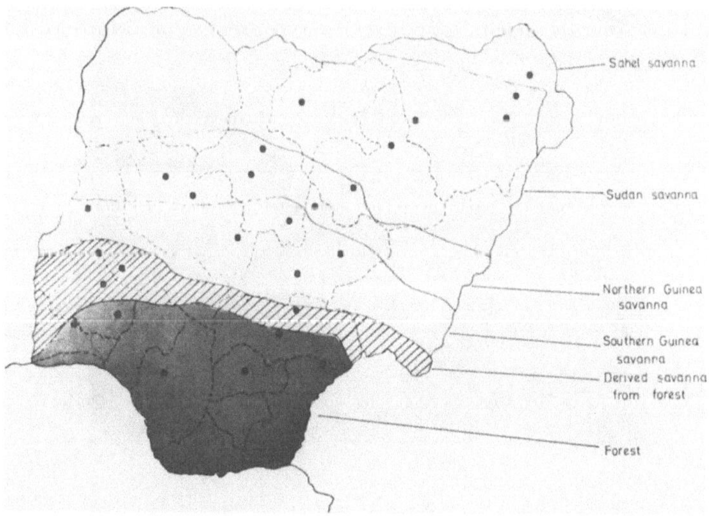

Fig. 2. Map of Nigeria showing the major sampling areas (●) and the different savanna zones.

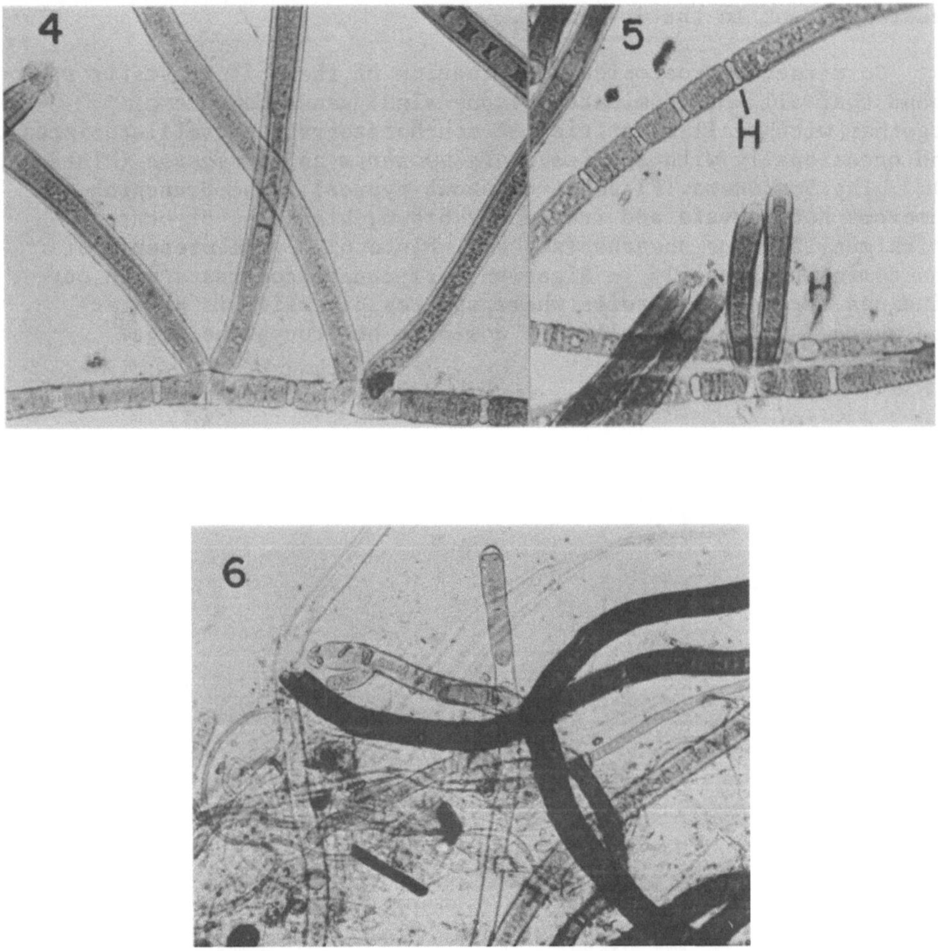

Plates 4-6. Light micrographs of a *Scytonema* sp from a Nigerian soil crust. Note false branching, heterocysts (H) and blackish sheaths of certain filaments (x335).

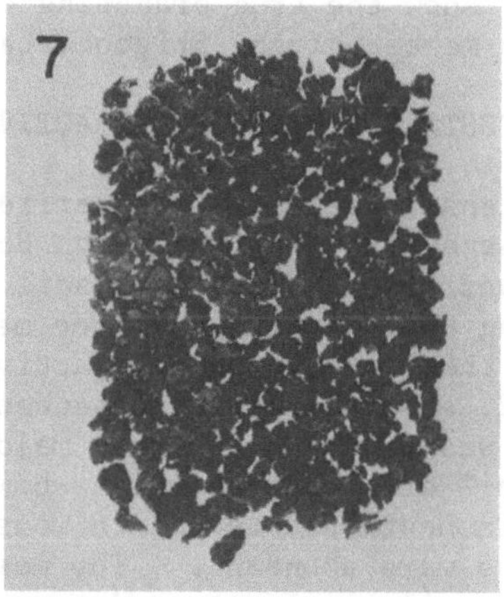

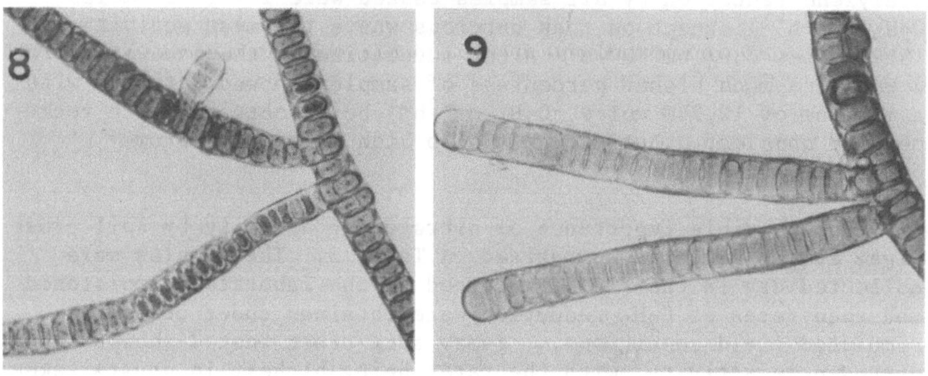

Plate 7. Colonies of *Stigonema panniforme* from Brazil (x 0.45).

Plates 8-9. Light micrographs of *Stigonema panniforme*. Note true branching (x 270).

Studies on Brazilian algae were concerned with one alga *Stigonema panniforme*. In parts of the Amazon region of Brazil near Manaus this alga develops in the open "Campina" areas among the rain forest. It is most abundant in the rainy season when it extensively covers the bare soil, but during the dry season it is less abundant, developing over the soil surface as blackish granules, a few mm in diameter (Plate 7). Light micrographs of the alga (Plates 8-9) show the true branching of this alga, which is a very distinctive member of the Stigonematales.

TESTS FOR NITROGENASE ACTIVITY

Tests for nitrogenase activity were carried out using the acetylene reduction assay (22). Studies on Scottish soils were carried out *in situ* and also by returning soil samples to the laboratory, incubating them in nitrogen-free medium for 4 weeks, and then re-assaying them for nitrogenase activity. The data, summarised in Table 2, show that light-dependent aerobic nitrogenase activity was detected in every major habitat type tested. However the % of each type which showed activity *in situ* ranged from only 3% in heathland to 80% in moist rock outcrops where blue-green algae were abundant. The mean rates of *in situ* acetylene reduction by all samples tested were low (< 2 nmoles C_2H_4 $cm^{-2}h^{-1}$) except on rock outcrops where the mean activity was 15.7 nmoles C_2H_4 $cm^{-2}h^{-1}$. After incubation in the laboratory for 4 weeks, a much higher percentage of samples showed activity with a maximum of 12,240 moles C_2H_4 $cm^{-2}h^{-1}$ being obtained with rock outcrop samples. Activity was also high in samples from permanent grassland.

The possible importance of nitrogenase activity by soil crust algae from Nigeria is summarised in Table 3. The samples were collected dry in the field, returned to the laboratory, moistened and then rates of C_2H_2 reduction were obtained under otherwise simulated field conditions. Every soil crust tested showed nitrogenase activity, with the rates being highest in crusts taken from the open areas in the forest (mean 11.2 nmoles C_2H_4 $cm^{-2}h^{-1}$) and minimal in crusts from the Sahel savanna (2.0 nmoles C_2H_4 cm^{-2} h^{-1}). On average these values are considerably higher than those found in Scottish soils, although the average rates obtained *in situ* on the rock outcrops in Scotland are higher than those of any of the Nigerian soil crust samples. Tests were also carried out in which the soil crusts were moistened with nitrogen-free culture medium and then maintained in the laboratory at 25°C and 3000 lux for 2 - 4 weeks before assaying for nitrogenase activity. Very much higher rates of nitrogenase activity were then obtained with a maximum value of 76.4 nmoles C_2H_4 $cm^{-2}h^{-1}$ being obtained in forest soil crust samples.

TABLE 2

Acetylene reduction rates by samples from various habitats in Scotland incubated *in situ* and in the laboratory

Soil type	No. of samples taken	% fixing *in situ*	% fixing after incubation for 4 weeks in the laboratory	Mean rates of C_2H_2 reduction *in situ* (nmoles $C_2H_4 cm^{-2} h^{-1}$)	Maximum rates of C_2H_2 reduction occurring in the laboratory (nmoles $C_2H_4 cm^{-2} h^{-1}$)
Arable	48	6	54	0.02	41
Bogland	8	12	12	0.3	8
Coniferous forest	39	8	8	0.2	292
Deciduous woodland	32	18	28	0.5	765
Freshwater marshes	30	35	68	0.5	387
Permanent grassland	54	22	45	1.2	3627
Heathland	33	3	3	0.3	342
River banks	48	31	55	1.1	1503
Rock outcrops	12	60	76	15.7	12240
Marine rocky shores	8	50	50	3.6	299
Sand dunes	18	22	72	1.3	*
Salt marshes	5	80	80	1.6	279

*not tested. The *in situ* tests were carried out during the months of May - September.

TABLE 3

Acetylene reduction by algal crust samples from various habitats in Nigeria

Area	Mean rate of C_2H_2 reduction (nmoles $cm^{-2}h^{-1}$)	Maximum rate of C_2H_2 reduction (nmoles $cm^{-2}h^{-1}$)
Sahel savanna	2.0	10.0
Sudan savanna	6.6	25.2
Northern Guinea savanna	5.7	29.7
Southern Guinea savanna	9.2	23.7
Derived savanna	10.6	24.5
Forest	11.2	76.4

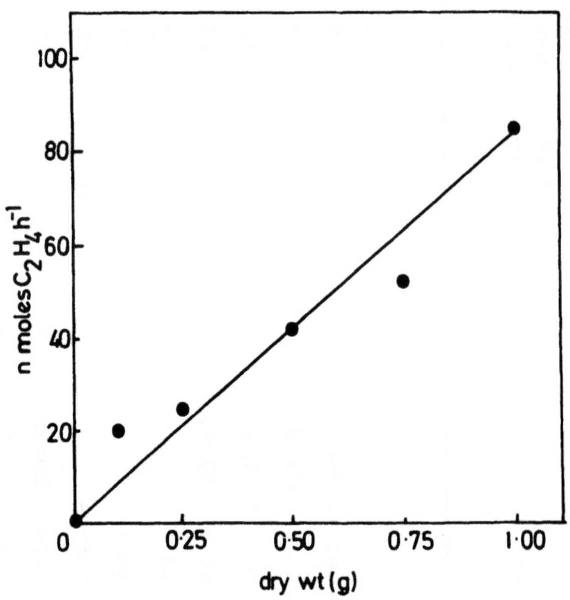

Fig. 3. Light-dependent acetylene reduction by *Stigonema panniforme* at 25°C and 3000 lux. The acetylene reduction assay period was 60 min.

Tests for nitrogenase activity by *Stigonema* were carried out in the laboratory by weighing out different quantities of algae collected during the dry season, wetting these and then testing for light-dependent nitrogenase activity 24 h later. As Fig. 3 shows nitrogenase activity increases with increase in algal biomass indicating that the alga is a N_2-fixing species. No data were obtained on the quantitative significance of this alga in the field.

ENVIRONMENTAL FACTORS AFFECTING NITROGENASE ACTIVITY BY ALGAE

In natural ecosystems N_2-fixing algae are subjected to a variety of extreme and often rapidly fluctuating environmental conditions (18,32). Among the environmental parameters which are important are pH, temperature, light, moisture, and nutrients other than nitrogen. The effect of variation in these on the nitrogenase activity of the various soil algae are presented below.

pH

Blue-green algae are characteristic of neutral and slightly alkaline soils; they are less common in acid soils and are rarely active in pure culture at pH levels below 5. Data on the effect of pH on algae from Scotland, Nigeria and Brazil are presented in Fig. 4. It is seen that all three types have a wide pH tolerance with an optimum near pH 8 and with good activity occurring at pH 10. Nitrogenase activity by the Scottish samples decreases markedly below pH 6. The tropical algae, on the other hand, show good activity even at pH 4. Such algae must possess an efficient pH buffering mechanism because *in vitro* the nitrogenase enzyme of cyanophytes is very susceptible to pH change outside the range 7.0 - 7.5 (10). The capacity of tropical algae to fix N_2 under acid conditions may be a factor contributing to their ecological success in many areas.

Temperature

Most blue-green algae grown in laboratory culture show temperature optima near 32.5 - 35°C, although exceptions occur (see 32), and field populations may show very different temperature optima. Data on the effect of temperature on soil cores from Scotland, soil crusts from Nigeria and *Stigonema* from the Amazon region are presented in Fig. 5.

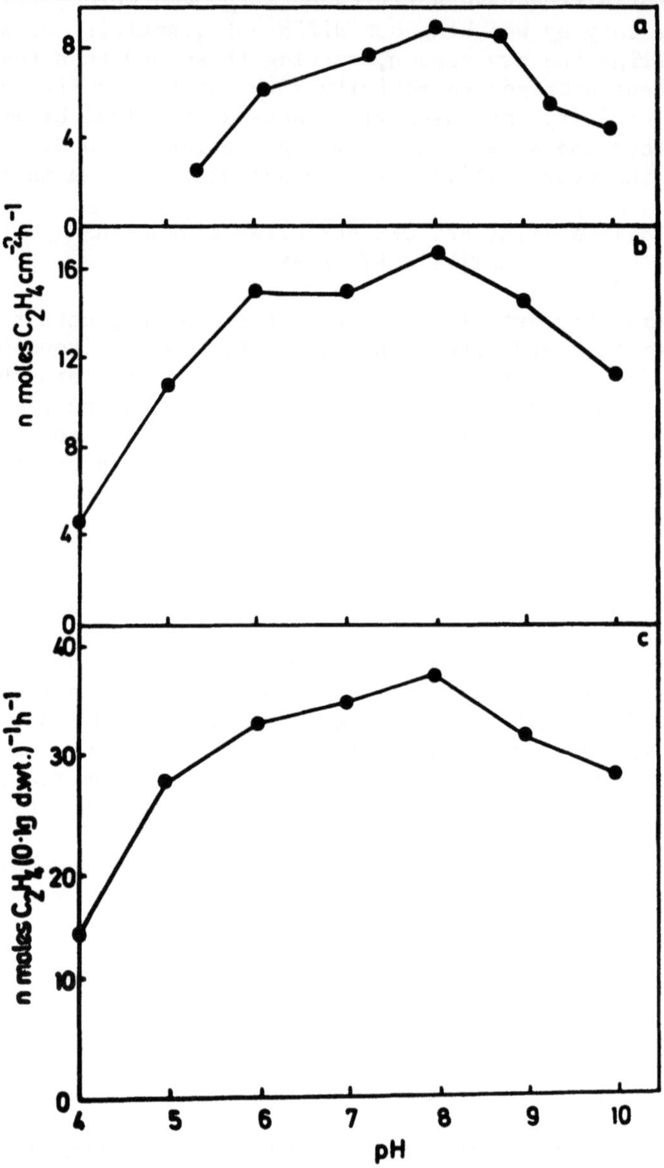

Fig. 4. The effect of pH on acetylene reduction by (a) a Scottish soil *Anabaena*, (b) *Scytonema* crusts and (c) *Stigonema panniforme*. The samples were incubated at the various pH levels for 24 h prior to the 60 min acetylene reduction assay. The pH of the medium did not change from the values given by more than 0.2 pH units during the experimental period.

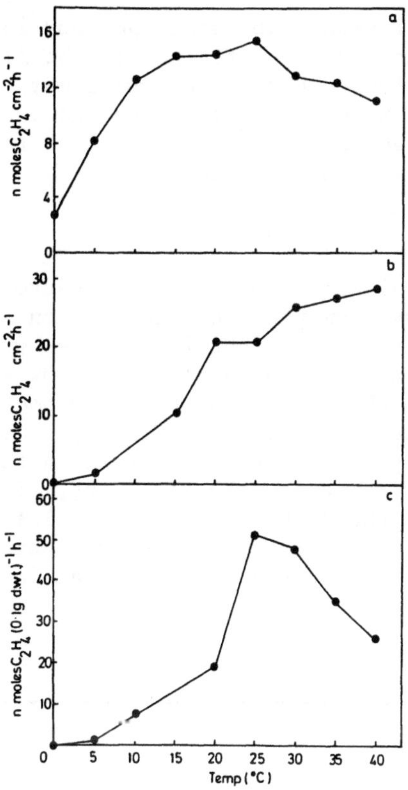

Fig. 5. The effect of temperature on acetylene reduction by (a) Scottish soil cores dominated by *Nostoc* and *Cylindrospermum*, (b) *Scytonema* crusts from Nigeria, and (c) *Stigonema panniforme* from Brazil.

The algae show rather different responses to temperature. Thus, while the Scottish samples reduced acetylene at 0°C, the tropical forms showed little activity even at 5°C. Activity by the Scottish samples, however, was still high at 40°C, although they had a temperature optimum of 15 - 25°C. The Nigerian samples, surprisingly, showed increased nitrogenase activity with increase in temperature to 40°C, while activity of *Stigonema* declined steeply above 30°C. The Scottish algae showed the greatest response to temperature increases from 0 - 10°C, with the corresponding ranges for Nigerian and Brazilian samples being 15 - 20°C and 20 - 25°C respectively. There is thus a general direct correlation between the temperature responses of the algae and the temperatures of the habitats from which they were collected, although the Scottish algae were able to reduce acetylene at temperatures which would seldom, if ever, be experienced in Scotland, except possibly in certain localised soil niches for short periods in summer. Alexander (1), for example, reported how in the Arctic, heat absorption by dark coloured mosses in summer may result in high temperatures and high rates of N_2-fixation then. The difference in response of the Nigerian and Brazilian algae to high temperatures is, however, unexpected.

Light

Many blue-green algae are obligate photoautotrophs receiving their necessary energy from photophosphorylation, some grow photoheterotrophically at low light intensities which do not support photoautotrophic growth, and others grow slowly in the dark (see 32). Data obtained with *Stigonema panniforme* show the type of result most usually obtained (Fig. 6). It is seen that on incubating this alga in the dark, there is a rapid decline in nitrogenase activity, compared with the light, with activity ceasing after 20 h in the dark. In soil samples from Scotland a rather similar pattern was observed, although some samples showed a low rate of dark nitrogenase activity for at least 24 h.

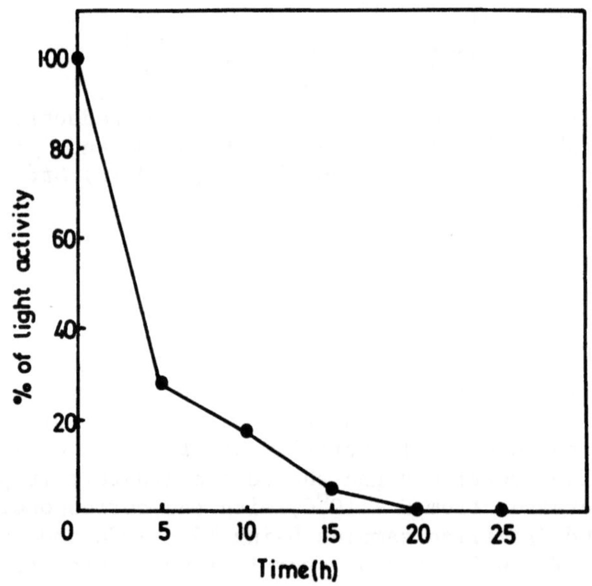

Fig. 6. Nitrogenase activity in the dark as % of nitrogenase activity in the light by *Stigonema panniforme*. The alga had been growing previously at 3000 lux and 25°C before being placed in the dark at 0 time.

Data were also obtained on the response of the tropical algae to increase in light intensity. It was found that both photosynthesis (O_2 evolution) and nitrogenase activity increased with increase in light intensity up to 80,000 lux probably because the dark-pigmented sheaths of these algae served as a light screen. This was likely to be the case particularly with the *Stigonema* which occurred on light coloured "Campina" soils. Data on the relative light transmission through algal mats of different thickness were then obtained by filtering different amounts of *Stigonema* onto Whatman GFC filters, clearing the filters in Cedar Wood oil and then measuring transmission of light of 665nm (the absorption maximum of chlorophyll *a*). As Fig. 7 shows, with increase in algal biomass there is, as expected, a decrease in light transmission with an algal layer less than 2 mm thick reducing light transmittance by approximately 80%. Thus in nature it is likely that these tropical algae adapt to the pertaining light intensity by varying their pigmentation and/or by aggregating together to cause self-shading, as Castenholz noted with some filamentous hot spring algae (5).

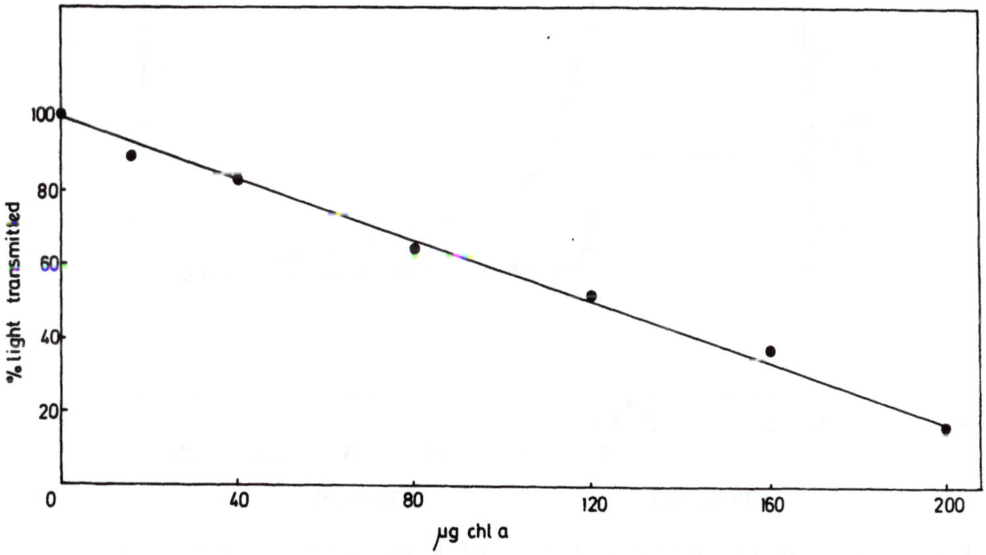

Fig. 7. Relative transmittance of 665 nm wavelength light through *Stigonema panniforme* mats. A value of 200 µg chl *a* corresponds approximately to an algal layer 2 mm thick. Light was supplied by a 100 watt incandescent bulb placed 15 cm from the algal mat.

Moisture

Desiccation or lack of moisture is a major factor affecting algal growth and nitrogenase activity by blue-green algae in temperate (12,13,25), tropical (26,37) and polar regions (7,14). However blue-green algae with their gel-like protoplasm and thick mucilaginous sheaths are able to absorb water extremely quickly when it is available and lose it much more slowly. Thus as Fig. 8 shows when air-dried *Stigonema panniforme* clumps are moistened they absorb several times their dry weight of water within 60 sec and take several hours to lose that absorbed water even on subsequent incubation at a relative humidity of 40%. Similar results were obtained with algae from Nigerian and Scottish soils.

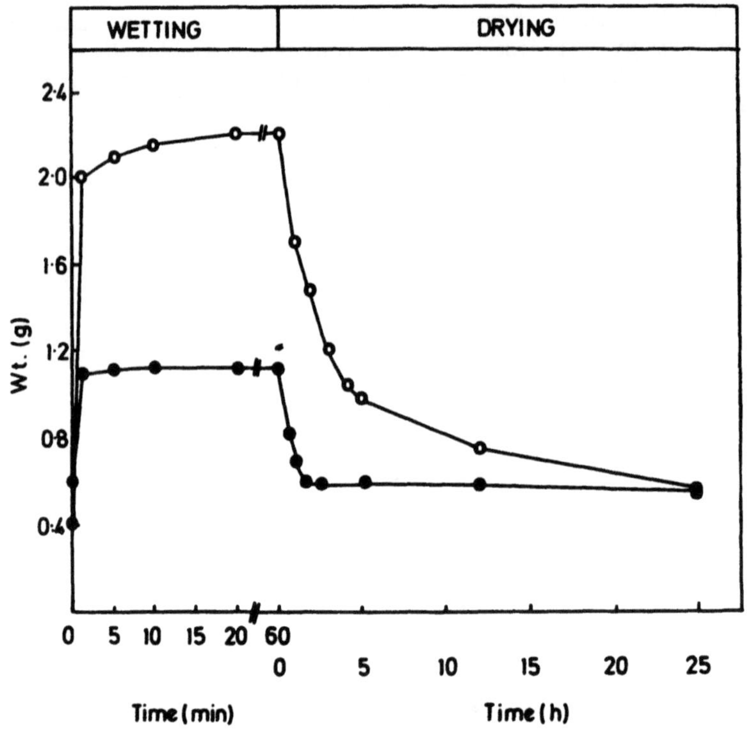

Fig. 8. Uptake and subsequent loss of water by *Scytonema* from Nigeria (●) and by *Stigonema panniforme* from Brazil (O). The samples were initially placed at a relative humidity of 40% until they equilibrated, they were then placed in water at 0 time and uptake of water measured by weighing of surface-dried samples at intervals thereafter. After 60 min they were removed from the water and the rate of water loss at a relative humidity of 40% monitored. The temperature throughout was 25°C.

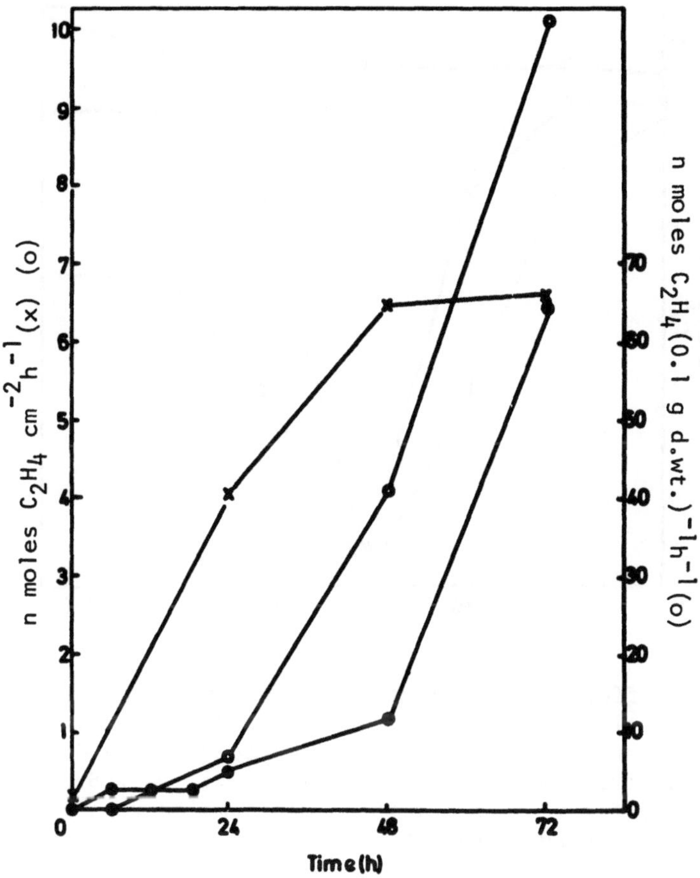

Fig. 9. Time course of acetylene reduction after rewetting air-dried samples of algae (*Nostoc* and *Anabaena*) from Scottish soils (**X**), *Scytonema* from Nigeria (**O**) and *Stigonema panniforme* from Brazil (●). The temperature was 25°C.

The results in Fig. 9 show further than on re-wetting air-dried and inactive algae, light-dependent nitrogenase activity restarts within 24 h even in the case of Nigerian and Brazilian samples which had been kept dry for several months before remoistening. Another factor which may be important, and which has been seldom considered in relation to moisture supply is the relative humidity of the atmosphere. Thus as Fig. 10 shows, Scottish soil algae retained under a relative humidity of 97.5% sustain an active nitrogenase when other factors are non-limiting, whereas algae exposed to lower relative humidities (87% and 75%) lose their activity within 60 h. Activity recovers just as quickly as it is lost however when the algae are subsequently returned to a high

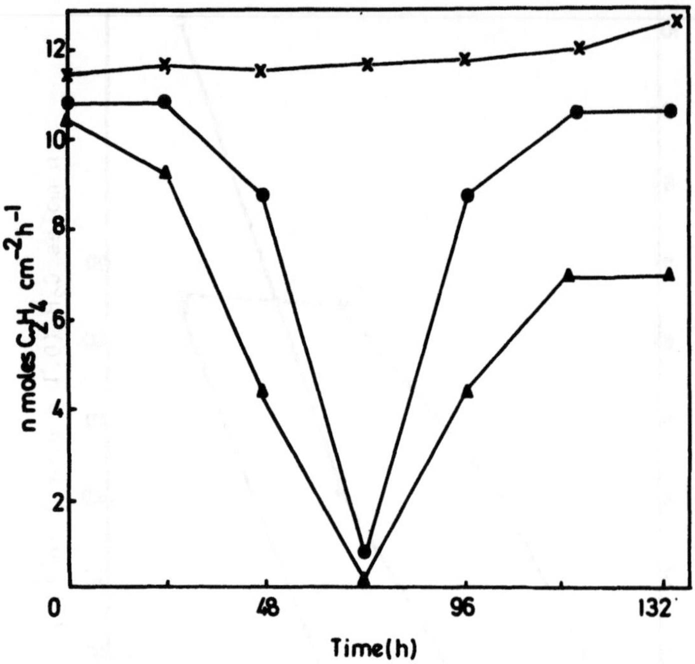

Fig. 10. The effect of different relative humidities on acetylene reduction by soil algae (*Nostoc* and *Anabaena*) from Scotland. The algae were incubated at the different relative humidities shown (X , 97.5; ● , 87.0; ▲ , 75.0) for 72 h, after which all samples were placed at a relative humidity of 97.5.

relative humidity. In natural ecosystems nitrogenase activity can be governed by soil moisture and/or the relative humidity of the atmosphere, and this is of importance in relation to the time at which field assays for nitrogenase activity are carried out. For example in studies on soil algae from Morocco we have found (W.D.P. Stewart and H.W. Pearson, unpublished) that *Nostoc* colonies showed nitrogenase activity in early morning when the relative humidity and soil moisture were high, but not around noon when the algae had become desiccated.

In connexion with the effect of moisture, it is of interest to note with soil crust algae from Nigeria, that on wetting them after a period of dryness, there is an immediate release of extracellular nitrogen, prior to the restart of nitrogenase activity. This production of extracellular nitrogen on change from one set of environmental conditions to another is similar in some respects to the findings of Jones and Stewart (16,17) for the marine *Calothrix scopulorum*.

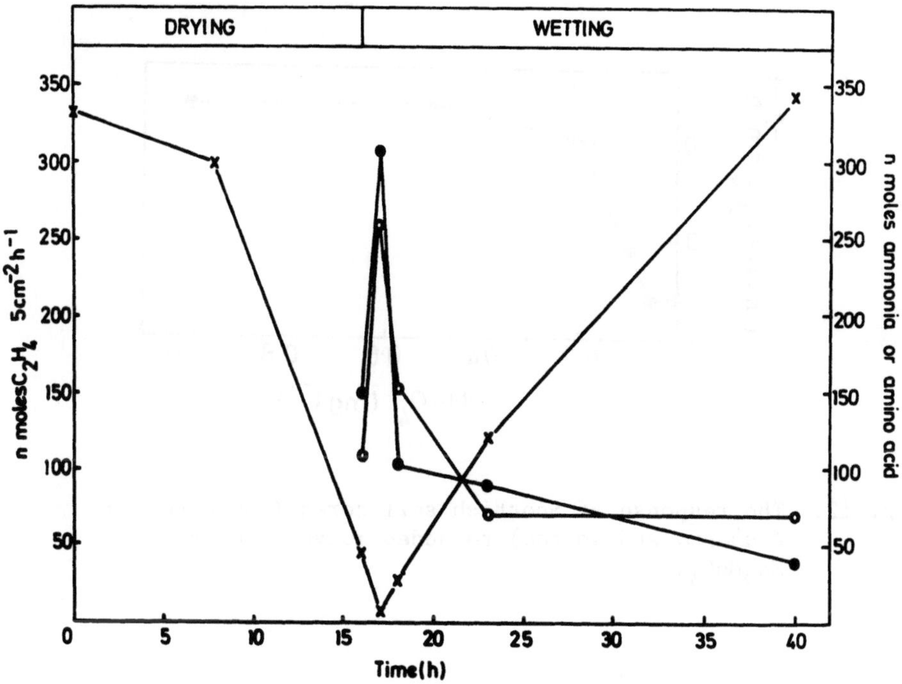

Fig. 11. The production of soluble nitrogen when dried *Scytonema* crusts from Nigeria are moistened X , acetylene reduction; O , soluble amino acids; ● , soluble ammonia (see 34 for methods). The experiment was carried out at 25°C.

Molybdenum

In tests on Scottish soils, data were also obtained on the responses of soil cores to molybdenum, an essential component of the nitrogenase enzyme (4). It was found that approximately one-quarter of the soils tested showed an increase in light-dependent nitrogenase activity when Mo concentrations as low as 0.1 p.p.m. were provided. In all cases 0.5 p.p.m. of added Mo was found to be saturating. Typical data are presented in Fig. 12. These data resemble those obtained by Wolfe (38) who found that the optimum molybdenum concentrations for N_2-fixation by *Anabaena cylindrica* was 0.2 p.p.m.

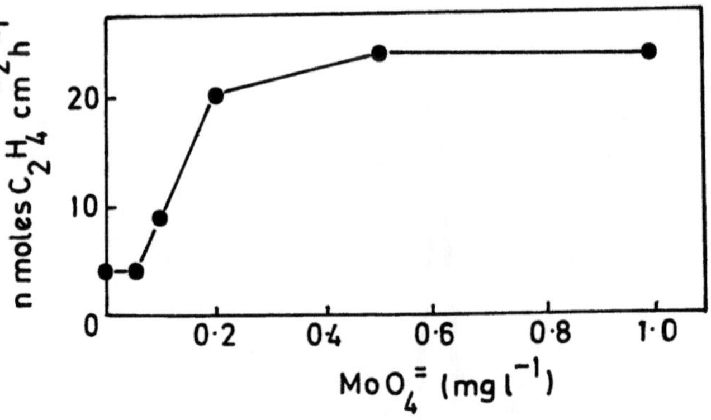

Fig. 12. The response of Scottish soil cores (dominated by *Anabaena* and *Nostoc*) to added molybdenum as Na_2MoO_4.

QUANTITATIVE SIGNIFICANCE OF N_2-FIXATION BY THE SOIL ALGAE

The quantitative significance of nitrogen fixation in any ecosystem depends on the total requirement of that ecosystem for nitrogen and on the relative contribution of biological nitrogen fixation, relative to other sources of nitrogen input. In tropical soils, for example, which are often characteristically poor in nitrogen, the input of a few kg nitrogen ha^{-1} ann^{-1} may be as important to the maintenance of the ecosystem, as are the additions of very much higher quantities of nitrogen by legumes in intensively cultivated agricultural land.

The data presented in this paper obtained using the C_2H_2 reduction assay, over short periods of time, provide information on whether or not nitrogenase activity is likely to occur in any habitat, but quantitative extrapolations of the data obtained to amounts of nitrogen fixed must be treated with extreme caution. Nevertheless it is important to obtain some indication of the order of magnitude of fixation in particular areas, and the following extrapolations attempt to do that. In general it is clear from our studies on Scottish soils that the overall input of biological nitrogen fixation by blue-green algae is very low, and assuming a 100-day period of fixation, each with activity occurring for a 16 h period and assuming a 3 : 1 ratio for the rates of C_2H_2 reduction : N_2 reduction then average rates of fixation calculated are of the order of 1 - 2 kg N ha^{-1} ann^{-1}. These values are very much lower than the values attributed to algae in the Broadbalk wilderness by Day *et al.* (6).

The data for Nigerian savanna soils are, on the whole, appreciably higher and assuming a 12 h period of fixation per day, a 3 : 1 ratio, as above, and 250 days of activity per year in the wet south-west of the country this could account for an annual input of combined nitrogen of approximately 3 g N $m^{-2}ann^{-1}$. In the dry north-east of the country, on the other hand, the input by algal crusts is probably around 0.3 g N $m^{-2}ann^{-1}$. According to Nye and Greenland (20) there is annual input in the soil-plant system of the tall-grass savanna of Nigeria of about 38 kg N $ha^{-1}ann^{-1}$ and in this connexion input of nitrogen from these crusts may be of considerable importance.

ACKNOWLEDGEMENTS

B.J. Harbott assisted in the study of the Scottish soil algae, which was carried out as part of the U.K. contribution to the International Biological Programme. We thank Gail Alexander for assistance with other aspects of the work.

REFERENCES

1. Alexander, V.A. (1975). In: Nitrogen Fixation by Free-Living Micro-organisms (ed. Stewart, W.D.P.) pp. 175-188. Cambridge University Press, Cambridge.

2. Allen, M.B. and Arnon, D.I. (1955). Pl. Physiol. Lancaster, 30, 366.

3. Becking, J.H. (1975). In: Proc. 1st Int. Symp. of Nitrogen Fixation (eds. Newton, W.E. and Nyman, C.J.) Vol. 2. pp. 556-580. Washington State University Press, Pullman, Washington, D.C.

4. Burns, R.C. and Hardy, R.W.F. (1975). Nitrogen Fixation in Bacteria and Higher Plants. Springer-Verlag, Berlin. 189 pp.

5. Castenholz, R.W. (1968). J. Phycol. 4, 132.

6. Day, J., Harris, D., Dart, P.J. and van Berkum, P. (1975). In: Nitrogen Fixation by Free-Living Micro-organisms (ed. Stewart, W.D.P.) pp. 71-84. Cambridge University Press, Cambridge.

7. Fogg, G.E. and Stewart, W.D.P. (1968). Br. Antarct. Surv. Bull., 15, 39.

8. Fogg, G.E., Stewart, W.D.P., Fay, P. and Walsby, A.E. (1973). The Blue-green Algae. 459 pp. Academic Press, London and New York.

9. Granhall, U. (1975). In: Nitrogen Fixation by Free-Living Micro-organisms (ed. Stewart, W.D.P.) pp. 189-198. Cambridge University Press, Cambridge.

10. Haystead, A. and Stewart, W.D.P. (1972). Arch. Microbiol. 82, 325.

11. Henriksson, E. (1971). Pl. Soil Sp. Vol., 415.

12. Henriksson, E., Henriksson, L.E. and Da Silva, E.J. (1975). In: Nitrogen Fixation by Free-Living Micro-organisms (ed. Stewart, W.D.P.) pp. 199-206. Cambridge University Press, Cambridge.

13. Hitch, C.J.B. and Stewart, W.D.P. (1973). New Phytol. 72, 509.

14. Horne, A.J. (1972). Br. Antarct. Surv. Bull., 27, 1.

15. Jones, M.J. and Wild, A. (1975). Tech. Comm. No. 55, Commonwealth Bureau of Soils, Harpenden.

16. Jones, K. and Stewart, W.D.P. (1969*a*). J. Mar. Biol. Ass. U.K., 49, 475.

17. Jones, K. and Stewart, W.D.P. (1969*b*). J. Mar. Biol. Ass. U.K., 49, 701.

18. Mague, T. (1977). In: A Treatise on Dinitrogen Fixation Section IV. Agronomy and Ecology (eds. Hardy, R.W.F. and Gibson, A.H.) Wiley Interscience, New York.

19. Moore, A.W. (1969). Bot. Rev., 35, 17.

20. Nye, P.H. and Greenland, D.J. (1960). Tech. Comm. No. 51, Commonwealth Bureau of Soils, Harpenden.

21. Peters, G.A. (1975). In: Proc 1st Int. Symp. on Nitrogen Fixation (eds. Newton, W.E. and Nyman, C.J.) Vol. 2, pp. 592-610. Washington State University Press, Washington, D.C.

22. Renaut, J., Sasson, A., Pearson, H.W. and Stewart, W.D.P. (1975). In: Nitrogen Fixation by Free-Living Micro-organisms (ed. Stewart, W.D.P.) pp. 229-246. Cambridge University Press, Cambridge.

23. Rippka, R., Neilson, A., Kunisawa, R. and Cohen-Bazire, G. (1971). Arch. Mikrobiol., 76, 341.

24. Rippka, R. and Waterbury, J.B. (1977). FEBS Letts. (submitted).

25. Shtina, E.A. (1972). In: Taxonomy and Biology of Blue-Green Algae (ed. Desikachary, T.V.) pp. 294-295. University of Mádras, Madras.

26. Singh, R.N. (1961). Role of Blue-green Algae in Nitrogen Economy of Indian Agriculture, 175 pp. Indian Council of Agricultural Research, New Delhi.

27. Singh, R.N. (1972). Physiology and Biochemistry of Nitrogen Fixation by Blue-green Algae. Final Technical Report, 1967-1972, 66 pp. Banaras Hindu University, Varanasi-5, India.

28. Stewart, W.D.P. (1967). Ann. Bot. N.S., 31, 385.

29. Stewart, W.D.P. (1971). Pl. Soil Sp. Vol., 377.

30. Stewart, W.D.P. (1973). A. Rev. Microbiol. 27, 283.

31. Stewart, W.D.P. (1975). In: Proc. 9th Europ. mar. Biol. Symp. (ed. Barnes, H.) pp. 637-660. Aberdeen University Press, Aberdeen.

32. Stewart, W.D.P. (1977). In: A Treatise on Dinitrogen Fixation, Section III Biology (eds. Hardy, R.W.F. and Silver, W.S.) pp. 63-123. Wiley Interscience, New York.

33. Stewart, W.D.P. and Lex, M. (1970). Arch. Mikrobiol. 73, 250

34. Stewart, W.D.P., Rowell, P. and Apte, S.K. In: Proc. 2nd International Symposium on Nitrogen Fixation (ed. Rodriguez-Barrueco, C.) Academic Press, London (in press).

35. Stewart, W.D.P., Rowell, P., Codd, G.A. and Apte, S.K. (1977). In: Proc. 4th Internat. Congress Photosynthesis, Reading, U.K. (in press).

36. Thuyet, T.O. and Tuan, D.T. (1973). In: Vietnamese Studies, Agricultural Problems, Vol. 4, pp. 119-127.

37. Venkataraman, G.S. (1975). In: Nitrogen Fixation by Free-Living Micro-organisms (ed. Stewart, W.D.P.) pp. 207-218. Cambridge University Press, Cambridge.

38. Wolfe, M. (1954). Ann. Bot. 18, 299.

CONTRIBUTION OF THE LEGUME-*Rhizobium* SYMBIOSIS TO THE ECOSYSTEM AND FOOD PRODUCTION

A.A. Franco

Programa Fixação Biológica de Nitrogênio
Convênio CNPq-EMBRAPA-UFRRJ, Km 47
23460 Seropedica, Rio de Janeiro, Brazil

Among the various biological systems which are able to fix atmospheric nitrogen, the symbiosis of Leguminoseae with *Rhizobium* seems to contribute most nitrogen to the ecosystem and to food production. Nitrogen fixation by the legumes amounts to 20% of the estimated biological N_2 fixed each year on earth, far more than any other single system, with figures similar to those of all the nitrogen fixed chemically by industry (56).

Leguminoseae have a worldwide distribution being one of the two flowering families with the highest numbers of species (5). Nodulated species are found in three of the four Leguminoseae sub-families. Nodulation is more common among Papilionoideae species followed very closely by Mimosoideae and Caesalpinoideae. Caesalpinoideae are more common in the tropics and present fewer nodulated species; no nodules were found in 80 species studied of tropical woody species of the Swartzoideae sub-family (67). It seems suprising that less nodulated species occur in the sub-families which are more common in the tropics. However insufficient work has as yet been carried out in the tropics, and most species examined here have been sampled from ecosystems in equilibrium.

Leguminous plants represent the only known crop that can be self sufficient in nitrogen nutrition and that may leave fixed nitrogen in residues in the soil. The contribution of legume N_2-fixation is also important in natural uncultivated systems. In tropical Africa and America the overall contribution of naturally growing legumes to the nitrogen economy is greater than that of cultivated species (23).

NITROGEN GAINS IN NATURAL SYSTEMS

Legumes are predominant in climatic savannas (Caatingas) and biotic savannas in Africa (9), providing most of the natural forage to feed cattle and goats during the dry season (43, 45). They are also predominant among the eight families found in edaphic savannas (Cerrado of Central Brazil, Chaparrados of Venezuela and Llanos of Colombia) (59). In the Amazon, forest legume trees are the most common species among the 30-40m high trees (20).

Nitrogen losses in climax savannas (10, 8, 37) and rain forests (7, 50) are usually minimal, although large amounts of nitrogen are cycling in the luxurious forest associations. In rain forests 250 kg N/ha is returned annually to the soil (4, 29) compared with 80 kg N/ha per year in dry savanna (4). Losses of nitrogen in these systems are small when compared with K, P, and Mg. In biotic savannas burning and leaching contributes considerable to loss of nitrogen. Greenland and Nye (29) estimated fixation of 23 to 58 kg N/ha annually for lowland forests, 17 to 45 kg for highland forests and 2 to 15 kg for savannas. Bonnier and Brakel (7) consider the legume symbiosis a product of nitrogen imbalance, and attribute the common lack of nodules in climax forests to the continuous recycling of nitrogen. The predominance of nodulated species on newer soils adjacent to rivers where equilibrium has not been reached, shows the importance of the legumes for natural ecosystems. The buildup of secondary forest is another example of subequilibrium. Mean gains of 40 to 50 kg N/ha annually were observed in a 50 year old secondary forest in Ghana and more than 100 kg N/ha in West Africa (19, 30).

GRAIN LEGUMES IN TROPICAL AGRICULTURE

Among all the major crops, soybean (*Glycine max*) presents the highest protein productivity e.g., 9.1 kg protein/ha daily compared with 4.5 for lima bean (*Phaseolus lunatus*), 2.7 for cowpea (*Vigna unguiculata*), 1.9 for rice (*Oryza sativa*), 1.6 for maize (*Zea mays*), and 0.6 for cassava (*Manihot esculenta*) (Table 1). One hectare of land cultivated with soybeans can produce enough protein to feed one person for 5500 days, on USA standards, while meat protein produced in the same area is only enough for 185 to 620 days.

Soybean production in sub-tropical and tropical regions is increasingly rapidly. In Brazil this crop is self sufficient in nitrogen and is currently competing with coffee as the most important export crop. Estimates of N_2 fixation by soybean vary between 40 and 206 kg N/ha annually (12, 34, 68), and reach up to 400 kg experimentally by increasing the CO_2 concentration in the

TABLE 1

Protein productivity of certain tropical food crops (42)

Crop	Estimated Yield (kg/ha)	Protein Content (%)	Crop duration (days)	Protein productivity (kg/ha/day)
Leguminous				
Soybean	2,800	38	95	9.1
Lima bean	3,200	25	115	4.5
Cowpea	1,800	25	80	3.3
Peanut	1,600	26	120	2.7
Winged bean	1,400	31	112	3.0
Chick pea	2,500	20	125	2.7
Mung-bean	900	24	75	1.9
Root Crops				
Sweet Potato	20,000	1.3	120	1.4
Potato	15,000	2.0	125	1.3
Cassava	20,000	1.2	220	0.6
Cereal				
Rice	5,000	7.5	140	1.9
Maize	4,000	9.5	120	1.6
Sorghum	3,500	10.1	110	1.5

air (36). Integration of acetylene reduction assays indicates that only 40% of the total plant nitrogen is derived from the symbiosis, while only 16% came from N_2-fixation when 150 kg N/ha was supplied as fertilizer (26). Using nodulating and nonnodulating isolines Pal (53) found values up to 83% for total plant nitrogen derived from the symbiosis and up to 32% in plants receiving 200 kg N/ha. However, more important than the overall participation of the symbiosis is the nitrogen derived from biological fixation at seed filling stage (26, 66). More than 200 kg N/ha was necessary for nonnodulating isolines to reach the yield of well-nodulated isolines (6, 54).

Peanut (Arachis hypogea) is second to soybean in the total world production of legume grain, mainly for oil production. Peanuts nodulate effectively with about 30% of the cowpea rhizobia

(2, 61). Field grown plants generally show good nodulation (61). However evaluating the effectiveness of this association has as yet received little attention. Yields of 5600 kg/ha have been observed on light soils (46), although the Brasilian average is 1308 kg/ha and the Indian only 652 kg/ha (21).

Dry beans (Phaseolus vulgaris) are the most important legume crop for human consumption in South America. Brazil is the largest bean producer but still needs to import to meet national needs. The bean plant is very sensitive to pests and diseases. The symbiosis of beans is especially sensitive to temperature, water stress and to soil acidity. The specific requirement of this plant for Mo in acid soils is discussed elsewhere in this symposium (24). Large differences among Rhizobium phaseoli strains have been observed in relation to tolerance to manganese toxicity and high temperatures (17, 22, 32). The host has specific Rhizobium requirements and inoculation is essential (7, 61), especially to guarantee early nodulation and nitrogen fixation (27).

In Table 2 the yields and estimated nitrogen fixation for some of the major grain legumes are presented. Cowpea, pigeon pea (Cajanus cajan) and Phaseolus aureus nodulate with cowpea type rhizobia. Chick pea (Cicer arietinum) is important for semi-arid regions and does have specific Rhizobium requirements.

Besides the species discussed, Rachie (57) listed several potentially important legumes: bambara ground nuts (Voandzeia subterranea), guar (Cyamopsis tetragonoloba), hyacinth beans (Lablab niger), rice beans (Phaseolus calcaratus), winged beans

TABLE 2

Yields and estimated N_2 fixation of the major tropical grain legumes in kg/ha

Legume	World Average in Developing Countries	Reported High Yields	References	Estimated N_2-fixed	References
Soybean	1407	4000	(11)	40-206	(12,34,68)
Bean	454	4310	(28)	-	-
Peanut	588	5500	(46)	49	(51)
Pigeon pea	531	5000	(58)	90-150	(12,70)
Cowpea	201	3400	(3)	90-354	(1,69,70)
Chick pea	636	3500	(58)	41-270	(51)
P. aureus	-	2500	(58)	224	(1)

(Psodocarpus tetragonolobus) and velvet beans (Mucuna pruriens). Okigbo (52) listed several other edible legumes used in Southern Nigeria and Pinchinat (55) described several species used in tropical America. Amongst these the potential of winged beans should be stressed; these are being studied in several parts of the world. Under experimental conditions yields of 4590 kg/ha were obtained (71).

FORAGE LEGUMES IN TROPICAL AGRICULTURE

The results obtained up to now have shown that tropical legumes can produce as much as clover and medics in temperate regions. In areas with pronounced dry seasons, the limitations imposed by water deficiency may be difficult to overcome. In humid tropical areas on the other hand, nitrogen fixation can be continuous all year round and higher rates of N_2-fixation than in temperate regions could be expected. However, for most of the areas under these conditions, other limitations may affect yields so that tolerance to low fertility and acid soils becomes more important. Stylosanthes spp. is a good example of a legume adapted to these conditions. Stylosanthes roots are heavily infected with mycorrhiza (48), and the plant is less sensitive to nutritional deficiencies and acid soils (14). However some of the best cultivars have problems with delayed nodulation and are very specific in rhizobial requirements (25, 44, 56).

Glycine wightii cultivars differ considerably, with tetraploid cvs. tending to nodulate easier and earlier than diploid cvs. (15). Under field conditions it is seldom well nodulated, and inoculation does not assist since nodulation problems are not due to Rhizobium specificity (12, 18, 41, 49) but rather to environmental factors such as phosphorus deficiency, manganese toxicity (16, 64) or high soil temperature (63).

Commercial Centrosema pubescens seed contains sparsely and profusely nodulating lines (62). This species is relatively specific in Rhizobium requirements, but is appropriate for low elevation pastures and has greater drought resistance than Pueraria phaseoloides. Nitrogen fixation of 520 kg/ha annually has been recorded for C. pubescens (47). Siratro (Macroptilium atropurpureum) nodulates profusely with cowpea type rhizobia and associates very well with several grasses. Siratro accounted for 25% of a hill pasture in association with Millinis minutiflora after 3 years of fertilization with phosphorus and micronutrients. With only phosphorus fertilization it almost disappeared (14).

Several other legumes are used for forage production, e.g., Desmodium, Galactia, Lablab, Indigofera, Lotononis, Cajanus, Arachis, Leucaena, etc. Special mention should be given to the forage legume tree Leucaena leucocaephala that has been used extensively in Australia (38). It grows in prolonged dry periods (38, 60) and furnishes high protein forage (30%) all year (39). If the highly specific Rhizobium requirements are met (8, 40, 49, 61) it has great potential for areas with long dry periods. Nitrogen incorporation from soil and biological fixation annually of 800 kg/ha has been obtained with L. latisiligua cv. Peru (31).

The use of legumes can improve considerably the protein yields of grass pastures, e.g., 113% yield increases of elephant grass (Pennisetum purpureum) associated with M. atropurpureum cv. siratro have been observed when compared with a pure stand of the grass (13). It was necessary to add 270 kg N/ha of Chloris gayana to equal the protein yields of the grass associated with S. guyanensis (33).

Several tropical legumes are highly specific in their requirements for rhizobia and these require special attention. Comprehensive soil fertility studies to define limiting factors and adaptation of the species under stressed conditions are major points to be considered to ascertain the full potential of N_2-fixation in tropical legumes.

REFERENCES

1. Agboala, A. A. and Fayemi, A. A. A. (1972) Agron. J. 12, 409.

2. Allen, O. N. and Allen, E. K. (1940) Bot. Gaz. 102, 121.

3. Anonymous (1974) World Crops, 26, 281.

4. Basilevic, N. I. and Rodin, L. E. (1966) Forest. Abstr., 27, 357.

5. Benson, L. (1957) Plant classification: Boston. D.C. Heath and Company.

6. Bhangoo, M. S. and Albritton, D. J. (1975) Ark. Farm Res. 24, 6.

7. Bonnier, C. and Brakel, J. (1969) Lutte Biologique contre la Faim, J. Duculat, S. A., Gembloux.

8. Campelo, A. B. and Campelo, C. R. (1970) Pesq. agropec. bras. 5, 333.

9. Campelo, A. B. and Döbereiner, J. (1977) In: A treatise on dinitrogen fixation Section IV: Agronomy and Ecology, R. W. F. Hardy and A. H. Gibson, John Wiley & Sons, New York, 191.

10. Coaldrake, J. E. (1962) Commonw. Bur. Pasture Field. Crops Bull. 46, 25.

11. Cooper, R. L. (1971) Agron. J. 63, 490.

12. Date, R. A. (1973) Soil Biol. Biochem. 5, 5.

13. De-Polli, H., Franco, A. A. and Almeida, D. L. (1973) Bol. Tech. Inst. Pesq. Exp. Agropec. Centro Sul, no 104.

14. De-Polli, H., Carvalho, S. R., Lemos, P. F. and Franco, A. A. (1977) Intern. symp. on Biol. nitrogen fixation in the tropics, Brasilia, Brazil.

15. Diatloff, A. and Ferguson, J. E. (1970) Trop. Grassl. 4, 223.

16. Diatloff, A. and Luck, P. E. (1972) Trop. Grassl. 6, 33.

17. Döbereiner, J. (1966) Plant Soil 24, 153.

18. Döbereiner, J. and Coser, A. C. (1970) X Int. Congr. Microbiol., Mexico, Abst. 220.

19. Dommergues, Y. (1963) Bois Far. Trop. 87, 9.

20. Ducke, A. and Black, G. A. (1953) Ann. Acad. Bras. Cienc. 25, 1.

21. F. A. O. (1974) Production Yearbook, Rome, Vol. 28, 1.

22. Fonseca, O. O. M. da and Franco, A. A. (1977) Unpublished results.

23. Foury, A. (1950) Les Cahiers de la Rech. Agron. (Robat, Morocco), 3, 25.

24. Franco, A. A. (1977) Micronutrient requirements of legume Rhizobium symbiosis in the tropics. Int. Symp. on the limitations and pot. of biol. nitrogen fixation in the tropics, Brasilia, Brazil.

25. Franco, A. A. and Vincent, J. M. (1976) Plant Soil 45, 27.

26. Franco, A. A., Fonseca, O. O. M. da, Marriel, I. E. (1977) Rev. Bras. Cienc. Solo (In Press).

27. Franco, A. A., Neyra, C. A. and Pereira, J. C. (1977) Inter. symp. on the limitations and pot. of biol. nitrogen fixation in the tropics, Brasilia, Brazil.

28. Graham, P. (1977) In: Sistema de produccion de frijol, CIAT, Programa de frijol, A-43.

29. Greenland, D. J. and Nye, P. H. (1959) J. Soil Sci. 9, 284.

30. Greenland, D. J. and Kowal, J. L. M. (1960) Plant Soil 12, 154.

31. Gomez, A. A. and Zandstra, H. G. (1976) In: Exploiting the legume-Rhizobium symbiosis in tropical agriculture, J. M. Vincent, A. S. Whitney and J. Bose (eds.), Univ. Hawaii Misc. Publ. 145, 81.

32. Guss, A. and Döbereiner, J. (1970) An. V Reun. Lat.-Am. Rhizobium, Rio de Janeiro, 193.

33. Haggar, R. J. (1971) J. Agr. Sci. (Cambridge) 77, 427.

34. Hardy, R. W. F., Burns, R. C., Herbet, R. R., Holsten, R. and Jackson, E. K. (1971) Plant Soil Special Vol. 561.

35. Harkness, K. A. (1967) New horizons in food production, Dept. Agric. Engineering, Ohio State Univ., Columbus (Mimeo).

36. Havelka, U. D. and Hardy, R. W. F. (1974) In: Proc. 1st Int. symposium on nitrogen fixation, W. E. Newton and C. J. Nyman (eds.), Washington State University Press, p. 456.

37. Hutjens, J. D. M. (1971) Plant Soil 34, 393.

38. Hutton, E. M. (1964) Commonw. Bur. Pasture Field Crops Bull. 47, p. 79.

39. Hutton, E. M. and Bonner, I. A. (1960) J. Aust. Inst. Agr. Sci. 26, 276.

40. Ishizawa, S. (1954) J. Sci. Soil Manure, Japan 24, 1.

41. Lopes, E. S., Lovadini, L. A. C., Gargantini, H., Miyasaka, S. and Leon, J. C. C. (1970) Ann. V. Reun. Lat.-Am. Rhizobium, Rio de Janeiro, p. 266.

42. Luse, R. A. and Okwuraiwe, P. E. (1975) In: Proc. IITA, Collaborators' meeting on grain legumes improvement, R. A. Luse and K. O. Rachie (eds.), IITA, Ibadan, Nigeria, 98.

43. Malate-Berliz, J. V. C. and Pereira, J. A. (1965) Proc. Int. Grassland Congr. IX, S. Paulo, Brazil, Vol. 2, 1299.

44. Mannetje, L.'t. (1969) Aust. J. Bot. 17, 553.

45. Marques, J. R. de A. (1965) I. Congr. Pan-Am. Conserv. Solo, S. Paulo, 777.

46. Marquette, J. (1966) Agron. Trop. (Paris) 21, 1148.

47. Moore, A. W. (1967) Emp. J. Exp. Agr. 30, 239.

48. Mosse, B. (1976) In: Exploiting the legume-Rhizobium symbiosis in tropical agriculture. J. M. Vincent, A. S. Whitney and J. Bose (eds.), Univ. Hawaii Misc. Publ. 145, 275.

49. Norris, D. O. (1967) Trop. Grassl. 1, 107.

50. Norris, D. O. (1969) Trop. Agr. (Trinidad), 43, 265.

51. Nutman, P. S. (1971) Sci. Prog. Oxf., 55.

52. Okigbo, B. N. (1976) In: Exploiting the legume-Rhizobium symbiosis in tropical agriculture, J. M. Vincent, A. S. Whitney and J. Bose (eds.), Univ. Hawaii Misc. publ. 145, 98.

53. Pal, U. R. (1975) Acta Agronomica Academiae Scientiarum Hungaricae 24, 430.

54. Pal, U. R. and Saxena, M. C. (1975) Expl. Agric. 11, 221.

55. Pinchinat, A. M. (1976) In: Exploiting the legume-Rhizobium symbiosis in tropical agriculture. J. M. Vincent, A. S. Whitney and J. Bose (eds.), Univ. Hawaii Misc. publ. 145, 171.

56. Quispel, A. (1974) In: The biology of nitrogen fixation, A. Quispel (ed.), North-Holland Publ, Co., Amsterdam, p. 1.

57. Rachie, K. O. (1973) Series Seminars 2E, CIAT, Columbia, 123.

58. Ramanujan, S. (1973) Proc. First IITA grain legume improvement workshop, Ibadan, Nigeria, 37.

59. Rizzini, C. T. (1963) Simp. sobre o Cerrado, Univ. S. Paulo, Brazil, p. 127.

60. Rocha, G. L. da, Werner, J. C., Mattos, H. B. and Pedreira, J. V. S. (1970) An. Sem. Metodol. Planej. Pesq. Legum. Trop., Rio de Janeiro, p. 1.

61. Sandmann, W. P. L. (1970) Rep. Grassl. Res. Stat. Marandellas, Rhodesia.

62. Serpa, A. and Cunha Filho, L. A. (1970) An. V. Reun. Lat.-Am. Rhizobium, Rio de Janeiro, p. 1.

63. Souto, S. M. and Döbereiner, J. (1968) Pesq. agropec. bras. 3, 215.

64. Souto, S. M. and Döbereiner, J. (1969) Peq. agropec. bras. 5, 59.

65. Souto, S. M., Coser, A. C. and Döbereiner, J. (1970) An. V. Reun. Lat.-Am. Rhizobium, Rio de Janeiro, p. 78.

66. Thibodeau, P. S. and Jaworski, E. G. (1975) Planta (Berl.) 127, 133.

67. Vincent, J. M. (1974) In: The biology of nitrogen fixation, A Quispel, North Holland Publ., Amsterdam, p. 265.

68. Weber, D. F., Coldwell, B. E., Sloger, C. and Vest, H. G. (1971) Plant Soil, Special Vol., 293.

69. Wetselaar, R. (1967) Aust. J. Exp. Anim. Husb. 7, 518.

70. Wetselaar, R., Jakobsen, P. and Chaplin, G. R. (1973) Soil Biol. Biochem. 5, 35.

71. Wong, K. C. (1975) In: Proc. of UMAGA/FAUM Conference on Malaysian food self sufficiency, Univ. Malaysia, Kuala Lumpur.

PLANT INFLUENCE IN SYMBIOTIC NITROGEN FIXATION

Joachim F.W. von Bülow

Departamento de Engenharia Agronômica
Universidade de Brasília
Brasília, D.F., Brazil

One of the preoccupations of people involved in problems of agriculture is to enable farmers to use the cheapest, most plentiful and most efficient sources of nitrogen in the most profitable way for crop production. If one assumes a sugar cane production of 80t/ha fresh weight with 2% N-content, plants must absorb 1600 Kg/ha N every year. Although sugar cane is a long-season C_4 species, more efficient in photosynthesis than most crops, maximum photosynthesis is only possible with adequate N absorption. A non-limiting N supply is possible with reasonable energy costs, if one thinks in terms of the amount of coal consumed by a modern NH_3 industrial plant. However, social and environmental problems caused by pollution will be increasingly difficult. And what about the nature of our present crop plants which evolved under domestication with marginal supplies of mineral nitrogen? This is a question to concern the agronomists and the plant breeders. Adaptive features of our crop plants with limited N-supply are: first, N-nutrition controls the rate of root/shoot growth; second, reduced N is remobilized from older leaves to the reproductive system; third, the balance of reproductive growth/vegetative growth is shifted to reproduction. Such traits may have been accentuated during selection by man for high growth rates and gigantisms. Whereas nutrient and water deficits may act as controlling limits to desired patterns in growth and development, the same genotypes in luxury environments respond with decreasing efficiency (excessive vegetative growth) of which there are many examples: cotton, cassava, maize, etc. (52).

To make crops more responsive to non-limited mineral fertilizer, Japanese and American plant breeders have selected dwarf wheat and rice cultivars, capable of more efficient use of huge amounts of

N-fertilizer. Basically, this is what the "Green Revolution" is all about. Since then, energy problems have arisen and pollution has grown extremely difficult to control. With food prices rising rapidly, millions of poor people are undernourished with all the negative and mostly irreversible consequences for human physical and intellectual capabilities!

Biological nitrogen fixation is contributing to produce food more cheaply and with fewer social and environmental difficulties. A great deal of knowledge has been accumulated about the symbiosis of legumes with N_2-fixing bacteria (60). It has been discovered and confirmed by $^{15}N_2$ incorporation, that in tropical environments grass-bacteria symbiotic associations are also able to fix atmospheric nitrogen (23, 19). However, very little has been done for raising biologically fixed nitrogen output. Genetic manipulations are underway trying to transfer nif genes from one organism to another. Improvement of the N_2-fixing ability through genetic screening of both plants and bacteria is the way to take advantage of existing variability and potentials of favourable host-bacteria combinations. New variability can be obtained through hybridization and mutation breeding.

This paper represents an attempt to review some of the recent work on legume and grass N_2 fixation which is of interest for the agronomist and plant breeder. Genetic improvement depends on the existence of genetic variability and the possibility for identifying superior genotypes. Many environmental factors and plant traits interfere with the selection work. It is hoped, that discussing the possibilities and difficulties, will stimulate interest and support for this important field of research.

VARIABILITY OF N_2 FIXATION

Genetic variability of N_2-fixation efficiency exists as demonstrated by inter and intra-specific differences. The most complete collection of germplasm for each species involved is one necessary condition for the selection of more efficient and, at the same time, agronomically desirable genotypes. New variability may be produced through mutation breeding, hybridization, and genetic manipulations. The plant breeder must use different methods according to whether species are cross or self pollinators, self compatible or self incompatible.

Inter-specific Differences

Legumes. Among Leguminosae cross-inoculation group specificity is a separating factor, but symbiotic "promiscuity" exists and is related to cross-pollination of host species (81, 38). Pea root nodules have been found which contain more than one Rhizobium species. Resistance of plants to infection by their specific bacteria has to be taken into account. Host genes causing resistance or ineffective symbiosis are mostly recessive (59, 60) and can be selected against, with elimination in case of selfing. Different ploidy levels affect nodulation of different species in different ways. The same is true with a number of nodules, which is controlled by major host genes in some species, but usually is a polygenic trait (59).

Grasses. Grasses in sub-tropical and tropical regions are mostly associated with Azospirillum lipoferum. Paspalum notatum is associated with Azotobacter paspali (25). Roots from forage grasses and grain crops have been sampled and assayed in different ways for nitrogenase activity (21). However the actual average amounts of N_2-fixed per ha are not yet defined. Early specific N_2-ase activity data measured as nmoles C_2H_4/(h x g of dry roots) (Table 1) show large apparent differences between species and locations (11, 23, 21).

TABLE 1

Nitrogenase activity (C_2H_2) in forage and grain Gramineae species. Values obtained from excised roots during a preliminary survey near Rio de Janeiro, Brazil (24).

Species	nmoles C_2H_4/(h x g dry roots)	
	minimum	maximum
Brachiaria mutica	150	750
B. rugulosa	5	150
Cynoden dactilon	17	269
Digitaria decumbens	21	404
Hyparrhenia rufa	20	30
Milinis minutiflora	13	41
Panicum maximum	20	299
Paspalum notatum.	2	283
Pennisetum purpureum	5	954
Zea mexicana (Jutiapa)	0	303
Z. mexicana (Bolsas)	0	717
Z. mays (cultivars)	74	1,260
Sorghum bicolor (cultivars)	9	155
Triticum aestivum (cultivars)	48 (average)	
Pennisetum americanum (cultivars)	0	215
Eleusine caracana (cultivars)	0	77

Inter and Intra-Variental Differences

Legumes. There are reports about varietal differences of N_2 fixation within nodules. The genetics of ineffective types has been studied mainly in Trifolium and Lotus species. Genetic knowledge of deficient or resistant host mutants (59, 60) is important for studying the nature of N_2-fixation and specific bacterial strain-host interactions, and will be of direct or indirect use in breeding experiments. For instance, the non-nodulating soybean mutant, no no, is now available in isogenic lines of several cultivars, and it serves as a useful check in fixation experiments (7, 51).

Breeding for improvement of N_2-fixation efficiency within heterogeneous populations should also be possible for quantitative traits such as: early or late nodulation, number of nodules, and efficiency of N_2-fixation within nodules. Using all root:shoot combinations of 8 soybean genotypes in reciprocal intervarietal grafting, it has been possible to prove significant effects of the shoot genotype, sampling date, and the interaction of shoot genotype with sampling date, number, fresh weight and acetylene reduction activity of nodules. Root genotypes were independent of shoot genotypes, except for nodule fresh weight. Root genotype effects on total nodule activity per plant is due to inherent differences in specific nodule activity (50). A significant effect of root genotypes on seed yield was related partly to the N_2-fixing ability (49). In Stylosanthes guyanensis, differences in seed isozymes among plant genotypes could be used to predict nodulation effectiveness of Rhizobium strains.

Grasses. Early observations were obtained on Paspalum notatum (25), and apparent differences in N_2-ase activity of field extracted roots were obtained among different genotypes within other forage grass species as well (22). Cultivars and hybrids of Digitaria decumbens, assayed from intact soil-plant cores, sampled from a field experiment, yielded mean estimates significantly different among genotypes, ranging from 234 to 970 g N fixed per ha x day (22, 17). Estimates of nitrogenase activity have been obtained from several grain crops, including Zea mays, Sorghum bicolor, Triticum aestivum and Pennisetum americanum. Apparent differences exist among races of Zea mexicana (Table 1) (11, 24, 22, 13).

A corn variety diallele cross experiment was grown without N-fertilizer during the hot and cool seasons of 1976, near Rio de Janeiro. N_2-ase activity (C_2H_2) of roots, excised during flowering and late grain-filling growth stages, was different for genotypes at the two growth stages. Average heterosis was 134% in relation to the variety average (100%). The relationship between grain yield and N_2-ase activity was non-significant: r = -0.27 (Table 2, Fig. 1) (13). A tentative mass and S_1-line selection

TABLE 2

N_2-ase (C_2H_2) activity averages on an individual plant basis for a 4-variety diallele cross experiment of corn (Zea mays L.) during the hot and cool seasons of 1976. Samplings were done at the silking and grain-filling stages, except for the 3rd planting where samplings were done only at the grain-filling stage (August 1976). Experimental field of the UFRRJ, Brasil (13 and unpublished).

Genotypes	Grain Yield (Kg/ha)	nmole of C_2H_4/(h x g dry roots)					
		Jan. 76	Feb. 76	June 76	July 76	August 76	Average
1 Piranão	1944	354	272	121	339	212	260
1 x 2	1625	286	262	118	258	167	218
1 x 6	1917	278	301	155	341	116	238
1 x 8	1403	495	413	127	363	207	321
2 UR - I	1944	88	191	100	165	64	122
2 x 6	1458	119	233	81	161	107	140
2 x 8	1082	358	238	72	244	92	201
6 Comp. Amplo	2069	128	217	63	231	141	156
6 x 8	1417	355	356	97	406	134	270
8 Cateto Nort.	792	420	153	110	383	218	257
Average	1565	288	264	104	289	146	218

for N_2-ase activity was carried out in two corn varieties. Evaluation experiment results show an increase of N_2-activity after one cycle of S_1-line selection and no response at all for the mass-selected variety. Data obtained at various levels of N-fertilizer indicate significant interactions among grain production and N_2-ase activity (Fig. 2, unpublished).

Inbreeding and testing for N_2-ase activity has been done within cultivar UR-I for several generations. Results of screening among S_1-lines in replicated plots during the hot season of 1974/75 near Rio de Janeiro have been published (11). Only one line, which showed extremely high N_2-ase (C_2H_2) activity, was significantly different from other lines of the same variety after an operationally feasible pre-screening process which admitted a reasonable number of lines, 17 among a total of 276, for more detailed sampling. Inbreeding and sampling of these lines has been done ever since.

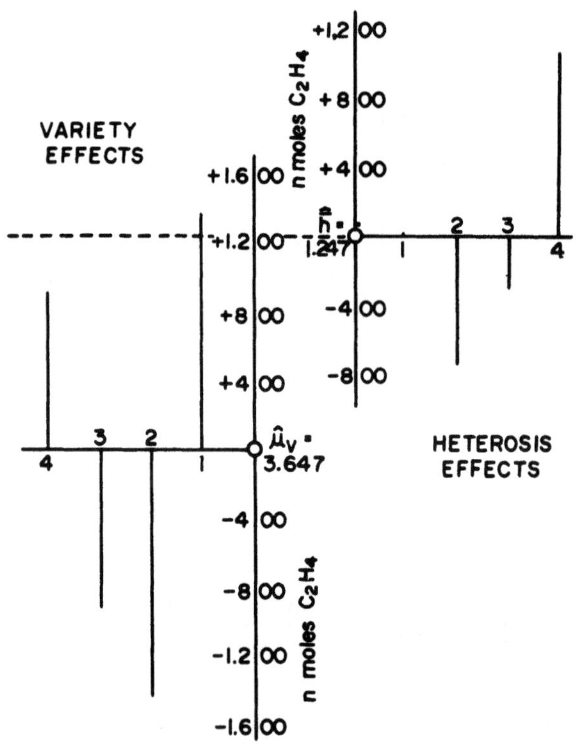

Fig. 1. Demonstration of variety and heterosis effects for N_2-ase activity (C_2H_2) in a 4-variety corn diallele cross experiment. Diallele analysis based on totals per genotype. Experiment grown during hot season, experimental field of the UFRRJ, Brasil (13).

TRAITS CORRELATED WITH N_2-FIXATION

The behavior of traits correlated with N_2-fixation, their genetic control and breeding value, would be extremely useful for the agronomist and breeder. Furthermore it may be easier to apply some selection index in addition to or instead of direct selection for N_2-fixation.

Photosynthetic Efficiency

Efficient photosynthesis and absorption of minerals from the soil ideally should satisfy completely all plant needs for organic synthesis and energy. Requirements of energy for symbiotic N_2-fixation are rather high and must be provided in addition.

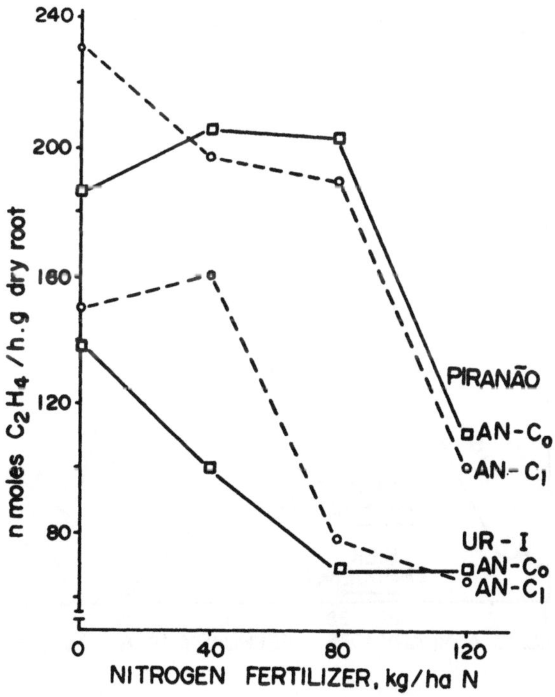

Fig. 2. Effects of 3 nitrogen fertilizer levels and 1 cycle of selection for higher N_2-ase activity on the N_2-ase activity of two corn varieties during the cool season of 1976. Experimental field of the UFRRJ, Brasil (unpublished).

Legumes. In a field experiment with two soybean cultivars, the acetylene reduction assay was performed at 7-10 day intervals, starting at flowering. At the end of flowering, treatments were: supplemental light, 25% shade, 50% depodding, 60% defoliation and control. The enhancing of the source:sink ratio (more light and depodding) maintained nodule activity well above the control in both varieties. Reduction of the ratio (shading and defoliation) decreased nodule activity below the control. In the control, the assimilate supply to nodules during podfilling was inadequate (49). Experiments with algae and different crop plants (64, 1, 44) provide further knowledge about the variability of the photosynthetic process.

Genotype differences for water-soluble carbohydrate accumulation in soybean plants were small among cultivars. There was an increase within leaves and petioles between 09:00 and 13:00 h daily. Soluble carbohydrates in plant parts decreased as seed developed, and may limit the seed-filling process (26). For apparent leaf photosynthesis and photosynthate translocation, differences were found among 16 soybean cultivars belonging to four different maturity groups. The order of cultivars was not consistent in growth stages, flowering, mid-pod fill and late pod-fill (2).

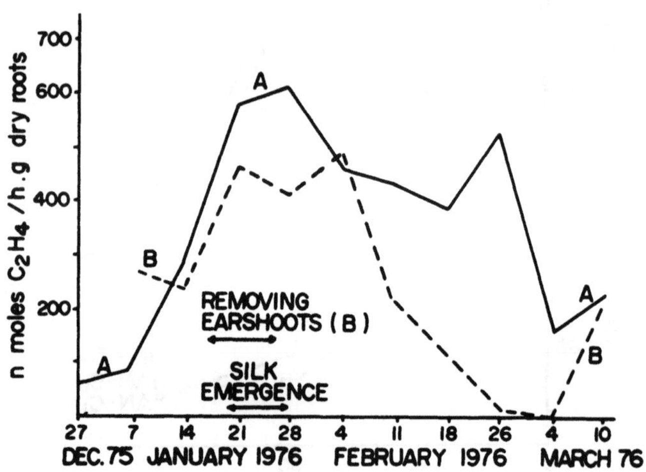

Fig. 3. N_2-ase activity (C_2H_2) of excised roots sampled weekly from a field of a brachytic corn variety, Piranão, during the vegetative and reproductive growth stages. In treatment "B" ear shoots were removed just before silking. Experimental field of the UFRRJ, Brasil (63).

Grasses. In corn, leaf senescence occurs during grain development (27), however corn possesses an intermediate sink, the culm, from which assimilates are translocated to the ear (grain filling) and to the root system (synthesis of organic compounds and N_2-fixation). Nitrogen nutrition is sometimes limiting to carbon dioxide assimilation in maize (9).

In a field experiment with the brachytic cultivar "Piranão," grown in the 1976/77 hot season near Rio de Janeiro, one treatment consisted of removing the earshoots just before silking. When ears were removed before silking, the N_2-ase activity decreased in relation to the control and no second peak, which normally would be expected at grain-filling, appeared. In treatments which were not sprayed with ammonium molybdate solution, it was found that N_2-fixation (C_2H_2) apparently is limited in a way different from *Glycine* or *Phaseolus*: N_2-ase activity increases or decreases within certain minimal needs of the plants' final photosynthetic sink, represented by the developing caryopses of the ear (Fig. 3).

Vegetative and Reproductive Growth Stages

Studying N_2-ase activity of a species at regular intervals during its whole life cycle provides data which may indicate the best possible moment for the plant breeder to take samples in the field. It should be desirable to sample at growth stages when activity of N_2-ase is highest. On the other hand, screening among seedlings would offer greater numbers for sampling, and work can be done in better controlled conditions. This would be advantageous if seedling efficiency correlates sufficiently well with the plants' overall efficiency of N_2-fixation.

Legumes. In soybeans many studies indicate that activity per plant increases towards flowering with a maximum near the end of flowering. There is a considerable decline during the early pod-filling, due to a decline in specific activity of nodules. This occurs immediately prior to when the growth rate of pods and seeds exceeds that of total plant tops. The cause is clearly an inadequate assimilate supply to nodules during pod-filling (49, 73). In common beans (*Phaseolus vulgaris*) N_2-ase (C_2H_2) activities increased rapidly from the 3-node vegetative to the early pod-filling stages and then decreased to zero at maturity. There were cultivars with similar total dry matter production but with a 2-3 fold N_2-ase activity. Furthermore, seed yields and total N uptake were positively related. Therefore it may be possible to increase both symbiotic N_2-fixation and seed yield through selection (80).

Grasses. One of our observations when first sampling maize roots in 1974, was that N_2-ase activity appeared to be quite low during vegetative growth but increased markedly at flowering.

Field trials carried out since then provided relatively detailed knowledge about growth stage variation of N_2-ase activity (C_2H_2) for corn and sorghum (Fig. 4) (6, 11, 63). There was a significant interaction between cultivars and growth stages (Table 2) (13).

The Root System

Obviously, roots are important in our discussion of plant influences upon N_2-fixation, because it is within roots, root nodules, or in the rhizosphere, that N_2-fixation is more important in crop yields.

Legumes. In soybeans, data indicate that only a small portion of the root system can be responsible for much of the water uptake (71). Increasing the number of roots by grafting caused a drastic increase in growth and seed yield. The shoot:root ratio based on dry weight of plant material remained constant at approximately 10:1 (69). Resistance of roots to infection by N_2-fixing bacteria is under genetic control. Much remains to be explained on how such resistance or nonnodulating genes actually function (59, 60).

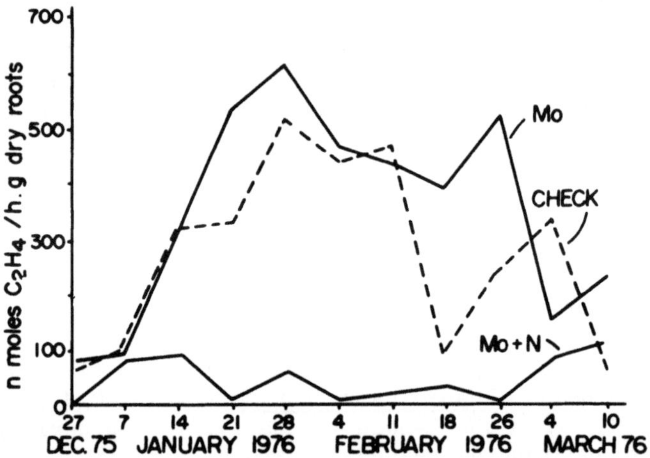

Fig. 4. N_2-ase activity (C_2H_2) during the growth stages of the bracyhytic corn variety Piranão at two levels of nitrogen fertilizer, 35 kg N/ha (-N) and 235 kg N/ha (+N), and with (+Mo) or without (-Mo) spraying with 1.6 kg/ha ammonium molybdate (63).

Grasses. The root system of *Gramineae* develops in a different way from that of *Leguminosae*. Its primary root (tap root) and seminal roots in some species constitute a temporary system, being substituted soon by the typical fasciculated root system. In the field, corn roots increase rapidly in length and fresh weight for 80 days after planting, remain constant for about 14 days, and then decline rapidly during the grain-filling stages. After tasseling, roots start dying, but net root length remains constant, because new roots are being formed. At later reproductive growth, net root length decreases rapidly (55). Because of obvious difficulties in studying the root system, very little information is available. However, good estimates are needed on the extent and weight of the root system, the distribution and thickness of roots, etc. Different methods have been applied for the difficult task (55, 42). In corn, most of the N_2-ase activity apparently takes place in the thicker segments of roots (21). Such root segments are found near the culm in the upper soil layer (28, 54). The total corn root weight apparently varies in about the same direction as does grain yield (56). Genotype differences of corn and sorghum root exudates of organic acids were found to be small (35) and probably are not important for N_2-ase activity within the roots.

Nutrient uptake efficiency. Efficiency of nutrient uptake may be different for different genotypes of legume or grass species. Efficiency varied due to different ploidy levels for different genotypes. In corn, efficiency of dry matter production per unit of P absorbed was greatest for later maturing hybrids (10). In peanuts, gain in yield due to iron chelate treatment ranged from 22% for three commercial cultivars which are inefficient in iron absorption on calcareous soils. The range for efficient genotypes was only 8% to 18%. The mean yield of untreated efficient cultivars equalled approximately that of treated inefficient cultivars (39). In Brazil, because huge areas of tropical soils are extremely poor in phosphorus and rich in free aluminum and exhibit "cerrado" type vegetation, crops are being screened for more efficient P absorption and resistance to high Al^{3+} levels in the soil. More efficient and resistant corn and sorghum lines have been found (65, 4, 30, 18). Zinc deficiency and excess phosphorus have a negative influence on N_2-fixation, directly through nodule nutrition or indirectly by affecting the host nutrition (18).

UNFAVOURABLE ENVIRONMENTAL CONDITIONS

When one talks about plant traits which influence N_2-fixation, the influences of the environment also must be considered, the phenotype being the product of environmental influences upon the genotype. Unfavourable environments may inhibit partially or completely a certain trait, and this is mainly true in regard to quantitatively inherited characteristics. In environmental conditions, which are favourable for maximum values of the trait being studied, gene expression is near its potential maximum. Therefore, selection should be done under the best possible set of environmental conditions required by the trait we want to improve. This, by the way, means that the plant breeder, before all, must be an excellent agronomist: he must know how to grow the crop and he must know how crops respond to different environments.

Oxygen Concentration and Temperature

The N_2-fixing nodule or root has O_2 requirements for the production of ATP to support N_2-ase activity. But, the internal partial pressure of O_2 (pO_2) in nodules must be maintained at a low level because of the oxygen sensitivity of N_2-ase (15). Combinations of sub-ambient pO_2 and low temperatures as well as low or high rhizosphere temperatures at ambient pO_2 are limiting to soybean N_2-fixation. When heating, the decrease is slow and reversibility is low (37). Growth of *Azospirillum lipoferum* at constant partial pressures of oxygen has been studied (62). This organism has a very poor oxygen protection mechanism and is sensitive to low temperatures. N_2-ase activity was maximal between 33 and 40°C with a pronounced drop below 33°C (21).

Nitrogen Fertilizer

According to present knowledge, maximum yields of some crops and forages cannot be achieved without fertilizing the crop with some kind of reduced nitrogen. However, N_2-fixation is inhibited by fixed nitrogen at levels such as occur in highly fertilized soils (38). In Brazil, such soils are rare. Therefore our problem is how to apply nitrogen fertilizer without inhibiting N_2-fixation. It appears that the so-called "Green Revolution" has been accompanied by a not very desirable "Fertilizer Revolution." What we want from now on might be called the "Biological Efficiency Revolution."

Legumes. Large amounts of N-fertilizer inhibit N_2-ase activity, but a moderate soil-N level has little effect on N_2-fixation of soybeans (8) and is important for initial growth, even in the presence of adequate inoculation (40). Urea and organic matter were much less detrimental to N_2-fixation than nitrate or NH_3-N sources (16, 74). Furthermore, considerable applied N is leached and contributes to pollution, whereas biologically fixed N apparently does not (47).

<u>Grasses</u>. N_2-ase inhibition by large amounts of N-fertilizer has been observed in maize (63, Fig. 4), sorghum (6), rice (5), and *Digitaria decumbens* (23). Two maize cultivars were tested at different fertilizer-N levels; one tolerated up to 80 kg N/ha while the other dropped its root N_2-ase activity sharply at that level (Fig. 2). Apparently the more tolerant genotypes also should be the most desirable for breeding purposes where adequate nitrition levels are required for successful selection for sufficiently high grain yields. Selection for better N_2-ase activity without considering adequate grain production may lead to negative results (Fig. 5).

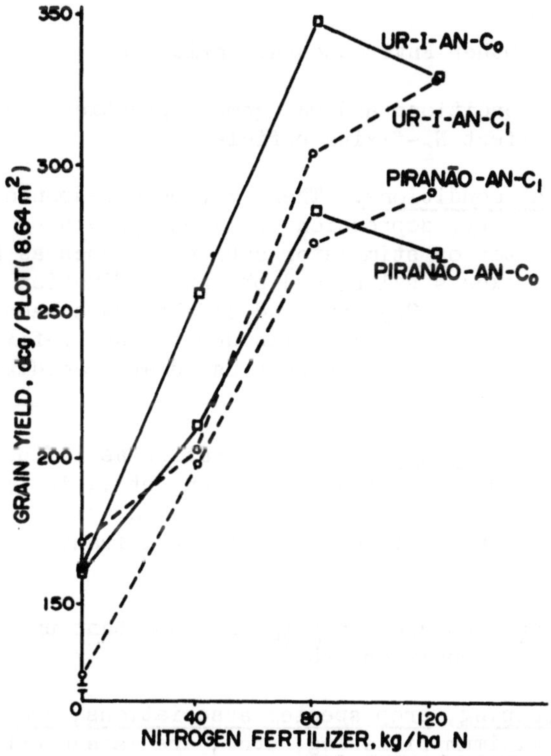

Fig. 5. Effect of 3 nitrogen fertilizer levels and 1 cycle of selection for higher N_2-ase activity on the N_2-ase activity of two corn varieties during the cool season of 1976. Experimental field of the UFRRJ, Brasil (unpublished).

Planting Density

In soybeans it has been observed that after flowering the percentage of red nodules declined, while the percentage of green, inactive nodules increased. After thinning normal strands down to only one plant per m^2, 2.5 times as much active nodule volume per plant was observed. Furthermore, the usual post-flowering activity decline in nodules was postponed for several weeks (78). At the same time, in the uppermost canopy level (20%) the weight per seed was 23.9% greater in the thinned plants while in the lowest level (20%) seed weight remained unchanged, as compared to unthinned plots (77). There are no results as far as I know, in regard to the influence of plant competition on N_2-ase activity in Gramineae.

Other Environmental Influences

A number of conditions and management practices in the field affect or might affect N_2-fixing efficiency:

Soil chemical conditions. There may be a response to aluminum and manganese toxicity, depression of legume growth by liming, presence of inhibitor or stimulator substances such as herbicides, etc., presence of excess salts, and phosphate fixation in latosolic soils (70, 57, 53, 79). Soybeans and alfalfa have been found to develop considerable acidity in soils due to the N_2-fixation process. A yield of 10t/(ha x year) produces a calculated acidity equivalent to 600 kg $CaCO_3$/ha (61).

Soil physical conditions. Soil aeration as influenced by cultivation, mulching, interrow compaction, etc., is important to root growth (36, 14, 72); soil moisture, evaporation, precipitation and drought conditions certainly are of significant importance (45, 31, 34).

Inoculum placement and quantity are important under temperature and moisture stress conditions (82).

Multiple cropping, crop species associations, crop rotation, with or without minimum tillering, etc., are management practices, or production systems, increasingly applied in modern agriculture. N_2-fixation must be studied in the context of these systems (20, 32, 47).

This review of environmental influences on N_2-ase activity, however brief, should give an idea of the multiplicity and complexity of factors to be taken into account not only in sampling for genetic selection but also in sampling during physiological and agronomical studies of N_2-fixation. Let us not forget: first, field research is of prime importance for agricultural production improvement; second, superior genotypes can express themselves fully only in the environment best suited for the trait under consideration; thirdly, with huge error mean squares; due to environmental variability, one usually cannot draw valid conclusions about any treatment effects or differences among genotypes. Such are some of the most important reasons why we must learn more about and be able to control environmental influences which modify the inherent capacity of N_2-fixation by plant-bacteria systems.

DISCUSSION AND CONCLUSIONS

Biological nitrogen fixation may be enhanced both by environmental control and genetic improvement. Interspecific variability can possibly be put to use in a few cases. Materials stored in germplasm banks should be screened for outstandingly efficient genotypes within species. More primitive populations, not yet submitted to selection under modern high nitrogen conditions, might be good sources of genes for better N_2-fixing ability. *Zea mexicana*, from which corn possibly has evolved, was sampled; however, N_2-ase (C_2H_2) activity was higher in corn. The most active corn S_1-lines were maintained by inbreeding through S_4, always testing and choosing the better N_2-fixing (C_2H_2) individuals. Results are extremely variable and repeatability has been low.

Tentative selection, using mass and among S_1-line selection and recombination methods, seems to indicate that the methodology for estimating the N_2-fixing capacity must be substantially improved in order to decrease the sampling error. It appears that selection among cultivars, races or ecotypes, as well as among families or inbred lines, is more likely to succeed than selection within cultivars or families. Genotypic differences and even heterotic effects in open-pollinated corn composites have been demonstrated, in spite of high experimental and sampling errors. Environmental influences in field tests were extremely high (coefficient of variability 40 to 50%). A great deal of variability is due to sampling and lab procedures. For example the sampling error mean square obtained when dividing the number of roots excised from one plant into two samples, represented about 40% of the experimental error. When sampling two plants, the sampling error mean square was equal (100%) to the experimental error mean square. Inadequate field or lab technique can not always be compensated for by increasing the number of replicates. At least five plants should be sampled within inbred lines and many more must be taken within half-sib families for selection purposes. Sampling should be done in a favourable environment for high N_2-ase activity. In the case of maize this requires low N-fertilization, up to 40kg N at planting, low initial soil-N, liming, and added molybdenum. High N-fertilization, 80 to 120kg N/ha, depending on the cultivar, inhibits N_2-ase activity almost completely.

There was no favourable correlation between N_2-ase activity (C_2H_2) and grain yields among ten corn composites and their crosses, contrary to what happens in soybeans and common beans. When selecting corn for better N_2-ase activity, without selecting also for higher grain yield, productivity can decrease significantly. Because of limitations imposed against taking the most adequate sample number in the field, and also to have better environmental control, it would be desirable to test plants in the greenhouse in the seedling growth stage. Corn seedlings have shown N_2-ase activity (C_2H_2); however, quantities of N_2 fixed were extremely small (68). Several thousand soybean seedlings could be tested in one month, and it was claimed that screening for better fixing genotypes is possible in spite of huge standard deviations (76).

Efficiency of molecular nitrogen fixation is not only closely related to the photosynthetic ability of the host plant, but it appears also that N_2-fixation efficiency, as a quantitatively controlled trait, presents similar or even more difficult problems to improvement through gene frequency as well as environmental modifications.

LITERATURE CITED

1. Alderfer, R. G. and Eagles, C. F. (1976) Bot. Gazette 137, 351.

2. Ashley, D. A., Bhagsari, A. S. and Boerma, H. R. (1976) Agron. Abstracts, 69.

3. Australia, Division of Tropical Agronomy, CSIRO. Annual Report 1974-1975, Brisbane.

4. Bahia Filho, A. F. C., França, G. E., Pitta, G. V. E., Magnavaca, R. and Perera, P. (1976) XI Reunião Bras. Milho e Sorgo. Piracicaba.

5. Balandreau, J., Rinaudo, G., Fares-Hamad, I. and Dommergues, Y. (1975) In: "Nitrogen Fixation by Free-living Microorganisms" by Stewart, W. D. P. (Ed.) Cambridge Univ. Press, pp. 57-70.

6. Berkum, P. van, Neyra, C. A. and Bülow, J. F. W. v. (1976) XI Reunião Brasileira Milho e Sorgo. Piracicaba.

7. Bernard, R. L. and Weiss, M. G. (1973) In: Caldwell, B. E. (Ed.) Soybeans. Amer. Soc. Agron., Madison, Wis. U.S.A., pp. 117-154.

8. Bhangoo, M. S. and Albritton, D. J. (1976) Agron. J. 68, 642.

9. Blackwood, G. C. and B. J. Miflin (1976) J. Exp. Botany 27, 735.

10. Bruetsch, T. F. and Estes, G. O. (1976) Agron. J. 68, 521.

11. Bülow, J. F. W. von and Döbereiner, J. (1975) Proc. Nat. Acad. Sci. U.S.A. 72, 2389.

12. Bülow, J. F. W. von, Mariel, I. E., Nery, M., Neves, M. C. P., Peres, J. R. R. and Döbereiner, J. (1975) XXVII Reun. Annual SBPC, Belo Horizonte.

13. Bülow, J. F. W. von, Podestá Filho, J. A. and Döbereiner, J. (1976) XI Reunião Bras. Milho e Sorgo, Piracicaba.

14. Chaudhary, M. R. and Prihar, S. S. (1974) Agron. J. 66, 350.

15. Criswell, J. G., Havelka, U. D., Quebedeaux, B. and Hardy, R. W. F. (1977) Crop Sci. 17, 39.

16. Criswell, J. G., Hume, D. V. and Tanner, J. W. (1976) Crop Sci. 16, 400.

17. Day, J. M., Neves, M. C. P. and Döbereiner, J. (1975) Soil Biol. Biochem. 7, 107.

18. Demeterio, J. L., Ellis, R., Jr. and Paulsen, G. M. (1972) Agron. J. 64, 566.

19. De-Polli, H., Matsui, E., Döbereiner, J. and Salati, E. (1976) Soil Biol. Biochem. 9, 119.

20. Desir, S. and Pinchinat, A. M. (1976) Turrialba 26, 237.

21. Döbereiner, J. (1977) Rev. Bras. Ciencias do Solo (in print).

22. Döbereiner, J. (1976) Inter. Symp. Genetic Control Diversity of Plants, Lahore, Pakistan.

23. Döbereiner, J. and Day, J. M. (1976) First Int. Symp. N_2-Fixation, 1974, Washington State University - Proceedings, pp. 518-538.

24. Döbereiner, J., Day, J. M. and Bülow, J. F. W. von (1975) Proc. II Int. Winter Wheat Conf., Zagreb, Yugoslavia, pp. 221-236.

25. Döbereiner, J., Day, J. M. and Dart, P. J. (1972) J. Gen. Microbiol. 71, 103.

26. Dunphy, E. J. and Hanway, J. J. (1976) Agron. J. 68, 697.

27. Feller, Urs K., Soong, Tai-Sen T., and Hageman, R. H. (1977) Pl. Phys. 59, 290.

28. Follett, R. F., Allmaras, R. R. and Reichman, G. A. (1974) Agron. J. 66, 288.

29. Poth, H. D. (1962) Agron. J. 54, 49.

30. Franca, G. E., Pitta, G. V. E., Bahia Filho, A. F. C., Magnavaca, R. and Pereira, P. (1976) XI Reunião Bras. Milho e Sorgo. Piracicaba.

31. Gallagher, J. N., Biscoe, P. V. and Hunter, B. (1976) Nature 264, 541.

32. Garcia, M. J. and Pinchinat, A. M. (1976) Turrialba 26(4), 409.

33. Gardner, C. O. and Eberhart, S. A. (1966) Biometrics 22, 439.

34. Garlipp, G. W. E. A. (1976) Agroplantae 8, 21.

35. Gaskins, M. H., Napoli, C. and Humbell, D. H. (1976) Agron. Abstr., 71.

36. Grable, A. R. (1966) Advan. Agron. 18, 57.

37. Hardy, R. W. F. and Criswell, J. G. (1976) Agron. Abstr, 72.

38. Hardy, R. W. F., Filner, P. and Hageman, R. H. (1975) In: "Crop Productivity-Research Imperatives" by Brown _et al._ (Eds.) Kettering Foundation, Yellow Springs, Ohio, pp. 133-176.

39. Hartzook, A., Karstadt, D., Naveh, M. and Feldman, S. (1974) Agron. J. 66, 114.

40. Hatfield, J. L., Egli, D. B., Leggett, J. E. and Peaslee, D. E. (1974) Agron. J. 66, 112.

41. Heenan, D. P. and Carter, O. G. (1976) Crop Sci. 16, 389.

42. Hignett, C. T. (1976) J. Australian Inst. Agric. Sci. 42, 127.

43. Hiranpradit, H., Foy, C. L. and Shear, G. M. (1972) Agron. J. 64, 274.

44. Ho, L. C. and Rees, A. R. (1977) The New Phytologist 78, 65.

45. Hume, D. J., Criswell, J. G. and Stevenson, K. R. (1976) Can. J. Pl. Science 56, 811.

46. Johnston, A. W. B. and Beringer, J. E. (1976) Nature 263, 502.

47. Jones, M. B., Street, J. E. and Williams, W. A. (1974) Agron. J. 66, 256.

48. Lahav, E., Harper, J. E. and Hageman, R. H. (1976) Crop. Sci. 16, 325.

49. Lawn, R. J. and Brun, W. A. (1974) Crop Sci. 14, 11.

50. Lawn, R. J., Fischer, K. S. and Brun, W. A. (1974) Crop Sci. 14, 17.

51. Lin, M. C. and Hadley, H. H. (1976) Crop Sci. 16, 321.

52. Loomis, R. S., Ng, E. and Hunt, W. F. (1976) In: "CO_2 Metabolism and Plant Productivity" Burris, R. H. and Black, C. C. (Eds.) Univ. Park Press, pp. 269-286.

53. Lucena, J. M. and Doli, J. D. (1976) Revista Comalfi 3, 241.

54. Mayaki, W. C., Stone, L. R. and Teare, I. D. (1976) Agron. J. 68, 532.

55. Mengel, D. B. and Barber, S. A. (1974) Agron. J. 66, 341.

56. Mengel, D. B. and Barber, S. A. (1974) Agron. J. 66, 399.

57. Munns, D. N. and Fox, R. L. (1976) Plant & Soil 45, 701.

58. Neyra, C. A., Pereira, P. A., Bülow, J. F. W. von and Döbereiner, J. (1976) XI Reunião Bras. Milho e Sorgo. Piracicaba.

59. Nutman, P. S. (1970) In: Anais do Seminário "Metodologia e Planejamento de Pesquisa com Leguminosas Tropicais" IPFACS Brasil. pp. 122-132.

60. Nutman, P. S. (Ed.) 1976. Symbiotic Nitrogen Fixation in Plants, Cambridge Univ. Press, London.

61. Nyatsanga, T. and Pierre, W. H. (1973) Agron. J. 65, 936.

62. Okon, Y., Houchins, J. P., Albrecht, S. L. and Burris, R. H. (1977) J. Gen. Microbiology 98, 87.

63. Pereira, P. A. A., Bülow, J. F. W. von and Neyra, C. A. (1977) Sociedade Brasileira Ciências Solo (in print).

64. Peterson, R. B., Friberg, E. E. and Burris, R. H. (1977) Plant Phys. 59, 74.

65. Pitta, G. V. E., França, G. E., Bahia Filho, A. F. C., Trevisan, W. L. and Schaffert, R. (1976) XI Reunião Bras. Milho e Sorgo. Piracicaba.

66. Quebedeaux, B. and Hardy, R. W. F. (1975) In: "CO_2 Metabolism and Plant Productivity" Burris, R. H. and Black, C. C. (Eds.) University Park Press, pp. 199-200.

67. Quebedeaux, B. Havelka, U. D., Livak, K. L. and Hardy, R. W. F. (1975) Plant Phys. 56, 761.

68. Ruschel, A. P. and Ruschel, R. (1974) Anais X Reunião Brasileira Milho e Sorgo, Sete Lagoas, M. G. 128-132.

69. Sanders, J. L. and Brown, D. A. (1976) Agron. J. 68, 713.

70. Sartain, J. B. and Kamprath, E. J. (1975) Agron. J. 67, 507.

71. Stone, L. R., Teare, I. D., Nickell, C. D. and Mayaki, W. C. (1976) Agron. J. 68, 677.

72. Veiga, Clóvis L. and Oliveira, Luiz B. (1976) Rev. Centro Ciencias Rurais 6, 295.

73. Vest, G., Weber, D. F., and Sloger, C. (1973) In: "Soybeans, Improvement, Production and Uses" Caldwell, B. E. (Ed.) ASA, Madison, U.S.A. pp. 353-390.

74. Vigue, J. T., Harper, J. E., Hageman, R. H. and Peters, D. B. (1977) Crop Sci. 17, 169.

75. Voorhees, W. B., Carlson, V. A. and Senst, C. G. (1976) Agron. J. 68, 976.

76. Wacek, T. J. and Brill, W. J. (1976) Crop. Sci. 16, 519.

77. Weil, R. R. and Ohlrogge, A. J. (1976) Agron. J. 68, 583.

78. Weil, R. R. and Ohlrogge, A. J. (1975) Agron. J. 67, 487.

79. Weir, C. C. (1977) Trop. Agric. 54, 87.

80. Westerman, D. T. and Kolar, J. J. (1976) Agron. Abstr. 79.

81 Wilson, J. K. (1939) Memoir 221, Cornell Univ., Ithaca, NY.

82. Wilson, D. O. (1975) Agron. J. 67, 76.

PLANT PHOTOSYNTHESIS

C.C. Black, R.H. Brown, and R.C. Moore

Biochemistry, Agronomy, and Botany Departments
University of Georgia, Athens, Georgia 30602, U.S.A.

Green plant photosynthesis has proven to be much more diversified in nature than anyone thought a decade ago. From a myoptic concentration on the reductive pentose phosphate (C_3) cycle a decade ago, plant biologists today are examining a great variety of plants from throughout the world with confident expectations of finding more diversity in photosynthesis. This exceptional interest in studying a variety of plants has its origins in basic discoveries about photosynthesis, photorespiration, nitrogen metabolism, and other aspects of plant biology, but it also originates from a stronger more universal awareness of our direct dependence upon plants for food, fiber, and fuel. Such basic knowledge about plants coupled with a keener awareness of the value of plants are the bases for this symposium. The aim of this contribution is to present the known diversity in plant photosynthesis and to consider some environmental pressures which regulate the efficiency of green plant photosynthesis.

The knowledge of diversity in the biochemistry of photosynthesis arose from work on sugarcane, a tropical crop plant (12), which quickly led to the discovery of the C_4-dicarboxylic acid (C_4) cycle of photosynthesis (11). The details of the biochemical diversity within C_4 photosynthesis are still being unraveled (5) but we now are aware of the diversity shown in Table 1. In addition we also know that C_4 photosynthesis is closely associated with definite physiological responses and leaf anatomy (1,10,11) and that diversity exists in these characteristics also (5,17).

As C_4 photosynthesis was being studied a new awareness arose (10) about a group of plants known to exhibit Crassulacean acid

TABLE 1. DIVERSITIES IN THE BIOCHEMISTRY OF C_4 PHOTOSYNTHESIS

Major BSC[a] Decarboxylase	Energetics of Decarboxylation in BSC	Major Substrate Moving From: MC to BSC	Major Substrate Moving From: BSC to MC[b]	Representative Plants
$NADP^+$ malic enzyme	Production of 1 NADPH/CO_2	Malate	Pyruvate	*Zea mays* *Digitaria sanguinalis*
NAD^+ malic enzyme	Production of 1 NADH/CO_2	Aspartate	Alanine/ Pyruvate	*Atriplex spongiosa* *Portulaca oleracea*
PEP carboxy-kinase	Consumption of 1 ATP/CO_2	Aspartate	PEP[c]	*Panicum maximum* *Sporobolus poiretii*

[a]MC = mesophyll cell; BSC = bundle sheath cell.

[b]3-PGA also moves into the mesophyll cells to support the synthesis of hexoses and starch in all three groups of plants.

[c]Nitrogen balance also must be maintained via an aminotransferase-type shuttle likely involving alanine. PEP also may be converted to pyruvate with the formation of ATP in BSC.

metabolism (CAM) (16). These plants also fix CO_2 to synthesize 4-carbon dicarboxylic acids but the synthesis occurs at night and an internal leaf decarboxylation occurs during the day. Thus a temporal separation of steps in CO_2 metabolism distinguishes CAM from other plant photosynthesis.

We have presented detailed evidence delineating another group of plants which are characterized as intermediate between C_3 and C_4 plants (2,4). *Panicum milioides* is the best studied C_3-C_4 intermediate plant, but one of us (RHB) visited Brazil and other areas of South America last year on a plant collection tour and found other intermediate species including *Panicum schenckii* and *P. decipiens*. So both the work on C_4 and C_3-C_4 intermediate plants has used tropical plants in gaining a more complete understanding of the diversity of plant photosynthesis. These results point to the need for concentrated work on photosynthesis, particularly in tropical plants.

Table 2 is a summarization of some characteristics of these plant groups. Additional characteristics were summarized earlier (1) for C_3, C_4, and CAM plants and some will be presented later in this manuscript. Note comparisons are made under defined physiological environments, usually at atmospheric O_2 and CO_2 levels with mature leaf tissue. Without dwelling on the details of these comparisons, which has been done before (1,5,10,11), we will conclude that different groups of plants exist with their unique characteristics. Now we will illustrate some of the diversity of plant photosynthesis by characterizing C_4 photosynthesis which will be integrated with some aspects of nitrogen metabolism in plants.

Initially C_4 plants were thought to be tropical plants possibly due to early work with sugarcane in combination with the realization that they have high temperature optima for leaf photosynthesis and efficiently utilize the intense light of the tropics. Figure 1 illustrates the general temperature response curves of C_4 leaf photosynthesis and shows the range of optimal temperatures found in specific C_4 plants. The temperature optimum ranges from 30°C in *Spartina* growing along the English coast to 47°C in *Tidestromia* growing in Death Valley California (Fig. 1). The light response of C_4 leaf photosynthesis has the general pattern shown for Bermudagrass in Figure 2 where full summer sunlight intensities are near the saturation intensity. But the C_4 plant *Tidestromia* gives no indication of saturation even at the high irradiations found in Death Valley (Fig. 2).

In comparison to the majority of C_3 plants, C_4 plants do have higher temperature and light intensity optima for photosynthesis and growth (1,10). But there are some C_3 plants which have light

TABLE 2. CHARACTERISTICS OF PLANT GROUPS DIFFERING IN PHOTOSYNTHESIS

	C_3	C_4	C_3-C_4 Intermediate	CAM
Initial Product of PS	3-PGA	Oxaloacetate	3-PGA	Dark - oxaloacetate Light - 3-PGA
PS CO_2 Compensation (ppm)	35 to 85	0 to 10	15 to 20	Diurnal variability 0 to 60
Response of leaf PS to: ∿ 2% O_2	30 to 40% increase	No effect	Intermediate between C_3 & C_4	No effect in short times (≤ 10 min.)
∿ 50% O_2	Inhibition	Slight inhibition		
Energetics of PS (CO_2:ATP:NADPH)	1: 3: 2	1: 5: 2	--	1: 6.5: 2
Leaf Fraction I Protein (% of soluble)	40 to 50	8 to 25	--	--
Leaf N for Maximum PS (% DM)	6.5 to 7.5	3 to 4.5	--	--
Physiological Quantum Requirement: Measured	12 to 22	12 to 18	--	--
Calculated	14	16	--	18
Cross Sectional Leaf Anatomy	Green MC colorless BSC	Highly developed green BSC and green MC	Green BSC and MC	Succulent, green MC 6 to 10 layers thick
Gm H_2O/gm DM	450 to 1000	250 to 350		25 to 125

PS = photosynthesis; MC = mesophyll cells; BSC = bundle sheath cells.

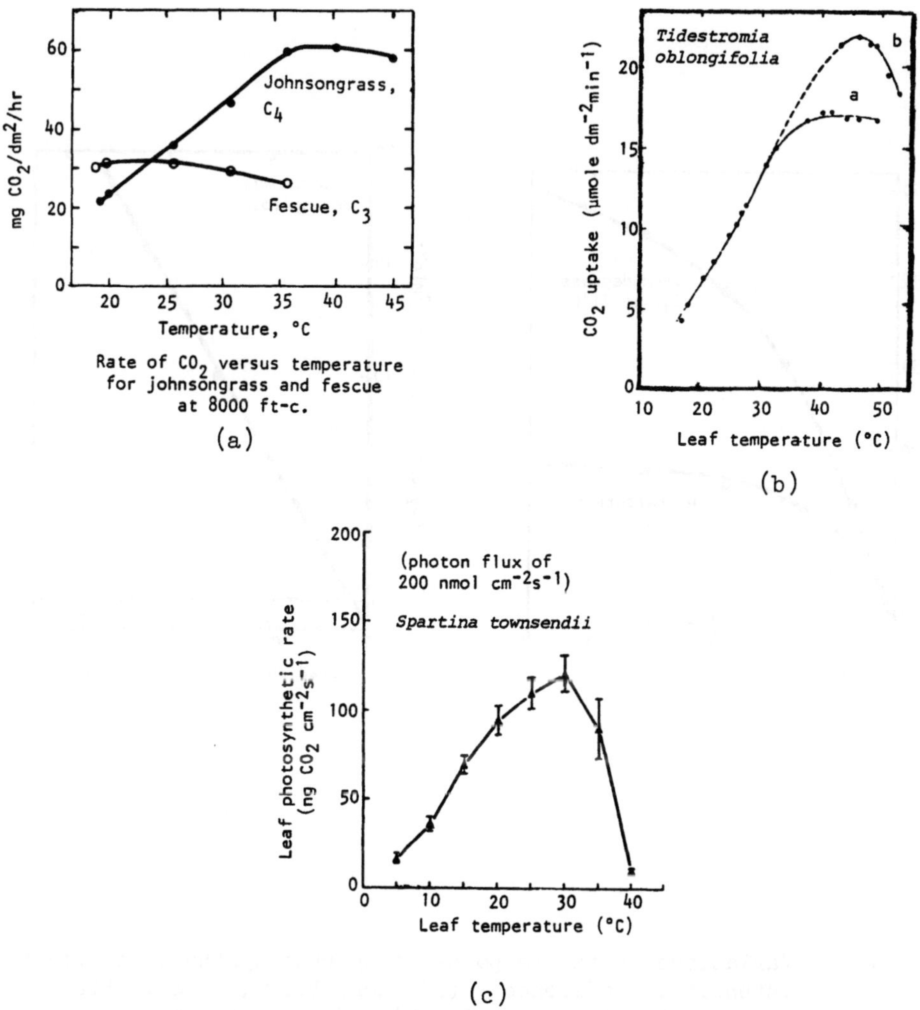

Figure 1. Variations in the response of C_4 photosynthesis to temperature. References: (a) Chen, Brown, Black. Weed Sci. 18, 399-403, 1970 (b) Björkman, Pearcy, Harrison, Mooney. Science 175, 786-9, 1972 (c) Long, Incoll, Woolhouse. Nature 257, 622-4, 1975.

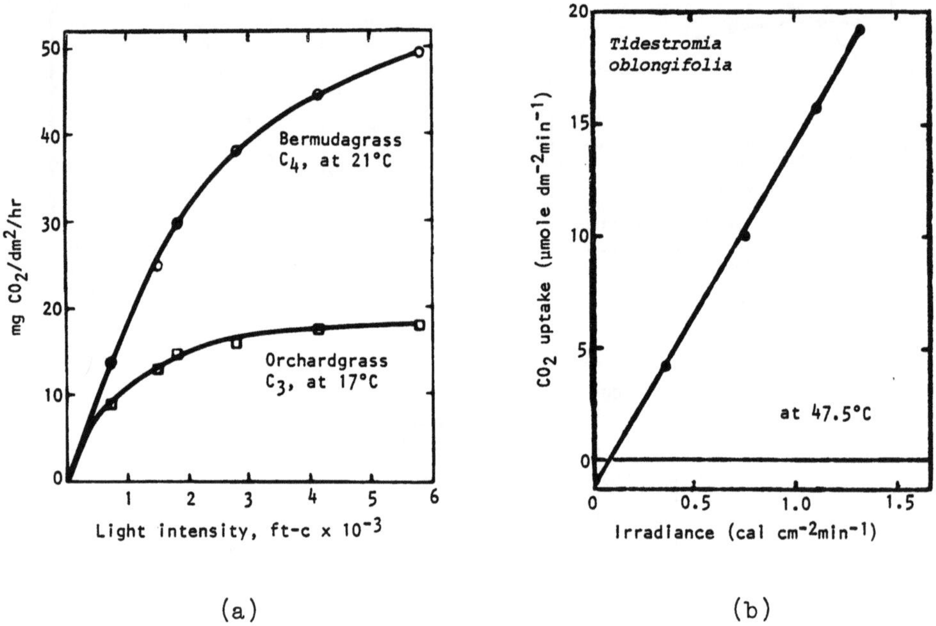

Figure 2. Variations in the response of C_4 photosynthesis to light intensity. References: (a) Chen, Brown, Black. Plant Physiol. 44, 649-54, 1969 (b) Björkman, Pearcy, Harrison, Mooney. Science 175, 786-9, 1972.

and/or temperature responses similar to C_4 plants, such as sunflower, *Typha, Cassava,* and *Camissonia* (1,14). The underlying biochemical bases for these unusual C_3 plants have not been elucidated.

Related to the light and temperature response curves of C_4 versus C_3 photosynthesis, we recently studied the quantum efficiency of plant photosynthesis (6). That is, rather than the efficiency of growth or yields, what is the minimum quantum requirement? The question was raised because the energetics of C_4 photosynthesis (Table 2) show an additional 2 ATP's are needed above that of C_3 plants. Also we questioned the contribution of photorespiration to the efficiency of photosynthesis. Photorespiration is influenced by temperature, O_2, and CO_2 levels (1,5, 10,19) which also strongly influence photosynthesis. The measured quantum requirements in physiological environments for C_3 photosynthesis vary from ∿ 12 to ∿ 22 (Table 2) and are strongly dependent upon temperature plus O_2 and CO_2 levels. In contrast, the physiological quantum requirement of C_4 plants is independent of these environmental variables (6). Leaves of C_4 plants do not exhibit photorespiration but the pathway is present (1,5). If photorespiration is active in C_4 leaves, how can the quantum requirement be independent of O_2, CO_2, and temperature? It has been suggested that the RuDP carboxylase/oxygenase in C_4 plant bundle sheath cells must be in an environment rich in CO_2 (9). Hatch has supported the concept that C_4 plants concentrate CO_2 in their bundle sheath cells to 15 to 60 μM or more (9). Assuming that the Km for RuDP carboxylase is 20 μM (9), the enzyme in C_4 bundle sheath cells would be near saturation with CO_2. But, since O_2 is a competitive inhibitor of CO_2, a dramatic response in RuDP carboxylase activity should occur as the O_2 concentration decreases from 21% to 2%. Such a decrease did not influence the C_4 quantum requirement (6). We do know, however, that changing O_2 concentrations changes the $^{14}CO_2$ fixation products made by C_4 plant leaves, showing that O_2 is affecting the carbon biochemistry of C_4 photosynthesis (8). We conclude that RuDP carboxylase/oxygenase is regulated in some unknown manner in C_4 leaves so that its activity is independent of atmospheric CO_2 and O_2 concentrations or that some other control phenomena are overriding the effects of environment in the C_4 leaf.

Net C_4 photosynthesis does not respond to O_2 changes near atmospheric but C_4 photosynthesis also is near saturation at atmospheric levels of CO_2 (Table 2, 10,17,19). Thus C_4 plants are efficient in making good use of air levels of CO_2 in contrast to C_3 or CAM plants which are not saturated by air CO_2 levels (5,19). It is the very active PEP carboxylase acting as a CO_2 trap in mesophyll cells which not only allows C_4 plants to suppress the external loss of CO_2 via photorespiration but also allows C_4 plants to efficiently remove CO_2 from air to keep RuDP carboxylase operating efficiently in bundle sheath cells. Thus there is little response of C_4 leaf photosynthesis to raising CO_2 levels in air.

Through their ability to effectively use air CO_2, C_4 plants also have restricted H_2O loss simultaneously so that their losses in transpiration are much less than C_3 plants. C_4 plants require about 50% less water than C_3 plants to produce dry matter (Table 2). At a given temperature the stomatal resistance is higher in a C_4 leaf while the internal CO_2 concentration is lower than in a C_3 leaf. Therefore, a C_4 plant loses less H_2O and has a steeper CO_2 gradient from the air to the leaf interior so that CO_2 fixation is efficient and water loss is minimized. CAM plants efficiently regulate H_2O loss (Table 2) via a night CO_2 fixation, high mesophyll resistance, and closed stomata during the day (5).

The biochemical pathway for CO_2 assimilation via a spatial separation of C_4 organic acid synthesis into mesophyll cells and the C_3 cycle into bundle sheath cells has been well documented in C_4 plants (1,5,10). The diversity of C_4 biochemistry (Table 1) also has been documented (5). So these characteristic features of C_4 photosynthesis will not be covered here.

Now we will consider the nitrogen use efficiency of C_4 plants since it is clear that C_4 plants have a unique pathway of leaf nitrogen assimilation and nitrogen is utilized more efficiently in C_4 leaves than in other groups of plants (Table 2, 3,15). During nitrogen assimilation, whether via NO_3^-, NH_4^+, or N_2 reduction, it is evident that plants supply a) a reductant which can be ferredoxin or a reduced pyridine nucleotide, b) ATP, and c) a carbon sketeton for the reduced nitrogen to attach to or carbon to be oxidized to form a or b. All of these components come from photosynthesis directly or, excepting ferredoxin, from photosynthetic storage products such as starch. So the close biochemical relationship between photosynthesis and nitrogen assimilation is easy to visualize. However, there are complex temporal as well as spatial separation problems involved. Photosynthesis is in green tissues and is only active during the day whereas nitrogen assimilation is in non-green and green tissues and may occur in the dark and/or light.

Recently the hypothesis has been developed that C_4 plants utilize nitrogen more efficiently than C_3 plants (3). The supporting data show about a 2-fold greater dry matter production per unit of leaf nitrogen (Table 3). In addition, leaf photosynthesis appears to be a near linear function of nitrogen concentration in C_3 and C_4 species, but there is nearly a 2-fold greater rate of photosynthesis per unit of leaf nitrogen with C_4 plants (*P. maximum*)

TABLE 3. GRASS DRY MATTER YIELDS AND NITROGEN CONTENT AT VARIOUS LEVELS OF NITROGEN FERTILIZATION

Nitrogen Applied	Dry Matter Yield		N Concentration in Forage	
	C_3	C_4	C_3	C_4
kg/ha	t/ha/yr		% DM	
112	3.8	8.3	2.52	2.13
224	5.8	11.4	2.77	2.26
448	7.2	16.1	3.25	2.75
896	6.9	17.5	3.50	3.00

C_3 = *Festuca arundinacea*; C_4 = *Cynodon dactylon*. Established sods grown for 4 years in a field experiment on the coastal plain of Virginia (3).

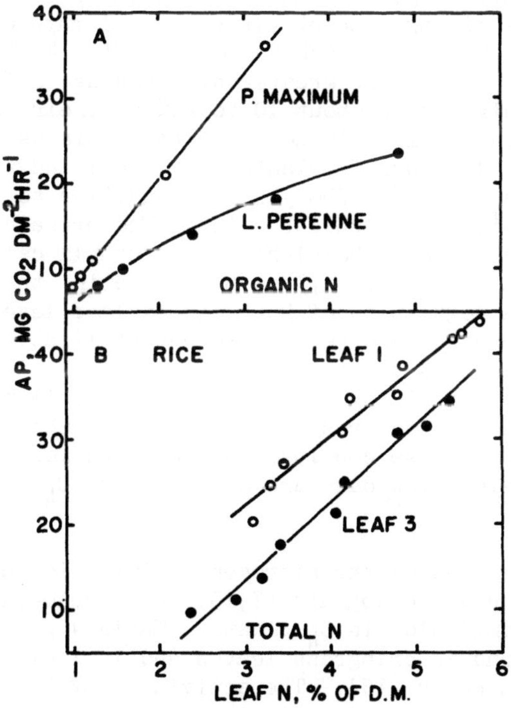

Figure 3. Relationships between leaf nitrogen concentration and apparent photosynthesis in *Panicum maximum* (C_4), *Lolium perenne* (C_3) and rice (C_3)(3).

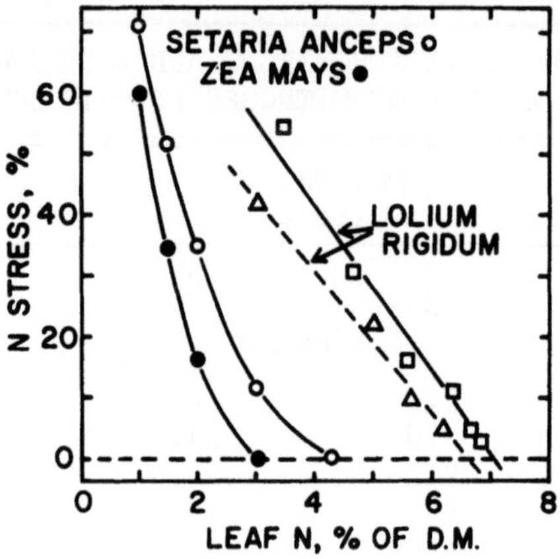

Figure 4. Relationships between leaf nitrogen content and nitrogen stress. The data sources and methods of calculation are given by Brown (3).

than with C_3 plants (Fig. 3). C_4 plants such as *Setaria* and *Zea* also require less nitrogen to obtain a maximal rate of dry matter accumulation than C_3 plants such as *Lolium* (Fig. 4, Table 2). Perhaps one reason for their greater nitrogen use efficiency is that C_4 plants only invest about 10 to 25% of their soluble leaf protein in RuDP carboxylase in contrast to C_3 plants which invest 40 to 50% (Table 2). Thus C_4 plants have regulated the amount of nitrogen invested in RuDP carboxylase in addition to sequestering the enzyme in their bundle sheath cells. The presence of the efficient CO_2 trap, PEP carboxylase, in mesophyll cells allows C_4 plants to use less RuDP carboxylase by regulating the CO_2 level around it. How a regulation of RuDP carboxylase level occurs is unknown but it results in a very efficient utilization of available soil nitrogen in dry matter production (Table 3). Therefore, C_4 plants can grow on nitrogen-poor soils; low soil nitrogen also may have been a selection pressure which led to the evolution of C_4 photosynthesis. These conclusions are based mostly on data from C_4 grasses since few data are available on C_4 dicot species (3).

We also have studied the nitrogen assimilation pathways in leaves of nutsedge and crabgrass (7,15) and developed schemes for NO_3^- and NH_4^+ assimilation in C_4 plants. Table 4 presents some enzyme levels found in crabgrass leaves and in isolated mesophyll and bundle sheath cells (15). The activities of NO_3^- assimilating enzymes in leaf extracts and in isolated bundle sheath and mesophyll cells are high enough to account for proposed *in vivo* rates of NO_3^- assimilation of 6 to 12 µmoles/mg chl/hr (15). These findings are the first reported where enzymatic activity met this criterion. Although some of the enzymes are found in varying

TABLE 4. LOCATION AND ACTIVITY OF NITROGEN ASSIMILATION ENZYMES IN CRABGRASS LEAVES

Enzyme	Whole Leaf	Mesophyll Cells	Bundle Sheath Strands	MC Plus BSS (1:1)
		µmoles/mg chl/hr		
Nitrate Reductase	5.1	8.3	N.D.	4.3
Nitrite Reductase	25.7	39.4	N.D.	20.2
Glutamate Dehydrogenase				
NADH-dependent	16.7	4.1	28.3	12.7
NADPH-dependent	6.3	1.1	11.1	7.1
Glutamine Synthetase	44.2	33.6	45.6	38.8
Glutamate Synthetase	10.1	6.5	13.9	8.0

N.D. = not detectable.

levels in both mesophyll and bundle sheath extracts, nitrate and nitrite reductases are localized exclusively in mesophyll cells, NADH- and NADPH-dependent glutamate dehydrogenase are primarily in the bundle sheath, while glutamine and glutamate synthetase are in both mesophyll and bundle sheath cells. Generally, enzyme activities in the cell extracts are about two-fold higher than that of the whole leaf extract (Table 4); this shows quantitative recoveries of enzymes from the isolated cell types since leaf chlorophyll is about equally distributed in crabgrass cell types.

Based on the clear compartmentation of NO_3^- and NO_2^- reductase into mesophyll cells, two schemes are proposed for nitrogen metabolism in crabgrass, depending on environmental conditions. In one scheme, under conditions of ammonium nutrition, ammonium is incorporated into glutamate in roots via glutamate dehydrogenase. Glutamine is transported in the transpiration stream to leaves where, in the bundle sheath, it is combined with α-ketoglutarate to form glutamate and ultimately transamination occurs to form other amino acids.

However, since most of the nitrogen absorbed by plants is in the form of NO_3^-, a scheme was devised for leaf NO_3^- metabolism. NO_3^- in the transpirational stream is transported across the bundle sheath to mesophyll cells for reduction, where nitrate and nitrite reductase are located. A full complement (13) of NO_3^- metabolism enzymes is present in mesophyll cells (Table 4), so NO_3^- assimilation into amino acids occurs there, thus eliminating a need for intercellular leaf transport of ammonium. At first, this scheme appears to be relatively inefficient in that NO_3^- must be transported across a green bundle sheath cell to a mesophyll cell. NO_3^- reduction in the bundle sheath could allow for greater efficiency by the elimination of this intercellular transport. But the strict enzyme compartmentation in crabgrass leaves eliminates this possibility. Since NO_3^- and NO_2^- reduction require large amounts of reductant and ATP, and CO_2 reduction via the C_3 cycle occurs in the bundle sheath, it may be that it is advantageous for the plant to have NO_3^- and NO_2^- reduction spatially separated from this CO_2 reduction. C_4 mesophyll cells are subject to higher irradiation than bundle sheath cells so that ATP, a reductant, and a carbon skeleton are readily available.

The effects of nitrogen assimilation intermediates on photosynthetic O_2 evolution also was investigated and data were obtained to support the scheme of NO_3^- assimilation in C_4 leaves. Only NO_2^- was active as a Hill oxidant with mesophyll cells, in agreement with the exclusive localization of nitrite reductase in mesophyll cells (Table 4).

Investigations with all other nitrogen intermediates (15) on the mesophyll cell Hill reaction showed that these intermediates do not serve as Hill oxidants nor as inhibitors of photosynthetic O_2 evolution mediated by phosphoenolpyruvate (18). NO_2^- was the only nitrogen assimilation intermediate that served as a presumed alternate electron acceptor to $NADP^+$ for ferredoxin and hence promoted O_2 evolution in the light in mesophyll cells. The reduction of NO_2^- and the concurrent O_2 evolution are both light dependent (Fig. 5) and do not require ATP, as evidenced by the fact that uncouplers of photosynthetic phorphorylation stimulated O_2 evolution.

Bundle sheath cells were inactive in all Hill reaction experiments using the accepted intermediates (13) of nitrogen assimilation (15).

Since NO_2^- was the only nitrogen metabolism intermediate which acted as a Hill oxidant in crabgrass mesophyll cells, one expects stoichiometric amounts of O_2 to be evolved per NO_2^- reduced. If NO_2^- is reduced in the chloroplast by ferredoxin in a classical non-cyclic electron transport pathway and thus acts as an alternate electron acceptor to NADPH for light reduced ferredoxin, then the reaction stoichiometry should be:

$$3H_2O = 6H^+ + 6e^- + 1.5\ O_2$$

$$NO_2^- + 6H^+ + 6e^- = NH_3 + H_2O + OH^-$$

Therefore, 1.5 moles of O_2 should be evolved per mole of NO_2^- reduced. We measured stoichiometries of 1.55 and 1.33 in separate experiments.

Since NO_2^- served as a Hill oxidant in crabgrass mesophyll cells, its effect on PEP mediated O_2 evolution (18) was noted and further work was undertaken to characterize the effect of light intensities of photosynthetic O_2 evolution in the presence of one to several Hill oxidants, including NO_2^-, PEP, and PGA (Fig. 5, 15). While increases in light intensity beyond 2×10^5 erg/cm^2/sec resulted in no increase in O_2 evolution for any one of the oxidants acting alone, increases of approximately 20% were noted at higher light intensities in the presence of both PEP and NO_2^-. Upon the addition of PGA, an additional increase was observed even at intensities up to 5×10^5 erg/cm^2/sec. These systems, containing 3 Hill oxidants, yielded a light response curve somewhat similar shaped to those observed with most whole C_4 leaves (Fig. 2).

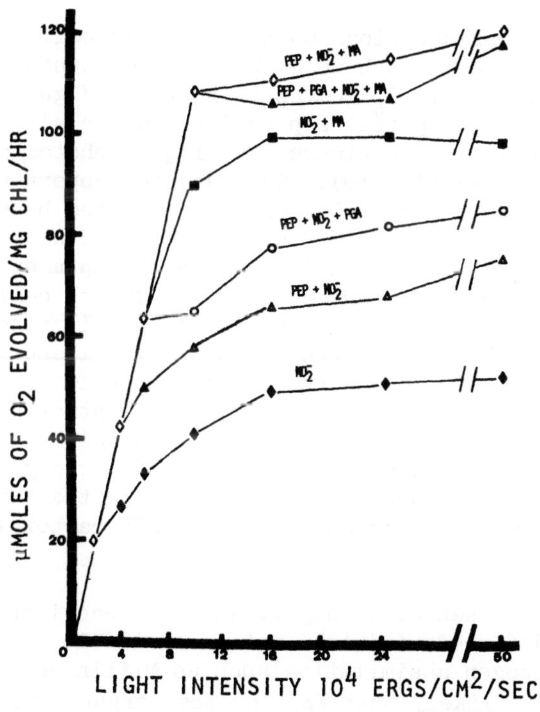

Figure 5. Influence of light intensity on the Hill reaction in isolated crabgrass mesophyll cells. The concentrations of additions were: PEP, 5mM; NO_2^-, 10 mM; PGA, 5 mM; and methylamine, (MA), 15mM.

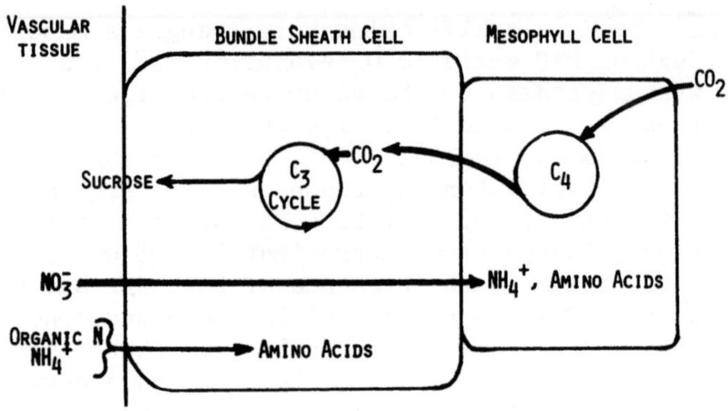

Figure 6. A simplified scheme for CO_2 and nitrogen assimilation in C_4 leaves.

From these considerations on nitrogen metabolism in C_4 plants, we can present a scheme combining photosynthesis with nitrogen assimilation in C_4 leaves as shown in Figure 6. Strong documentation for the spatial separations in the simplified pathway of CO_2 assimilation in Figure 6 during C_4 photosynthesis has been given previously (1,5,10). The data to support the spatial separation of nitrogen assimilation are presented here (7,15). Thus, in light, portions of CO_2 assimilation and nitrogen assimilation are spatially separated in C_4 leaves. Apparently NO_3^- enters and passes across a green C_4 bundle sheath cell which does not compete with an adjacent green mesophyll cell since it can not metabolize NO_3^- or NO_2^-. Thus the transpirational and water diffusion streams will move NO_3^- to the mesophyll cells for reduction. This might be thought of as a pulling or concentrating pathway for NO_3^- in mesophyll cells just as CO_2 is effectively pulled into the bundle sheath cells though the mesophyll PEP carboxylase. So nitrate and nitrate reductases act as nitrogen traps in the C_4 mesophyll cell in an analogous fashion to PEP carboxylase acting as a CO_2 trap.

In a temporal sense, at night, similar reactions might occur but the reduced ferredoxin would not be available to operate schemes for nitrogen assimilation such as Miflin and Lea propose (13). It is more likely that NH_4^+ or some organic form of nitrogen is translocated at night and incorporated into amino acids either in the bundle sheath cells or mesophyll cells without the use of a strong reductant such as ferredoxin.

Previously we emphasized the unique nature of photosynthesis in the green C_4 mesophyll cell in that it lacks RuDP carboxylase which comprises 40 to 50% of the chloroplast protein in all other green cells. In addition, a C_4 mesophyll cell will reduce oxaloacetate and 3-PGA, so it has a unique protein content and a unique pathway of carbon assimilation which only uses the activation and reduction portions of the C_3 cycle (5). We now wish to emphasize the unique nature of nitrogen metabolism in bundle sheath cells of C_4 plants. Clearly C_4 bundle sheath chloroplasts lack nitrite reductase and the cells lack nitrate reductase. So the ability to couple NO_3^- assimilation to chloroplast reactions is absent. These features show the unique protein composition and activity of C_4 bundle sheath cells among green nitrogen assimilating cells.

In brief, C_4 plants do utilize their available nitrogen more efficiently than C_3 plants (Table 3) in producing dry matter and in fixing atmospheric CO_2 (Fig. 3). We propose that the basis for the efficient nitrogen and CO_2 utilization is the spatial separation of biochemical activities as shown in Figure 6 such that a division of labor occurs in adjacent cells which is greatly beneficial to the intact plant.

REFERENCES

1. Black, C.C. (1973) Ann. Rev. Plant Physiol. 24, 253.

2. Brown, R.H. (1976) CO_2 Metabolism and Plant Productivity, eds. R.H. Burris and C.C. Black, pp. 311-325, University Park Press, Baltimore-London-Tokyo.

3. Brown, R.H. (1977) Crop Sci., In Submission.

4. Brown, R.H. and Brown, W.V. (1975) Crop Sci. 15, 681.

5. Burris, R.H. and Black, C.C. eds. (1976) CO_2 Metabolism and Plant Productivity, pp. 1-431, University Park Press, Baltimore-London-Tokyo.

6. Campbell, W.H. and Black, C.C. (1977) Proc. Natl. Acad. Sci. USA, In Submission.

7. Chen, T.M., Dittrich, P., Campbell, W.H. and Black, C.C. (1974) Arch. Biochem. Biophys. 163, 246.

8. Foster, A. and Black, C.C. (1977) Plant and Cell Physiol. Special Issue #3, pp. 325-340.

9. Hatch, M.D. (1976) CO_2 Metabolism and Plant Productivity, eds. R.H. Burris and C.C. Black, pp. 59-81, University Park Press, Baltimore-London-Tokyo.

10. Hatch, M.D., Osmond, C.B. and Slayter, R.O. eds. (1971) Photosynthesis and Photorespiration, pp. 1-565, Wiley-Interscience, New York.

11. Hatch, M.D. and Slack, C.R. (1970) Ann. Rev. Plant Physiol. 21, 141.

12. Kortschak, H.P., Hartt, C.E. and Burr, G.O. (1965) Plant Physiol. 40, 209.

13. Miflin, B.J. and Lea, P.J. (1976) Phytochem. 15, 873.

14. Mooney, H.A., Ehleringer, J. and Berry, J.A. (1976) Science 194, 322.

15. Moore, R.C. (1977) M.S. Thesis, University of Georgia, Athens, Georgia. In Preparation.

16. Ranson, S.L. and Thomas, M. (1960) Ann. Rev. Plant Physiol. 11, 81.

17. Ray, T.B. and Black, C.C. (1977) Encyclopedia of Plant Physiol. Series 2, eds. E. Latzko and M. Gibbs. In Press.

18. Salin, M.L., Campbell, W.H. and Black, C.C. (1973) Proc. Natl. Acad. Sci. USA, 70, 3730.

19. Zelitch, I. (1971) Photosynthesis, Photorespiration, and Plant Productivity. Academic Press, New York.

INTERACTIONS OF PLANT PHOTOSYNTHESIS WITH DINITROGEN FIXATION AND NITRATE ASSIMILATION

Carlos A. Neyra

Programa Fixacão Biológica de Nitrogênio

Convênio CNPq-EMBRAPA-UFRRJ, Km 47
Seropédica 23460, Rio de Janeiro, Brazil

INTRODUCTION

The success of efforts to increase world food production in order to meet the ever-increasing needs relies in part on the ability of plants to obtain larger amounts of nitrogen. Nitrate assimilation and biological N_2-fixation are the two major sources of N for plant growth and protein yield and maximization of both processes under field conditions should lead to a greater input of N into crop plants. At moderate levels of combined N in the soil, the plants can benefit from the simultaneous or complementary operation of both nitrate assimilation and biological N_2-fixation as evidenced by the seasonal profiles of N_2-ase and nitrate reductase activities obtained for certain crops such as soybeans (22, 51), Phaseolus vulgaris (42), sorghum (54) and corn (41).

Because of the reductive nature of these two nitrogen assimilation processes, energy must be supplied into the system. Solar radiation and consequently photosynthesis are the ultimate sources of energy to reduce either N_2 or nitrate to the level of NH_4^+. Current evidence suggests that plants use similar amounts of energy for the assimilation of N by symbiotic N_2-fixation or assimilation of NO_3^- (19, 21, 36). However, detailed energy budgets with regard to these two processes are not available for most plants.

Plant cells cannot reduce N_2 directly to NH_4^+ because they lack the enzyme nitrogenase, but they benefit from the association with certain N_2-fixing bacteria (8, 49). The best known association of agronomic importance is the legume-Rhizobium symbiosis and recently a number of reports have shown the existence and operation in nature of associations between grasses and bacteria, which are able to bring about N_2-fixation of economic importance (7, 16, 17, 41, 46).

Regardless of the type of plant-bacterial association, the site of reduction of dinitrogen to ammonia is far removed from the site of photosynthetic production, and the link between these two processes is established by means of translocation of photosynthetic products.

PHOTOSYNTHESIS AND N_2-FIXATION

The amount of photosynthate available is considered to be one of the major factors controlling the rate at which N_2-fixation proceeds (20, 24, 44). The best demonstration that production of photosynthate is a major limiting factor for N_2-fixation in field-grown soybeans was obtained from CO_2-enrichment experiments on the soybean canopy, which resulted in a net increase of the amount of N_2-fixed (24). Conversely, total N_2-fixed was decreased under conditions of elevated pO_2 in the aerial part of soybeans (44). The major effect of CO_2 enrichment was attributed to an increase in net production of photosynthate made possible by a decrease in photorespiration produced by an elevated CO_2/O_2 ratio. The increase of pO_2 resulted in a smaller CO_2/O_2 ratio thus favoring photorespiration and decreasing availability of photosynthate (44).

Many of the tropical grasses able to support significant N_2-ase activity possesses the photosynthetic C-4 pathway (13) and are thereby able to attain larger photosynthetic rates and minimal losses of carbon due to photorespiration (11); this may result in a greater availability of photosynthate for growth and N_2-fixation. In grasses, most of the N_2-ase activity computed over a 24 hour period occurs during the light period and this may reflect the dependence of nitrogenase activity in grasses upon available photosynthate, as is the case for legumes (1, 15, 41).

A large proportion of the photosynthate produced in legumes appears to pass through the nodules. Minchin and Pate (36) reported that as much as 32% of the total photosynthate was translocated to the nodules where 5% was used for nodule growth, 12% in respiration and 15% returned to the shoot via the xylem as amino compounds generated by N_2-fixation. The photosynthates transported to the nodules are metabolized rather rapidly (29) thus providing the reductant and ATP required for the reduction of N_2 to NH_4^+ and also the C-skeletons needed for amino acid formation (6, 21). Sucrose is considered the major form of carbohydrate arriving in the nodules, and it is rapidly converted into glucose and fructose by the vigorous invertase present in nodule cells (6, 29); thus hexoses appear to be the main substrate for the bacteroids within nodule cells (6).

Carbon skeletons for N_2-fixation also could be produced by dark CO_2 fixation (12, 30). Christeller *et al*. (12) have shown in lupin nodules that both *in vivo* CO_2 fixation activity and *in vitro* phosphoenolpyruvate (PEP) carboxylase activity were significantly correlated with nodule acetylene reduction activity during nodule development. They proposed that the oxaloacetate (OAA) formed from PEP could be used as an acceptor for ammonia assimilation into amino acids (aspartate and asparagine). Lawrie and Wheeler (30) showed dark CO_2 fixation in broad-bean nodules and high levels of PEP carboxylase, malate dehydrogenase and malic enzyme in nodule extracts. They proposed that the reduction of OAA to malate is an important route for the metabolism of carbon fixed within the nodules.

Recent results indicate that dark CO_2 fixation and malate metabolism may also play a significant role in the supply of carbon or energy for N_2-fixation in grass-bacterial associations (41). Addition of malate or bicarbonate during preincubation of isolated sorghum roots doubled nitrogenase activity (53), but addition of glucose had no effect (Table 1). We have also demonstrated stimulation of nitrogenase activity by malate or bicarbonate in isolated maize roots (2). In addition, it has been shown that malate is one of the best substrates for growth and N_2-fixation by *Azospirillum* spp. (14), the bacteria proposed to be responsible for most of the observed N_2-ase activity in several forage grasses, sorghum and maize roots (7, 16, 41).

TABLE 1
Nitrogenase activity in isolated sorghum roots pretreated with bicarbonate, malate or glucose *

Pretreatment	Concentration	Nitrogenase Activity
	(mM)	(nmoles C_2H_4 hr^{-1} g dry wt^{-1})
Water	--	110
Bicarbonate	5	223
Malate	5	214
	50	186
Glucose	50	111

*Root samples taken from field-grown plants at flowering and pretreated for 4 hrs in the different solutions at the beginning of preincubation (16 hrs). All solutions were adjusted to pH 6.5 except for bicarbonate (pH 8.4) (Van Berkum and Neyra, ref. 53).

PHOTOSYNTHESIS AND NITRATE ASSIMILATION

The existence of a relationship between photosynthesis and nitrate assimilation has been discussed in a number of reports (4, 5, 28, 33, 39) with special emphasis on the generation of reductant for NO_3^- and NO_2^- reduction.

Canvin and Atkins (10) demonstrated for a variety of plant leaves that the assimilation of $^{15}NO_3^-$ into amino-^{15}N is strictly dependent on light. Similarly, Magalhães _et al_. (33) demonstrated that NO_2 reduction by isolated chloroplasts was strictly dependent on light and proceeded in the absence of any added enzyme or electron donor. _De novo_ amino N synthesis was also shown to be stoichiometrically associated with NO_2^- reduction in the light (33).

The energy required for the reduction of NO_3^- to NO_2^- has been proposed to arise from photosynthesis either directly from light via a chloroplast electron transport system (9, 18, 34) or indirectly via oxidation of sugars (28) or malate (40).

With the exception of certain photosynthetic prokaryotes, such as the blue-green alga _Anacystis_ nidulans (9), a direct involvement of light in NO_3^- reduction through a chloroplast reaction is somewhat improbable. First, the enzyme nitrate reductase in higher plants is localized in the cytoplasm (47, 48) and not in the chloroplasts. Second, nitrate reductase in leaf tissue is NADH dependent and not NADPH dependent (3, 52). In chloroplasts, NADPH is the predominant form of reduced pyridine nucleotide in the light (43). Third, the migration of reduced pyridine nucleotides in and out of the chloroplasts is very restricted (25). The problem imposed by the impermeability of the chloroplast membrane to pyridine nucleotides can be avoided by the oxidation of photosynthetic intermediates with the concomitant production of NADH in the cytoplasm (27, 28, 37, 50).

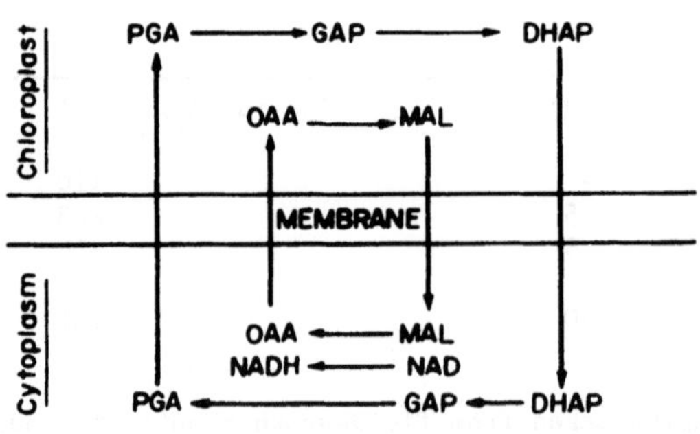

Fig. 1. Two different shuttles to produce NADH in the cytoplasm. For abbreviations, see text.

Two different shuttles have been proposed to operate in the transfer of reducing equivalents out of the chloroplasts (Fig. 1): the phosphoglycerate/dihydroxyacetone-P (PGA/DHAP) shuttle (outer arrows) generating NADH via a cytoplasmic NAD-glyceraldehyde phosphate dehydrogenase (GPD) (50) and the malate/oxaloacetate (Mal/OAA) shuttle (internal arrows) generating NADH via a cytoplasmic NAD-malate dehydrogenase (MDH) (26, 27).

The operation of the PGA/DHAP shuttle as a source of NADH for nitrate reduction was demonstrated both _in vitro_ and _in vivo_ by Klepper _et al_. (28). Evidence has also been presented for the operation of the MAL/OAA shuttle in the generation of NADH for nitrate reduction by corn leaves (37, 40). The results in Table 2 demonstrate that nitrate reductase is almost restricted to the mesophyll cells and that malate oxidation (malate + NAD assay) can be coupled to nitrate reduction in the mesophyll cell fraction. This demonstrates that the mesophyll cells contain not only nitrate reductase but also adequate levels of malate dehydrogenase activities to couple the oxidation of malate to nitrate reduction. Furthermore, the operation of both shuttles (PGA/DHAP and MAL/OAA) has been proposed to operate in corn leaves in association with the C-4 pathway of photosynthesis (37, 38). The mesophyll cells are considered the major site for nitrate assimilation in corn leaves, because they contain NO_3^- and all the required enzyme complements: nitrate reductase (NR), nitrite reductase (NiR), NAD-malate dehydrogenase (MDH), NAD-glyceraldehyde phosphate dehydrogenase (GPD), glutamine synthetase (GS) and glutamate synthase (GOGAT) (37, 38, 45).

According to the sequence of reactions proposed below (Fig. 2) for nitrate assimilation in corn leaves, NO_3^- is reduced in the cytoplasm while NO_2^- is reduced in the chloroplasts of the mesophyll cells. Photosynthetically produced malate (mesophyll

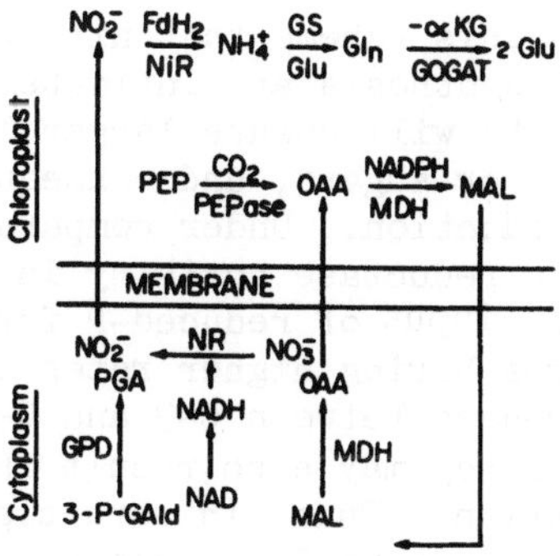

Fig. 2. Nitrate assimilation in the mesophyll cells of corn leaves. For abbreviations, see text.

TABLE 2

In vitro NR assayed directly with NADH or with malate + NAD coupled assay in mesophyll and bundle-sheath cell fractions*

Leaf Fraction	ENZYME ACTIVITIES	
	NADH	MAL +NAD
	(μmoles NO_2^- produced hr^{-1} mg $protein^{-1}$)	
Mesophyll	0.79	0.60
Bundle Sheath	0.084	0.066

*Leaf samples were obtained from 12-day old XL-81 corn seedlings cultured in vermiculite with full strength Hoagland solution (Neyra and Hageman, see ref. 38).

chloroplasts) and sugar intermediates (bundle sheath chloroplasts) are the prime sources of energy for nitrate reduction. Then, oxidation of malate or glyceraldehyde 3-P provides two alternative ways of generating NADH for nitrate reduction in the cytoplasm of the mesophyll cells. The reduction of NO_2^- to NH_4^+ occurring in the chloroplasts, uses energy arising from the light reactions in photosynthesis through electron transport, which provides reduced ferredoxin, the ultimate electron donor for nitrite reductase (32, 39). Ammonia can be primarily incorporated into glutamine via glutamine synthetase in the mesophyll chloroplasts (38, 45). The transfer of N from glutamine to the amino position of glutamate can occur via glutamate synthase (31, 35, 45). Further transfer of N from glutamate to OAA to form aspartate, could be expected from this model, thus providing a reasonable explanation for the appearance of aspartate during C-4 photosynthesis (11, 23).

It is evident from this scheme that nitrate assimilation and the C-4 pathway of photosynthesis are intimately related. A higher photosynthetic rate will ensure larger levels of transport metabolites, like malate or sugars, and consequently more energy available for nitrate reduction. Under comparable conditions of NO_3^- content and nitrate reductase activity in leaf tissues, it can be expected that the input of reduced-N for the plant will be higher in those leaves having higher rates of photosynthesis. In addition, the differences between C-3 and C-4 plants in terms of photosynthetic efficiency may also result in differences in the efficiency of N utilization. Thus, in corn plants (C-4) each gram of N incorporated produced about 100 grams of dry matter (Goic and Neyra, unpublished) while in soybeans (C-3) only 36 grams of dry matter was produced for each gram of N incorporated (Franco, personal communication).

REFERENCES

1. Balandreau, J. (1975). Thesis, D.Sc., Univ. Nancy, France.

2. Baldani, J. I., Pereira, P. A. A., Neyra, C. A. and Döbereiner, J. (1977). Int. Symp. on Limitations and Potentials of Biological Nitrogen Fixation in the Tropics. Poster No. 4.

3. Beevers, L., Fleshers, D. and Hageman, R. H. (1964). Biochim. Biophys. Acta. 89, 453.

4. Beevers, L. and Hageman, R. H. (1969). Ann. Rev. Plant Physiol. 20, 495.

5. Beevers, L. and Hageman, R. H. (1972). Photophysiology (A. C. Giese, ed.), Vol. VII, Acad. Press, New York, pp. 85-113.

6. Bergersen, F. J. (1977). A Treatise on Dinitrogen Fixation. (R. F. W. Hardy and W. S. Silver, eds.), John Wiley and Sons, New York, pp. 519-555.

7. Bülow, J. F. W. von and Döbereiner, J. (1975). Proc. Natl. Acad. Sci. U.S.A. 72, 2389.

8. Burns, R. C. and Hardy, R. W. F. (1975). Nitrogen Fixation in Bacteria and Higher Plants, Springer-Verlag, New York, pp. 14-38.

9. Candau, P., Monzano, C. and Losada, M. (1976). Nature 262, 715.

10. Canvin, D. T. and Atkins, C. A. (1974). Planta (Berl.) 116, 207.

11. Chollet, R. and Ogren, W. (1975). The Botanical Rev. (A. Cronquist, ed.), The New York Botanical Garden, New York, pp. 137-179.

12. Christeller, J. T., Laing, W. A. and Sutton, W. D. (1977). Plant Physiol. (in press).

13. Day, J. M., Neves, M. C. P. and Döbereiner, J. (1975). Soil Biol. Biochem. 7, 107.

14. Day, J. M. and Döbereiner, J. (1976). Soil Biol. Biochem. 8, 45.

15. Döbereiner, J. and Day, J. M. (1975). Nitrogen Fixation by Free-Living Microorganisms (W. D. P. Stewart, ed.), Cambridge University Press, Cambridge, pp. 39-56.

16. Döbereiner, J. and Day. J. M. (1976). Proc. 1st Int. Symp. on Nitrogen Fixation (W. E. Newton and C. J. Nyman, eds.), Washington State University Press, Pullman, p. 518.

17. Döbereiner, J., Day, J. M. and Dart, P. J. (1972). J. Gen. Microbiol. 71, 103.

18. Evans, H. J. and Nason, A. (1953). Plant Physiol. 28, 233.

19. Gibson, A. H. (1976). Proc. Soil Microbiol. and Plant Nutrition (Somiplan) Symp. (W. J. Broughton and C. K. Kohn, eds.), University of Malaya Press (in press).

20. Gibson, A. H. (1976). Proc. 1st Int. Symp. on Nitrogen Fixation (W. E. Newton and C. I. Nyman, eds.), Washington State University Press, Pullman, pp. 400-429.

21. Gibson, A. H. (1974). Mechanisms of Regulation of Plant Growth (R. L. Bieleski, A. R. Ferguson and M. M. Cresswell, eds.), The Royal Society of New Zealand, Wellington, pp. 13-22.

22. Harper, J. E. and Hageman, R. H. (1972). Plant Physiol. 49, 146.

23. Hatch, M. D. and Slack, C. R. (1970). Ann. Rev. Plant Physiol. 21, 141.

24. Havelka, U. D. and Hardy, R. W. F. (1976). Proc. 1st Int. Symp. on Nitrogen Fixation (W. E. Newton and C. J. Nyman, eds.), Washington State University Press, Pullman, pp. 456-475.

25. Heber, U. and Santarius, K. A. (1965). Biochim. Biophys. Acta 109, 390.

26. Heber, U. and Krause, G. H. (1971). Photosynthesis and Photorespiration (M.D. Hatch, C. B. Osmond and R. O. Slatyer, eds.), John Wiley Interscience, New York, pp. 218-225.

27. Heber, U. (1974). Ann. Rev. Plant Physiol. 25, 393.

28. Klepper, L. D., Flesher, D. and Hageman, R. H. (1971). Plant Physiol. 48, 580.

29. Lawrie, A. C. and Wheeler, C. T. (1975). New Phytol. 74, 429.

30. Lawrie, A. C. and Wheeler, C. T. (1975). New Phytol. 74, 437.

31. Lea, P. J. and Mifflin, B. J. (1974). Nature 251, 614.

32. Losada, M., Paneque, A., Ramirez, J. M. and Campo, F. F. del. (1963). Biochem. Biophys. Res. Comm. 10, 298.

33. Magalhães, A. C., Neyra, C. A. and Hageman, R. H. (1973). Plant Physiol. 53, 411.

34. Mendel, J. L. and Visser, D. W. (1951). Arch. Biochem. Biophys. 32, 158.

35. Mifflin, B. J. (1974). Plant Physiol. 54, 550.

36. Minchin, F. R. and Pate, J. S. (1973). J. Exp. Bot. 24, 259.

37. Neyra, C. A. (1974). Ph.D. Thesis, University of Illinois, Urbana, Champaign.

38. Neyra, C. A. and Hageman, R. H. (1974). Plant Physiol. 53, s-363.

39. Neyra, C. A. and Hageman, R. H. (1974). Plant Physiol. 54, 480.

40. Neyra, C. A. and Hageman, R. H. (1976). Plant Physiol. 58, 726.

41. Neyra, C. A. and Döbereiner, J. (1977). Adv. Agron. (N. C. Brady, ed.) (in press).

42. Neyra, C. A., Franco, A. A. and Pereira, J. A. (1977). Plant Physiol. 59, s-690.

43. Ogren, W. L. and Krogman, D. W. (1965). J. Biol. Chem. 240, 4603.

44. Quebedeaux, B., Havelka, U. D., Livak, K. L. and Hardy, R. W. F. (1975). Plant Physiol. 56, 761.

45. Rathnam, C. K. M. and Edwards, G. E. (1976). Plant Physiol. 57, 881.

46. Rinaudo, G., Balandreau, J., Dommergues, Y. (1971). Plant Soil (special volume) pp. 471-479.

47. Ritenour, G. L., Joy, K. W., Bunning, J. and Hageman, R. H. (1967). Plant Physiol. 42, 233.

48. Schrader, L. E., Ritenour, G. L., Eilrich, G. L. and Hageman, R. H. (1968). Plant Physiol. 43, 930.

49. Stewart, W. D. P. (1977). AMBIO 6, 166.

50. Stocking, C. R. and Larson, S. (1969). Biochem. Biophys. Res. Comm. 37, 278.

51. Thibodeau, P. S. and Jaworski, E. G. (1975). Planta (Berl.) 127, 133.

52. Wells, G. N. and Hageman, R. H. (1974). Plant Physiol. 54, 136.

53. Van Berkum, P. and Neyra, C. A. (1976). Plant Physiol. 57, s-533.

54. Van Berkum, P., Neyra, C. A. and Bülow, J. F. W. von (1976). Reunião Brasileira de Milho e Sorgo, Piracicaba, São Paulo.

SOME ASPECTS OF THE *ALNUS*-TYPE ROOT NODULE SYMBIOSIS

C. Miguel, A. Cañizo, A. Costa and C. Rodriguez-Barrueco

Centro de Edafología y Biología Aplicada

C.S.I.C. Salamanca, Spain

INTRODUCTION

The present chapter should perhaps review the progress made on the physiology of nitrogen fixation by non-leguminous angiosperms. But, at this stage there is not much interest in doing so, as many reviews on the subject now are available (4, 8). Therefore, we will try to give a new insight into this field, and information will be restricted to some of the knowledge acquired in our laboratory in recent times. In order to avoid ambiguity, whenever the term non-leguminous is employed we shall be referring to plants such as those given in Table 1. Thus, any other examples which do not fall within the "Alnus type" root-nodulated plants, such as listed by Bond, will be ignored here, although these other associations are very interesting.

LATE REPORTS ON ALNUS-TYPE ROOT NODULATED PLANTS

It is of particular interest that the number of genera and species of this type of plants reported as nodule-bearing is increasing gradually. The figures of 166 and 14 for the number of species and genera, respectively, have been communicated recently to the authors by Bond. Those figures mean a slight increase for the number of genera, but they mean a 50% increase in the number of species over the last ten years. There is apparently some reluctance to admit the genus *Arctostaphylos* among the non-leguminous nodule-bearing plants, because only one report is available on the existence of root nodules in that plant in Alaska (1). Further reports of nodulation of that genus both in Europe and elsewhere (9) were always negative. The authors have had the same experience from a search made in some parts of Central and Northeast Spain. No nodules were found in any of the plants checked at those sites.

The genus Colletia is included in Table 1, mainly from the results reported by Bond in his last IBP (International Biological Program) survey. That work refers to some species mainly Colletia paradoxa and C. armata; a new species is now reported as C. spinosissima (20). Locally called "curro," C. spinosissima is well nodulated and very widespread in Argentina. A. H. Merzari (personal communication to the authors) from the Faculty of Agronomics of the University of Buenos Aires mentions that the beneficial effects of Colletia in Argentinian lands has long been recognized by the Araucans Indians. The same Argentinian authors also report nodulation in four new species of Discaria, another member of the Rhamnaceae: D. americana, D. serratifolia, D. trinervis and D. nana. Another species of Discaria, D. articulata (Phil.) Miers, is reported to bear no nodules.

TABLE 1

Non-leguminous root-nodulated plants bearing Alnus-type root-nodules

Genus	Family	Order
Alnus	Betulaceae	Fagales
Myrica	Myricaceae	Myricales
Casuarina	Casuarinaceae	Casuarinales
Coriaria	Coriariaceae	Coriariales
Dryas Purshia Cercocarpus	Rosaceae	Rosales
Elaeagnus Hippophaë Shepherdia	Elaeagnaceae	Rhamnales
Ceanothus Discaria Colletia	Rhamnaceae	
Arctostaphylos	Ericaceae	Ericales

HOST PLANT-ENDOPHYTE COMPATIBILITY IN NON-LEGUMES

Two points are of interest on the problem of cross-inoculation among non-legumes: (a) the induction of nodulation, and (b) the induction of nitrogenase activity within the nodules formed. Certainly the induction of nodulation is a step prior to the establishment of symbiotic activity. The following experiments were intended to obtain evidence regarding the cross-infectivity at the levels: (a) between species from genera belonging to different families, (b) between species from genera of the same family, and (c) between species from one genus. In the present report we will remark on the nitrogen fixation by these crosses.

Viability of case (b) is shown by the three members of the Elaeagnaceae, *Elaeagnus*, *Hippophaë* and *Shepherdia*. It also is known that cross-infectivity between members of one genus (case c) is possible, though the nitrogen-fixing capability of the nodules formed may range from none to normal depending among other factors such as the geographical origin of the partners employed. However, reports on cross-inoculation between species of different families is in a controversial state. The negative results obtained, not many are available (2, 13), on cross-inoculation trials between two species, mainly *Alnus glutinosa* and *Myrica gale*, typical members of their respective genera, together with the different morphology shown by their respective endophytes *in tissue*, lead to the conclusion that their endophytic symbionts are distinctly different. Moreover, those results and a few more obtained with other plants are the basis for a classification of the non-legume endophytes in different species of the genus *Frankia* within the Actinomycetales, and are the basis for recognizing ten species of actinomycetes corresponding to their host plant (3).

In this type of experiment, crushed-nodules or habitat soil inocula must substitute for pure cultures of the respective endophytes. Therefore, any cross-inoculation studies should be regarded with caution. In this regard we call attention to previous work (23, 25) on inoculation trials between *Alnus glutinosa* and *Myrica gale* employing soil inocula. The high frequency of nodulation in both species by any one soil was interpreted as revealing the presence of the respective endophytes in the soil concerned. This view was supported by the fact that alder and bog myrtle were fairly widely distributed in the soils sampled. Similar results in experiments in water culture with crushed-nodule inocula were explained in terms of the *Myrica* endophyte being present as a contaminant on the surface of the alder nodules used for the preparation of the inoculum. The experiment on the infectivity of soils on alder and bog myrtle has been investigated further. A wider range of soils (one hundred samples) from places where no *Myrica* plants have been recorded, were tested for infectivity on both alder and Myrica gale (5). The results are summarized in Table 2. If we accept that both endophytes are one and the same, the Table indicates that alder plants are more susceptible to nodulation. However, it should be added that Myrica plants always show less growth and require more strict conditions to survive the drought and low environmental humidity prevailing in the author's greenhouse. The experimental design of this investigation only allows a tentative statement, but taken in conjunction with the outcome of previous experiments, the results add weight to the suggestion that the endophytes are the same. There is, however, the difficulty that the phenomenon would mean that the appearance presented by the endophyte within the nodule must be dependent on the nature of the host, for it is known that the endophytes of Alnus and Myrica nodules differ greatly in morphology as well as in structure (15).

Alternatively, the results obtained could be explained on the basis of both endophytes being able to set up an effective symbiosis with M. gale. A third explanation, that the M. gale endophyte is so widespread that it is present in the many soils of different origins.

TABLE 2

Capacity of soils to nodulate A. glutinosa and M. gale expressed as percentage of plants with nodules and statistical significance. From Bermúdez de Castro, Miguel and Rodrìguez-Barrueco, 1976.

Soil origin	Percentage of nodulation A. glutinosa (1)	Percentage of nodulation M. gale (2)	Percentage of nodulation in both hosts	Correlation coefficient P > 99% between (1) and (2)
All the soils	84	69	69	0.815 highly significant
Soils where alder was present	100	91.4	91.4	0.588 highly significant
Soils where alder was not present	44.8	13.7	13.7	0.872 highly significant

tested in the present experiment appears less probable. The above results may be supported by the finding that a crushed alder induces highly effective nodulation on M. faya, a more restricted species of Myrica (21). The inability to find even one soil which would cause nodulation on M. gale without nodulating alder supports the above hypothesis, unless of course it can be shown that the M. gale endophyte can enter the roots of A. glutinosa. With respect to this, we present results on the susceptibility of Alnus roots to the crushed-nodule inoculum from Myrica. Indeed, all the 10 plants of alder become nodulated when inoculated with a crushed-nodule inoculum from greenhouse-grown M. gale. Fixation of nitrogen by the nodules thus formed on alder was normal. Similar results were reported by (18) using Comptonia peregrina (M. asplenifolia) inoculum on alder plants. M. faya nodules also could induce nodulation on alder. Thus seven out of seven plants of Alnus glutinosa became nodulated when inoculated with a crushed-nodule inoculum from M. faya. However no nitrogen fixation was detected in the nodules formed in this instance.

The positive results obtained with Alnus and Myrica, although open to criticism, may serve as a basis for further investigation. The responses have been obtained reportedly as illustrated by the crosses shown in Table 3. The Table gives data in respect to nodulation ability at the inter-family, intra-family and intra-genus levels.

Results from the Tables indicate: First, that the geographical adaptation of the endophytes of Coriaria myrtifolia and C. japonica (7) towards their respective host plants, leading to a mutual incompatibility, is not shown between the species C. myrtifolia and C. nepalensis (Fig. 1), a newly described nitrogen-fixer (10); the latter are geographically distant being restricted to the Mediterranean and Himalaya area, respectively (14). The results obtained with the other combinations show that the inoculum from Hippophaë could induce nodulation in Myrica gale and M. faya, Coriaria myrtifolia and Elaeagnus angustifolia whereas it did not nodulate Alnus glutinosa. Whenever nodules were formed nitrogen-fixation proceeded normally. On the contrary Hippophaë plants did not nodulate when inoculated with a M. gale, M. faya and C. myrtifolia inoculum. The roots of the Hippophaë plant appear to be particularly hard for the other endophytes to enter; the endophyte from Elaeagnus, another member of the Elaeagnaceae, is an exception. The results might indicate that the host here controls the entrance of the microorganism. The Table also gives pH values at which inoculation is carried out. The negative result for the combination A. glutinosa x H. rhamnoides was given negative in the reverse direction, i.e. H. rhamnoides x A. glutinosa. The same negative result was shown by A. glutinosa x C. myrtifolia in both ways. However, the combination Alnus x

TABLE 3

Compatibility of host plant-endophytes in some non-leguminous angiosperms

Host \ Inoculum	Alnus glutinosa	Myrica gale	Myrica faya	Coriaria myrtifolia	Coriaria nepalensis	Hippophaë rhamnoides	Elaeagnus angustifolia
Alnus glutinosa		Inf.Eff. pH 5.4	Inf. -Eff. pH 5.4	-Inf. -Eff. pH 5.4		-Inf. -Eff. pH 5.4-7.0	Inf. Eff. pH 5.4
Myrica gale	Inf. Eff. pH 5.4						
Myrica faya	Inf. Eff. pH 6.3	Inf.±Eff. pH 6.3		Inf. Eff. pH 6.2		Inf. ?Eff. pH 6,3	
Coriaria myrtifolia	-Inf.-Eff. pH 5.4-7	Inf. Eff. pH 5.4-7	Inf. Eff. pH 7.2		Inf. Eff. pH 7.2	Inf. Eff. pH 7.5	
Coriaria nepalensis				Inf. Eff. pH 7.2			
Hippophaë rhamnoides	-Inf.-Eff. pH 5.4-7	-Inf. -Eff. pH 5.4-7	-Inf. -Eff. pH 6.3	-Inf. -Eff. pH 7.2			Inf. Eff. pH 6.3
Elaeagnus angustifolia	Inf. Eff. pH 6.3					Inf. Eff. pH 6.3	

Inf. = infective
Eff. = effective, nitrogen-fixing nodules formed.

Figure 1. From left to right: Coriaria nepalensis plants inoculated with its own nodule-extract, non-inoculated controls and inoculated with a C. myrtifolia nodule-extract (x 1/14).

Elaeagnus worked well, despite the adverse result noted for Hippophaë. Myrica faya plants were susceptible to all the inocula, forming effective nodules in all instances; however, an inoculum prepared from nodules of M. faya failed to induce nodules on Hippophaë and could only induce ineffective nodules on A. glutinosa.

In the light of the results presented in the previous table we suggest that the morphological diversity of the microorganisms at the host species level may be determined primarily by the environment provided in the nodule cell. In the same manner, the effectivity in nitrogen fixation by the nodules formed will be controlled by the host conditions. One finds a surprising number of controversial results published and communicated personally on this subject. The explanation for the results obtained is often poorly defined. In the authors' laboratory even Alnus glutinosa plants failed on occasions to nodulate when inoculated with a nodule extract from corresponding plants. The same has always been true for Coriaria species, although now in our laboratory we easily obtain nodules in C. myrtifolia and C. nepalensis growing in water culture when inoculated with their respective nodule-extracts. In the above results on nodule induction the environmental factors, such as pH, and nutritional requirements of the specific plant, may play a significant role. It seems that the number of different strains of Frankia which may exist must still be established with certainty.

THE EFFECT OF PLANT GROWTH SUBSTANCES ON NODULE INDUCTION IN *Alnus glutinosa*

The development of non-leguminous root-nodules, as well as those of legumes, follows successive stages, each of which may be controlled in a specific way. On the whole it seems that phytohormones are involved in nodule initiation, a concept which is based on the hormone theory of new plant organ formation. This hormone hypothesis is supported by the requirement of some factors like auxin and cytokinin for the induction of repeated division of polyploid cells. It is also known that Rhizobium species synthesize and excrete those hormones in vitro. There is also a specific inhibition of the proliferation stage in nodule initiation when abscissic acid is added exogenously (19). The present work contributes to this aspect of nodule initiation and development with two experiments. Experiment A in which three phytohormones, mainly GA_3 (gibberellin), 2IP (cytokinin) and IAA (auxin), were added in different concentrations and exogenously to alder plants inoculated with a crushed nodule extract. Experiment B was performed to test the relative effect of two auxins, IAA and NAA, and of the auxin precursor tryptophane on the infectivity of an alder-nodule inoculum.

The results of Experiment A are given in Table 4 and are graphically expressed in Figure 2. The lowest concentration of cytokinin employed (10^{-9}M) significantly favored the formation of nodules relative to the other levels. However, nodule dry weight was greater at the highest level of cytokinin where nodules were less numerous. There was a highly significant difference between the 10^{-6}M GA_3 level and the other level tested, including the 0 control. The nodules formed always were very small, and plant development was affected with GA_3 was added, particularly at the highest level.

Considering separately the effect of each substance in Experiment B (Table 5, Fig. 3), it is seen that a highly significant difference exists between the number of nodules formed at any of the levels of auxins (IAA or NAA) and tryptophane relative to the controls. However, no significant difference was found at most levels between the effect caused by the various auxins levels tested and those of tryptophane. The highest level of NAA (10^{-6}M) was toxic and affected both nodule number and plant dry weight drastically.

TABLE 4

Effect of IAA, 2IP and AG_3 on the nodule-inducing capacity of a crushed alder-nodule inoculum. Age of plants at harvest: 6 weeks. Means of 4 plants.

Treatment	IAA		2IP		AG_3	
	Number of nodules	Dry wt. per plant, mg.	Number of nodules	Dry wt. per plant, mg.	Number of nodules	Dry wt. per plant, mg.
0	7 (5-8)	261.1 (202-359)	7 (5-8)	261.1 (202-359)	7 (5-8)	261.1 (202-359)
10^{-9}M	-	-	10 (8-13)	373.7 (309-431)	-	-
10^{-8}M	20 (16-26)	407.4 (304-494)	8 (6-11)	295.8 (148-589)	5 (3-8)	91.1 (34 -234)
10^{-7}M	17 (12-23)	342.5 (212-606)	-	-	-	-
10^{-6}M	15 (13-19)	281.9 (161-524)	4 (3-6)	455.6 (327-659)	97 (73-109)	52.6 (41 - 66)

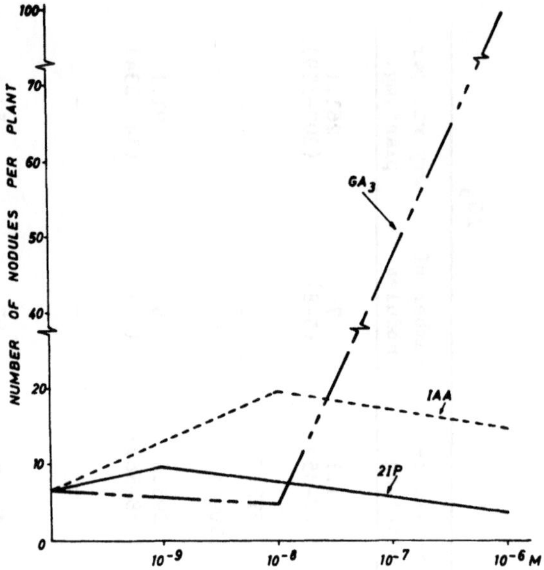

Figure 2. Effect of IAA, 2IP and GA_3 on the nodule-inducing capacity of a crushed nodule inoculum from Alnus glutinosa.

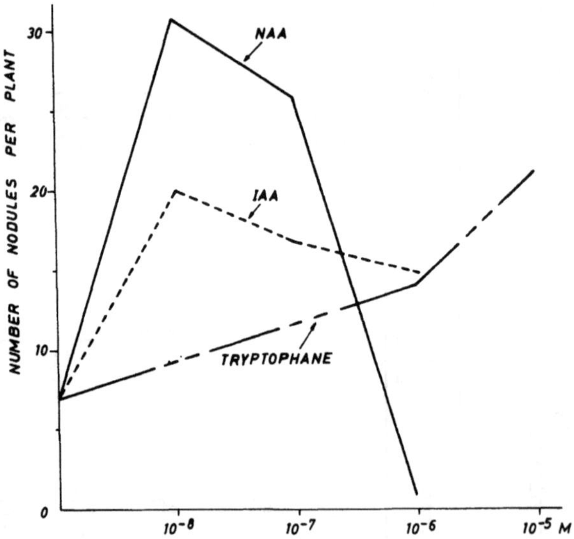

Figure 3. Effect of IAA, NAA and Tryptophane on nodule induction in Alnus glutinosa inoculated with a crushed-nodule inoculum.

TABLE 5

Effect of IAA, NAA and Tryptophane on nodule induction in *Alnus glutinosa*. Age of plants: 6 weeks. Mean values of 4 plants.

Treatment	IAA		NAA		TRYPT	
	Number of nodules	Dry wt. per plant, mg.	Number of nodules	Dry wt. per plant, mg.	Number of nodules	Dry wt. per plant, mg.
0	7 (5-8)	261.1 (202-359)	7 (5-8)	261.1 (202-359)	7 (5-8)	261.1 (202-359)
10^{-8}M	20 (16-26)	407.4 (304-494)	31 (27-40)	323.0 (283-390)	-	-
10^{-7}M	17 (12-23)	342.5 (212-606)	26 (21-34)	360.5 (274-513)	-	-
10^{-6}M	15 (13-19)	281.9 (161-524)	1 (0-1)	28.4 (22-41)	14 (10-16)	474.2 (389-686)
10^{-5}M	-	-	-	-	21 (14-27)	324.8 (220-482)

The possible role of the phytohormones in nodule development has been studied by others (11 16, 17), and in our own laboratory. We have detected endogenous gibberellin, cytokinin and auxin activity in root-nodules of the alder. If the nodules are capable of synthesis of the respective phytohormones mentioned, the relative contribution by the endophyte may be of importance. Certainly the role of the cytokinins may be related to the directed translocation of assimilates from the nodules to the shoots. Cytokinins have also been reported to affect photosynthetic rates, as well as to promote chlorophyll synthesis and chloroplast enzyme formation (26). Gibberellins and cytokinins have been detected in the different plant organs of the alder and have been found in highest concentration in the nodules. The nodules also were particularly rich in auxins. These substances now have been shown to increase nodule numbers at all levels tested. Further investigation on the effect of those substances, endogenous and exogenous, is required (6, 24). We may add that the cytokinins excreted by Rhizobium in vitro (22) are likely to play a part in nodule initiation. The results obtained here add support to that suggestion, since cytokinin addition at the level 10^{-9}M significantly increased the number of nodules relative to the cytokinin-free control. Therefore, a shift of hormone balances during root-nodule initiation and development is suggested. Auxins would affect favorably the first infection stages and cytokinins would contribute primarily to nodule development.

The gibberellin levels tested did not produce results that are readily interpreted, as all affected plant growth drastically. Therefore nothing positive can be concluded from the results obtained for the gibberellin levels tested for their effect on nodule formation. However, the finding that tryptophane increases nodule number, and that its transformation to IAA can be favored by gibberellin (12) suggests that the level of gibberellin supplied with the alder nodule extract might suffice to transform the tryptophane added.

REFERENCES

1. Allen, E.K., Allen, O.N., and Klebesadel, L.J. (1964) Science in Alaska, pp. 54-63. G. Dahlgren, Anchorage, Alaska.

2. Becking, J.H. (1970). Plant and Soil, 32, 611.

3. Becking, J.H. (1974). Bergey's Manual of Determinative Bacteriology, 701-706. R.E. Buchanan and N.E. Gibbons. The Williams and Wilkins Co. Baltimore.

4. Becking, J.H. (1975). The development and Function of Roots. pp.507-566. Third Cabot Symposium. Academic Press, London, N.Y.

5. Bermúdez de Castro, F., Miguel, C., and Rodríguez-Barrueco,C. (1976) Ann. Microbiol. (Inst. Pasteur), 127A, 307.

6. Bermúdez de Castro, F., Cañizo, A., Costa, A. Miguel,C. and Rodríguez-Barrueco, C. (1977). Recent developments in Nitrogen Fixation. W.E. Newton, J.R. Postgate and C. Rodríguez-Barrueco. Academic Press. London. In Press.

7. Bond, G. (1962). Nature, 193, 1103.

8. Bond, G. (1974). The Biology of Nitrogen Fixation, pp.342-378. A. Quispel. North Holland Pub. Co. Amsterdam.

9. Bond, G. (1976). Symbiotic Nitrogen Fixation in Plants,pp.443-474. P.S. Nutman. IBP. Cambridge University Press, Cambridge.

10. Cañizo, A. and Rodriguez-Barrueco, C. (in press). Revue d'Ecologie et Biologie du Sol.

11. Dullaart, J. (1970a). J. exp. Bot. 21, 1975.

12. Dullaart, J. and Duba,L.I. (1970b). Acta Bot. Neerl. 19,877.

13. Fletcher, W.W. (1953). Ph.D. Thesis. University of Glasgow.

14. Good, R. (1930). New Phytol. 29, 170.

15. Hawker,L.E. and Fraymouth, J. (1951). J. gen.Microbiol. 5,369.

16. Henson,I.E. and Wheeler, C.T. (1977a). New Phytol. 78, 373.

17. Henson,I.E. and Wheeler, C.T. (1977b). J.exp. Bot. 28, 205.

18. Lalonde,M., Knowles, R. and Fortin, J.A. (1975).Can. J.Microbiol. 21,1901.

19. Libbenga, K.R. and Bogers, R.J. (1974). The Biology of Nitrogen Fixation. pp. 430-472. A. Quispel. North Holland Pub.Co. Amsterdam.

20. Medan, D. and Tortosa, R. (1976). Boletin de la Sociedad Argentina de Botánica, 17, 323.

21. Miguel,C. and Rodríguez-Barrueco, C. (1974). Plant and Soil,41,521.

22. Phillips, D.A. and Torrey, J.G. (1970).Physiol. Plant.23,1057.

23. Rodríguez-Barrueco, C. (1968). J. gen. Microbiol., 52, 189.

24. Rodríguez-Barrueco,C. and Bermúdez de Castro, F. (1973). Physiol. Plant. 29, 277.

25. Rodríguez-Barrueco, C. and Bond, G. (1976) Symbiotic Nitrogen Fixation in Plants, pp. 561-565. P.S. Nutman. IBP.Cambridge University Press. Cambridge.

26. Wareing, P.F., Khalifa, M.M. and Treharne, K.J. (1968) Nature 220, 453

LEGUMES AND ACID SOILS

C.S. Andrew

C.S.I.R.O.
Cunningham Laboratory, Mill Rd., St. Lucia, Queensland, 4067, Australia

ABSTRACT

This paper discusses the prime effects of acid soil nutritional factors, other than those associated with the micro-plant nutrients, on legume-Rhizobium symbiosis in tropical legumes, with particular emphasis on pasture species. Factors discussed are pH, aluminium and manganese excess, and deficiencies of calcium, and phosphorus.

Discussion centres on the effect of each factor on nodule initiation, the efficiency of the legume-Rhizobium symbiosis, and on plant metabolism and growth. The responses of different species of legumes are also covered.

The relevant importance and inter-relationships of the above factors are governed largely by the geology and the inherent soil forming processes; thus low pH is usually associated with low calcium and phosphorus availability, but the degree of aluminium and/or manganese excess is influenced by the soil parent material and organic matter associated with the physical conditions of the soil. Soil pH has a marked effect on initiation of nodules and a lesser effect on efficiency of symbiosis. Calcium improves nodulation primarily at the intermediate soil pH range of 5.0 to 6.0; however, at low and high pH, the effect of the hydrogen ion concentration is dominant over that of calcium supply with respect to nodulation. Calcium is also very important in the efficiency of symbiosis and general plant growth. Aluminium excess is detrimental to nodule initiation, efficient symbiosis and plant growth, especially the roots. Manganese excess, unlike aluminium, does not affect nodule initiation or efficiency of symbiosis but causes a marked reduction in plant growth. Phosphorus deficiency minimises nodule initiation, and also lowers the efficiency of the symbiosis and reduces plant growth.

The paper also discusses certain aspects of nutritional diagnosis of the above factors.

INTRODUCTION

The importance of leguminous plants in agriculture is well recognised; indeed much of the soil nitrogen may have been mobilised per medium of legumes (36). The prime attribute of a leguminous plant is that of forming a symbiosis with Rhizobium and the resultant mobilisation of nitrogen from the air into the plant component parts. The quantity of nitrogen mobilised is quite often greater than that necessary to promote optimum dry matter production of the legume, and indeed this attribute may have been selected for during agricultural development.

There are several factors that may affect nitrogen production by legumes:

1) initiation and development of nodules,
2) efficiency of fixation of the legume-Rhizobium symbiosis,
3) nutrition of the host legume.

The inorganic plant nutrients, thirteen in number, have varying roles in relation to the above factors. Thus, certain nutrients have a single role e.g. initiation of nodulation, others have a multiple role, e.g. initiation of nodulation and efficiency of symbiosis. Indeed the effectiveness of the symbiosis, and of plant growth, results from a system in which adverse nutritional factors are absent or minimal, and essential nutrients are in sufficiency and in the correct balance.

Acid soils usually have some inherent adverse concentrations of nutrients and elements, coupled with some related nutrient deficiencies. The principal effects of soil acidity may be resolved into hydrogen ion concentration, deficiencies of calcium, phosphorus and molybdenum, and excessive quantities of aluminium and manganese (38, 91). This paper discusses each of the above factors (excluding molybdenum, see paper by Franco) in relation to nodule initiation and development, the efficiency of the legume-Rhizobium symbiosis, and the general effect on plant growth, with particular reference to tropical pasture legumes.

Soil pH

Early agriculturists in Europe and the U.S.A. recognised the importance of medium soil pH values (pH 6.0 - 7.0) for the optimum growth of such legumes as Medicago sativa, Trifolium repens and many other temperate species. Soils of low pH (pH 4.0 - 5.5) were known to be generally infertile. However, from the 1940's onwards several research workers indicated that the effect of low pH (pH 4.0 - 5.5) on legume growth was not significant provided all essential nutrients including nitrogen, were in sufficiency (12,69), and no toxic concentrations of elements were present. Such findings were obtained from investigations involving supplied nitrogen. Subsequent research has shown that pH plays an important role in the efficient functioning of a legume grown in the absence of applied nitrogen. In this connection, important factors are: a) Rhizobium survival and growth in the soil, b) root infection and nodule initiation, c) legume-Rhizobium symbiotic efficiency, and d) nutrition of the host plant.

Species of Rhizobium are known to differ in their tolerance of soil acidity (102), e.g. the slow growing types such as R. japonicum and R. lupini are more acid tolerant than the fast growing types such as R. meliloti. Within the latter group, Vincent (103) showed differences, e.g. R. leguminosarum was more tolerant than R. trifolii > R. phaseoli > M. meliloti. Differences between certain sero-groups have also been shown in R. japonicum (23). Few data exist for tropical legume rhizobia in this regard.

Most nodulation measurements that have been related to pH have been obtained from soil experiments. In such cases the observations may be confounded by factors other than pH per se., e.g. aluminium and manganese excess. Despite such problems, several workers have been able to characterise species with reference to pH requirements, e.g. alfalfa 6.2, red clover 6.0, sweetclover 6.5, cowpeas 5.0, lespedeza 4.5, soybean 6.0 (95). McIlvane and Pohlman (63) quoted optimum values for the above species at 6.7, 5.9, 6.6, 5.3, 5.8 and 6.0 respectively. Norris (80,81) claimed that tropical legumes in general were more tolerant of soil acidity than temperate legumes; however, more recent investigations have revealed that within both tropical and temperate legume groups, differences between species in their reaction to pH exist (3,28,48,53,73,75).

Following development and advancements in culture techniques, especially those related to water and sand culture investigations, researchers have been able to isolate the effect of pH from that of calcium, aluminium and manganese as well as other confounding factors. However, solutions having low nutrient concentrations present certain problems; firstly, solutions are poorly buffered, especially at low phosphorus and base concentrations, and secondly, relative fluctuations in nutrient concentration are large. These problems are mitigated by large volume solutions and/or frequent solution change in flowing cultures (2,14,27). Solutions used in early attempts to investigate the effect of pH on legume growth, contained relatively high concentrations of nitrogen. In such investigations most legumes, including M. sativa, showed only minor reductions in growth when the pH was lowered from 8.0 to 4.0 (3,12,35,69). A group of tropical and temperate pasture legumes was investigated by Andrew (3). Results from that experiment show that the growth of M. sativa, Macroptilium lathyroides, Desmodium uncinatum, Lotononis bainesii, Stylosanthes humilis, Glycine wightii, M. truncatula, M. scutellata, Trifolium repens, T. semipilosum, and T. rueppellianum was reduced only slightly by a reduction in pH from 6.0 to 4.0, when grown in the presence of applied nitrogen. However, the Medicago species and G. wightii were affected more than the other species. The growth of Stylosanthes humilis was greater at pH 4.0 that at pH 5.0 and 6.0. Examples of these results are shown in Figure 1.

In contrast to the above, plants wholly dependant on nitrogen fixation as a source of N demonstrated marked differences. For example, M. sativa failed to nodulate at pH 4.0 and 4.5, and required a pH of 6.0 to achieve 100% nodulation. In contrast, M. lathyroides and L. bainesii plants were effectively nodulated over the range of pH 4.0 to 6.0, whereas, the nodulation of D. uncinatum and S. humilis was reduced at pH 4.0 below that at pH 4.5

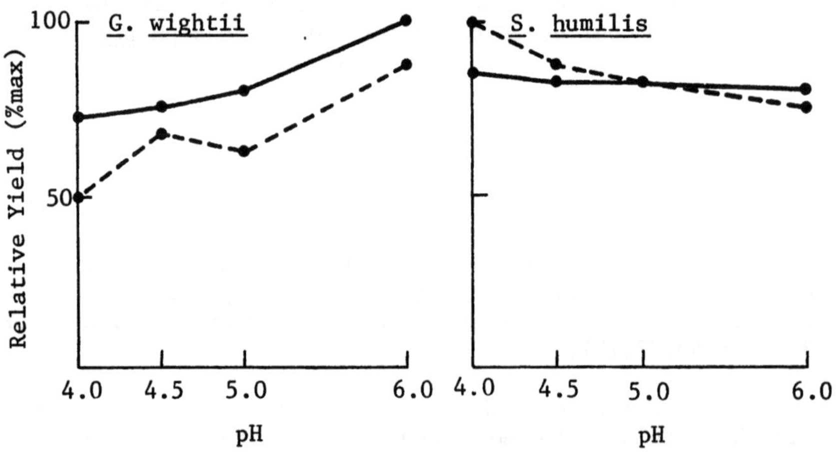

Figure 1. Effect of pH and calcium substrate concentration on the relative DM yield of two legumes grown in the presence of added nitrogen. ------ 0.125 mM calcium; ——— 2.0 mM calcium.

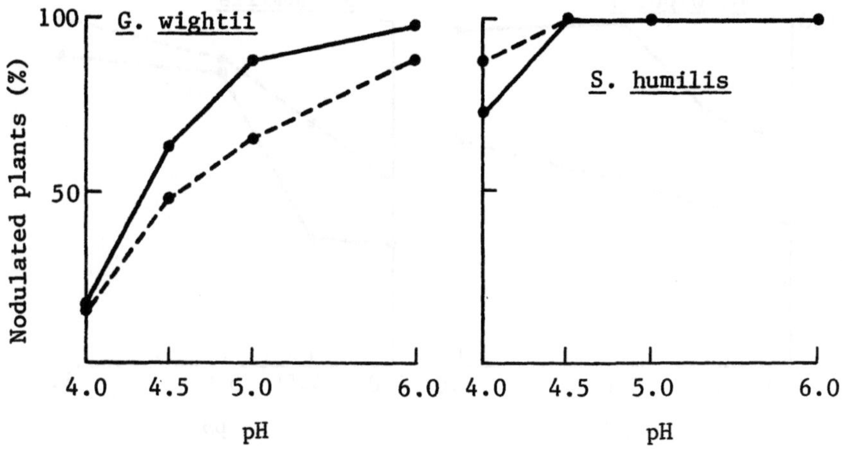

Figure 2. Effect of pH and calcium substrate concentration on percentage of legumes nodulated. ------- 0.125 mM calcium; ——— 2.0 mM calcium.

and above. The nodulation pattern of T. semipilosum and T. rueppellianum, (species introduced to Australia from high altitudes in the tropics), responded similarly to D. uncinatum (3). Examples of these results are shown in Figure 2. In fertile soils, significant quantities of available soil nitrogen are present and this may modify the response of legumes to pH, however, it is considered that the symbiotic effect is all important in the efficient functioning of the system.

The role of pH in the infection process is not clearly understood. Various hypotheses have been put forward, viz:- morphology of root hairs (71); colonisation of roots by rhizobia at sites of infection (55); structure of the mucigel layer (50); association of calcium (affected by pH) with cell walls and cell membranes, or enzyme activity (72,73).

Until recent times it was considered that once nodulated, legumes became self sufficient for nitrogen and produced optimum dry matter (41,54,70). However, it has now been shown that nodulated plants grown at pH levels below the optimum were less productive in dry matter than those grown at optimum pH (3,15,76). Figure 3 shows the effect of pH on the dry matter of some nodulated plants. The latter results are more in keeping with soil experiment results quoted earlier in this paper. Jensen (42) showed that maximum nitrogen concentrations in plant tops, coincided with

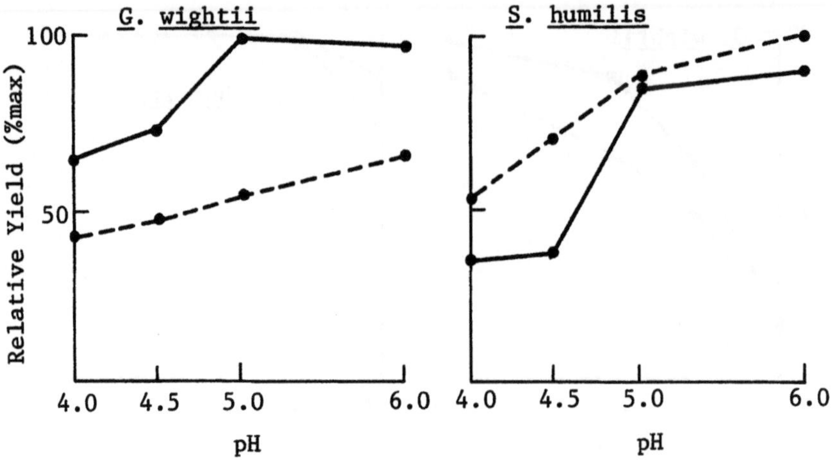

Figure 3. Effect of pH and calcium substrate concentrations on the relative DM yield of nodulated plants.
------- 0.125 mM calcium; ——— 2.0 mM calcium.

maxiumum dry matter production. For example, *M. sativa* fixed approximately twice as much nitrogen at pH 7.0 - 7.3 than at pH 4.9 - 5.2. The data of Andrew (3), Bear and Wallace (16), McNeur (62) and Andrew and Norris (7), support these observations.

Munns (74) has collated data from several publications to show the relative acid tolerance of thirty agronomic legume species in relation to degree of nodulation and growth index. Such a compilation, while being an approximation, is a valuable guide for agronomic use.

Calcium and its interaction with pH

Low available calcium supply is a corollary of soil acidity. In the latter case, soil acidity *per se* represents excess concentrations of hydrogen ion, essentially an adverse factor, whereas in the former case, low or deficiency concentrations of calcium pertain. In experiments in which both pH and calcium treatments exist, and interactions between the two sets of treatments also occur, the relative effect of pH usually dominates that of calcium. The largest effect of calcium occurs at the intermediate pH range, although this may be conditioned by the concentrations of calcium used in the experiment. Data from the investigation of Munns (72) indicate that nodulation of *M. sativa* was inhibited at calcium concentrations less than 0.2 mM, regardless of pH. Data obtained by Andrew (3) for 11 species, and by Loneragan and Dowling (54) for T. subterraneum, show that nodulation was not inhibited at low calcium concentrations provided the pH was optimum for both nodulation and growth. The experiment of Andrew (3) included a low (0.125 mM) and a high calcium treatment (2.0 mM). Species that were very sensitive to low pH (4.0 - 4.5 inclusive) did not nodulate or had poor nodulation irrespective of calcium treatment; likewise at pH 6.0 calcium treatment had little effect, but at pH 5.0 and 5.5 the high calcium treatment enhanced nodulation (see Figure 2 for examples). Calcium is an important nutrient in the initiation and efficient functioning of nodules (54,69,82,89, 93). The interaction of calcium with pH is explained in part by the effect of the hydrogen ion in reducing the uptake of calcium (6,13,16,27,96). Further, it has been shown that the concentration of calcium in legume roots is low, relative to that in the legume tops, and that enhanced calcium nutrition favours increased concentration of calcium in the legume tops without unduly affecting calcium concentration in the roots (6). It is suggested that the main factors operating are concerned with the absorption of calcium by the roots, and the translocation of calcium from the roots to the tops. Indeed, ratios of calcium concentrations in tops to those in roots may well represent a form of nutritional assessment.

The above discussion implies a strict interaction of calcium and pH; however, the concentrations of calcium in non-nodulated plants can in some instances be greater than those in nodulated plants, particularly at low pH treatment (e.g. 4.0 and 4.5), (6). Munns (72) stressed the importance of calcium at the time of nodule initiation. In the experiments of Andrew and of Munns, non-nodulated plants were acutely nitrogen-deficient and no new roots were developed; this implies that they failed to nodulate subsequently because of the severe shortage of nitrogen and their inability to produce new roots and thus new sites for infection of rhizobia, and that this allowed accumulation of calcium in the plant during a non-growth phase.

A review of data from the literature relative to effects of pH and calcium on legume performance reveals:- (a) legume species differ in their ability to nodulate at varying substrate pH values and calcium concentrations, (b) optimum growth of nodulated plants and maximum nitrogen production is dependent on satisfactory pH and calcium supply, (c) the effects of pH and calcium on plants grown in the presence of applied nitrogen are minor compared with those on plants grown in the absence of applied nitrogen. In all these respects the species have similar relationships but to varying degrees. Differences in the responses of the species cannot be wholly related to their mineral composition, e.g. S. humilis and L. bainisii were both very tolerant of low pH levels, however the former accumulated calcium to high levels, whereas the latter had relatively low calcium concentrations in its tissues (6). Despite this difference between the two species, both were adversely affected by high calcium concentrations (2.0 mM) in the substrate. Certain species of Stylosanthes are known to be adapted to relatively infertile soils and to tolerate acid soil conditions (31,39). In most cases positive responses of Stylosanthes to lime addition when grown in acid soil have not occurred beyond pH 5.5 (18,31,76). However, the reduction in dry matter production of S. humilis and L. bainesii at high calcium substrate concentration observed by Andrew (3) and further supported by Munns (74), has no simple explanation. In both cases the effect was not explained by micro-plant nutrient status, or by magnesium and potassium deficiencies. Calcium concentrations in S. humilis are usually high, even at low calcium substrate treatment; excessive amounts of calcium in plants grown with high calcium supply may well represent partial toxicities or at least inbalances of nutrients. This argument cannot be used in the case of L. bainesii because the calcium concentration in this species is relatively low, irrespective of the source of the plant material, laboratory or field grown.

Aluminium

Aluminium constitutes a large proportion of most soils, however the concentration of aluminium in the soil solution is neglible, except in acid soils, especially those with relatively large amounts of organic matter. In addition, aluminium can occupy varying proportions of the exchange sites in soil. The deleterious nature of aluminium in agriculture is well recognised (56,57,59,70,77,90,104). The effect of aluminium on the growth of different species within a genus, and varieties within a species, has been studied also (30,40,58,83).

Many of the laboratory experiments accomplished, have encompassed relatively high concentrations of aluminium in solution, e.g. up to 40-50 ppm of Al added. Such concentrations have little in common with those found in soil solutions, where concentrations rarely exceed 4 ppm. In the latter case, relatively high concentrations of aluminium are associated with low pH and low concentrations of available phosphorus and calcium. Under such circumstances the aluminium concentrations in solution are relatively stable. In early laboratory experiments, especially in those that were based on standard Hoagland solution or its equivalent, pH was usually 5.0 - 6.0 and the phosphorus concentrations were of the order of 1.0 to 2.0 mM (≡ 31 to 62 ppm P). Under such systems the concentrations of aluminium, added as treatments, were most unstable and precipitation of salts occurred. Results from these experiments bear little relationship to those in which the solution pH, aluminium, phosphorus and calcium concentrations have been kept at a minimum. For example, in investigations carried out by Munns (70) and Andrew et.al., (11), pH values were maintained at 4.0 to 4.2 and phosphorus concentrations were maintained at 1 to 2 ppm, and in some instances, 0.3 ppm. Even under these conditions it was necessary to frequently monitor and adjust the aluminium concentration to the required treatment levels. Suggestions for the cause of reductions in aluminium concentrations in solution culture are:- precipitation in the solution, absorption by roots, adsorption on specific exchange sites on the roots and adherance to root and root structures (59). Despite the differences in the details of the experiments conducted recently with those of earlier periods, the relative differences in the tolerance of species to aluminium excess are near parallel for the two systems. Difficulties arise however in attempting to relate the results of one experiment to those of another. Further problems arise with respect to the study of the effect of aluminium on nodule initiation, and efficiency of symbiosis, particularly for those species that fail to nodulate at pH levels of 4.0 to 4.5.

Species comparative experiments using legumes are few in number and are mostly concerned with temperate species, viz:- two Trifolium species (37), M. sativa and T. subterraneum (70), five temperate legumes (57), several M. sativa varieties (83). More recently five tropical and six temperate (including 2 African Trifolium species) legume species were compared by Andrew et.al., (11). All of the foregoing experiments were accomplished in solution culture having nitrogen added. The addition of aluminium up to 2.0 ppm in solution had no significant depressing effect on the dry matter production of M. lathyroides, L. bainesii, or S. humilis. The growth of D. uncinatum was depressed significantly, but only by the 2.0 ppm aluminium treatment, whereas the growth of G. wightii and M. sativa was severely depressed by all additions of aluminium. M. truncatula and M. scutellata responded similarly to M. sativa,growth of T. repens and T. semipilosum was reduced at the 2.0 ppm treatment but there was no significant effect on T. rueppellianum. The other interesting feature of the yield data from that experiment was a significant yield increase by the tolerant species, especially the tropical species, except for G. wightii, when grown in the presence of 0.5 ppm aluminium, compared with that obtained from the control treatment (Figure 4). Currently Miss Carvalho, (personal communication), taking advantage of a genus that generally nodulates at low pH, is studying the reaction of six Stylosanthes species to aluminium in the presence and absence of applied nitrogen. In the absence of applied nitrogen S. humilis grown in sand culture with four aluminium solution

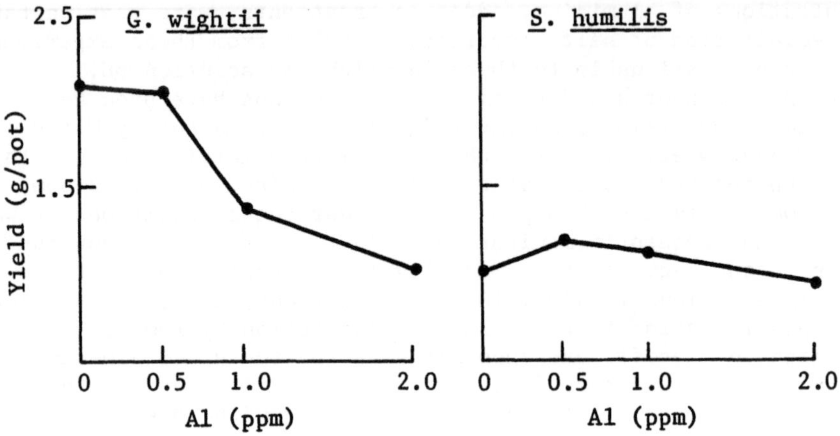

Figure 4. Effect of aluminium on the DM yield of tops of two tropical legumes.

treatments (0, 25, 75, 125 μM Al) showed a depression in yield at the third and fourth treatments and an increase in yield at the second treatment relative to that from the control. In that experiment weight of nodules was also reduced at the third and fourth treatments but the plant top to nodule weight ratios did not vary (Figure 5). Furthermore, the nitrogen concentration in the plant tops of S. humilis was not altered by the aluminium treatments. However, in S. scabra and S. viscosa nodule numbers and nitrogen concentrations were reduced. Other results from that program, including the results of transfer experiments and efficiency of symbiosis in fixing nitrogen (as judged by chemical composition and acetylene reduction techniques) will be reported in Miss Carvalho's Ph.D. thesis.

The reasons for differential tolerance of aluminium by species are not accurately known. Several suggestions have been made, viz: (a) organic acids produced by plants may act as chelating agents and prevent the precipitation of aluminium (45), (b) aluminium organic acid complexes might provide a detoxification system in resistant species (21), (c) identification of certain stable aluminium-organic acid complexes in plants, (97), (d) ability of plants to alter pH of substrate to varying degrees, (30). In another study involving twelve crop plants the latter argument was not verified (61). Differences in species response to aluminium excess have been related to cation capacity of plant roots (105) but this explanation has been questioned on the grounds that the

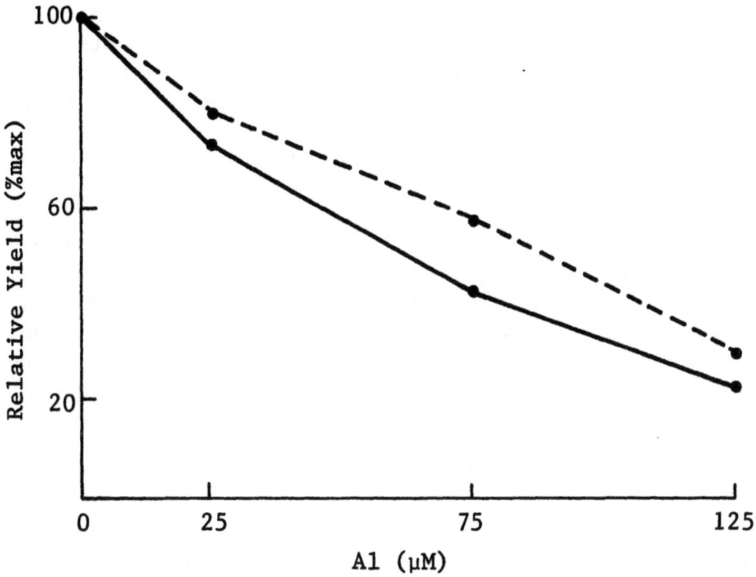

Figure 5. Effect of aluminium of relative DM yield of tops and nodules of S. humilis.
——— tops; -------nodules

carboxyl groups of the pectin in the cell wall, which are primarily responsible for cation exchange capacity, have little influence on ion uptake (21). Cation data from the investigation by Andrew et al. (11) also negates the explanation relative to cation exchange capacity. They showed that L. bainesii and S. humilis had contrasting potassium and calcium concentrations and yet both species were equally tolerant of aluminium excess.

Relative to aluminium concentrations in the plant roots, those in plant tops are very low. The concentration of aluminium in plant tops is not considered a useful guide for assessment of degree of toxicity or as a useful index of the degree of tolerance, due primarily to dust and soil contamination. However, there are suggestions that the tops of tolerant species usually have a lower aluminium concentration than sensitive species (11). This implies that uptake and translocation may be important factors, and this immediately involves the complementary effects of other plant nutrients. Two nutrients are considered here, namely calcium and phosphorus.

A prime effect of aluminium on nutrient composition of plant tops is a substantial reduction in calcium concentration (11,29,44, 57,59,70). Chemical composition data obtained from roots of plants grown in the presence of aluminium shows that the initial effect is a lowering of the root calcium concentration relative to that in the tops (11). This implies that aluminium has a prime effect on the uptake of calcium and its subsequent translocation. It has been shown that aluminium inhibited cell division in the adventitious roots of *Allium cepa* and *Agrostis stolonifera*, and the cessation of root elongation was closely correlated with the disappearance of mitotic figures (19,20). Many investigators have reported on the effect of aluminium in plant root morphology, colour of roots, and the development of mucilagenous material on the roots and have suggested that the prime effect is on the roots rather than on the plant tops. The data of Andrew et al. (11) and Andrew and Vanden Berg (10), show that the effect of Al on dry matter production of plant tops was greater than on the roots, and that the root weight ratios of affected plants was increased.

The effect of aluminium on the phosphorus concentrations in plant tops varies with the species (11,29,57,70,87,107). Such reductions are usually accompanied by increases in the phosphorus concentrations of roots. These have been attributed to, (a) internal precipitation of phosphorus and aluminium in or on the root (107), (b) increased uptake of P by the roots (19,90), (c) a dual effect, firstly a fixation of phosphorus by an adsorption-precipitation reaction at the cell surface or in the free space, and secondly, a reaction within the cell, possibly within the mitochondria (21). In contrast to the above effects of aluminium on phosphorus, several workers have indicated an enhancement of phosphorus uptake in plant tops by aluminium. Viets (101) working with whole plants and excised roots showed that in healthy plants, short term treatments with certain poly-valent cations including aluminium, stimulated the uptake of phosphorus. Similarly increases have been reported for *P. vulgaris* (86) and for a group of four relatively tolerant tropical legumes (10,11). A study of the effect of aluminium (plants pre-treated and grown in varying Al concentrations for 35 days) on the subsequent short time uptake and translocation of phosphorus in six legume species, using a radioactive labelled phosphorus treatment (10) showed that aluminium pre-treatment resulted in a progressive increase in phosphorus sorption by the plant roots over the full range of treatments. However, in the case of *M. sativa*, the 2.0 ppm treatment reduced the sorption below that of the remaining treatments. The phosphorus uptake into the tops of *M. lathyroides*, *D. uncinatum*, *L. bainesii*, and *S. humilis* was increased by aluminium pre-treatment, but that of the tops of *G. wightii* and *M. sativa* was reduced (Figure 6). Phosphorus translocation efficiencies established for the six species were generally reduced by aluminium pre-treatment, particularly in the most sensitive species *G. wightii* and *M. sativa*.

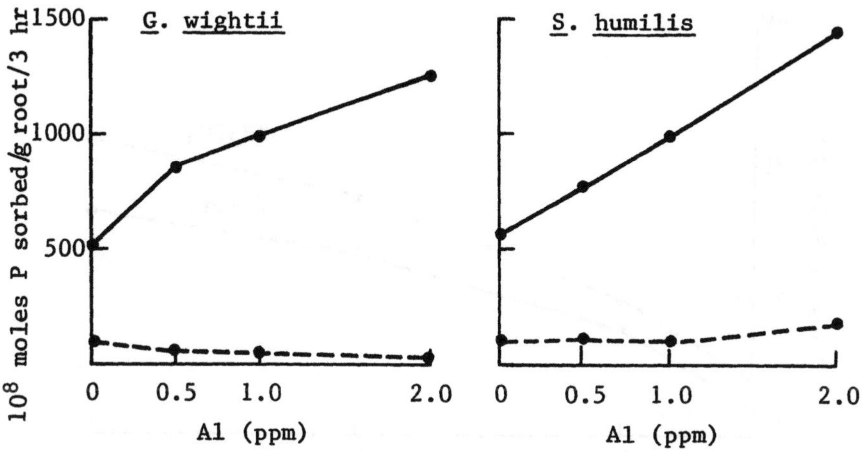

Figure 6. Effect of aluminium on phosphorus sorption by two legumes. ——— roots, ------- tops.

Results of phosphorus uptake studies using healthy excised roots of M. sativa, and in which aluminium treatments were imposed during the uptake phase, showed that the aluminium treatment enhanced phosphorus uptake over a ten minute period (10). During the foregoing experiment it was observed that roots which had sorbed phosphorus in the presence of added aluminium lost P-32 activity to the desorption solution, whereas roots untreated with aluminium did not. Over a 30 minute desorption period, 30% and 25% of the P-32 in roots was desorbed or removed by potassium dihydrogen phosphate and water respectively (Figure 7). These results pose the question as to whether the loss of phosphorus from roots affected by aluminium excess is due solely to desorption or whether some of the loss may be due to mass flow. The "Viets effect" (101) applies to aluminium as well as calcium and magnesium. The two latter ions used at moderate concentrations are not damaging to plant tissue or to its performance, and it is also known that calcium has some effect on cell membrane permeability (100); cells become leaky in a calcium - deficient environment. The relative effects of aluminium upon phosphorus absorption, while contrasting, may be due in part to the same phenomenon. Thus when aluminium is used at low concentrations and for short periods, growth of some plant species may be enhanced and phosphate sorption increased, but if used at excessively high concentrations for particular species, growth is impaired and phosphate sorption reduced.

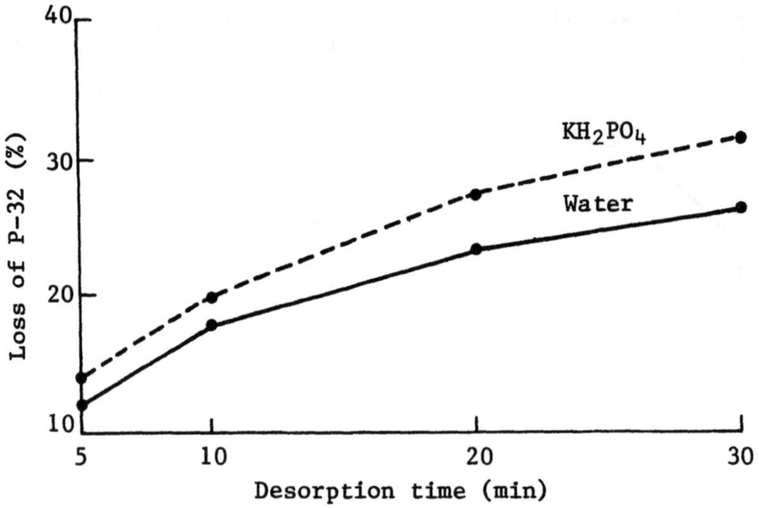

Figure 7. Loss of P-32 from aluminium-treated excised roots (8 ppm aluminium) of M. sativa during post rinsing with water or $1x10^{-5}M\ KH_2PO_4$ for varying time periods.

The visual symptoms of aluminium excess on roots are in some respects similar to those of calcium deficiency. This poses the question as to the inter-relationship of pH, calcium status, and aluminium substrate concentration. For example, in most acid soils, low pH is highly correlated with low available calcium and high available aluminium resulting in complex nutrition interactions. Thus, low pH reduces available soil calcium and also reduces uptake of calcium by the root, and the increased aluminium supply further restricts the uptake of calcium. The roots become calcium deficient, interactions of phosphate and aluminium occur in and/or on the root surface, and apparent increases in phosphate uptake takes place; but because of the damage to the roots and root cells, the normal energetic functioning of the roots is impaired and mass flow actions dominate. The net result of aluminium, therefore, is a plant with a damaged root system and reduced top growth, and having low concentrations of calcium, especially in the roots, and also low phosphorus in plant tops but an apparent increase in roots. Associated with the latter effect there appears to be a loss of normal plant nutrient uptake functions particularly those associated with the energetics of the plant system. Within the groups of legumes, differences in response to aluminium and other associated factors occur; it is therefore essential that species, and indeed varieties, be characterised with respect to the above. In agricultural systems, recourse should be had to the breeding of plants with improved tolerance of aluminium; this is particularly so in the case of pastures where the cost of soil amelioration to overcome toxicity, in part or in full, may not be economic or practical.

Manganese excess

Many reports of manganese toxicity in plants were made during the middle of the twentieth century (38,51,52,85). Results accrued during that period showed that plants differed in their response to high levels of soil available manganese, and in particular, legumes were more sensitive to the problem than non-legumes. However, only a few of the investigations were directed towards a comparison of legumes in this regard, and then only with temperate legumes. A comparison of five legumes was made by Morris and Pierre (68); the order of tolerance obtained was Lespedeza > sweet clover > soybean = cowpea > peanut. The tolerance of *P. vulgaris*, *Vicia sativa*, and *M. sativa* was found to be less than that for *T. pratense*, *T. repens* and *V. faba* (51). Subsequent investigations covered tropical legumes (4,94).

Manganese excess is generally associated with soils of low pH however exceptions occur depending on the soil parent material, the physical condition of the soil, organic matter content, and soil water status. Soil conditions that check microbial oxidation of manganese or its compounds cause soluble manganese to increase (acidity, sterilisation, air drying, water logging). Attempts to diagnose soil for manganese are based on the measurement of soil exchangeable manganese, readily reducible manganese, and water soluble manganese (33,49,99). However, it is accepted that the water soluble portion is in apparent equilibrium with the exchangeable manganous form and that the latter is controlled by the balance of oxidation-reduction conditions. It is therefore not surprising that manganese toxicity exerts its greatest influence at the germination and early seedling stage of growth. Morris (66) measured the manganese concentration in water soluble extracts (soil:water ≈ 1:2) of 24 acid soils, and obtained a range of 0 - 6.3 ppm Mn, the concentrations being primarily related to soil pH. Concentrations of manganese from 0 to 40 ppm in solution were used in water culture experiments by Andrew and Hegarty (4) in a study involving eight tropical legumes and *M. sativa*. Marked differences in dry matter response were obtained (Table 1). In the above solution culture experiment, the plants were grown in the presence of added nitrogen. Results of a soil culture experiment (94) in which excess quantities of

Table 1

Multiple range tests for linear coefficients of the regression of plant yield on solution manganese concentration for nine tropical legumes.

Rank	Regression Coefficients	Species	Grouping 5%	Grouping 1%
1	-0.0023	Centrosema pubescens	1-3	1-3
2	-0.0038	Stylosanthes humilis		
3	-0.0039	Lotononis bainesii		
4	-0.0066	Macroptilium lathyroides	4-6	4-6
5	-0.0077	Leucaena leucocephala		
6	-0.0080	Desmodium uncinatum		6-7
7	-0.0102	Medicago sativa	7-7	
8	-0.0128	Glycine wightii	8-8	8-8
9	-0.0159	Macroptilium atropurpureum	9-9	9-9

L.S.D. of coeff. at 5% level = 0.0012
L.S.D. of coeff. at 1% level = 0.0016

Data from Andrew and Hegarty (4).

manganese sulphate were added to a light textured soil, also showed marked differences in responses of five species, viz:- C. pubescens > G. javanica var. SP-1 > S. gracilis > G. javanica var Tinaroo > M. atropurpureum var Siratro. The relative order of response obtained above is similar to that shown in Table 1, despite the fact that a different experimental technique was used in each experiment. In the solution culture experiment, with nitrogen added, the concentration of nitrogen in the plant tops increased as a result of manganese treatment, presumably due to reduced plant yield and concentration of nitrogen in the plant tops (4). In the soil experiment, the data showed that large applications of manganese to the soil reduced plant growth, nodule numbers and weight, but increased nitrogen concentrations in the plant tops in all but one species. Undoubtedly, some soil nitrogen was available to the plants in that experiment, and this related to low plant yield caused by manganese treatment, would explain the observed increase in nitrogen in the plant tops.

A sand culture technique (2) was developed to study the effect of manganese (0 to 40 ppm Mn) on a group of eleven bred lines of M. atropurpureum grown in the presence (nitrogen group) and absence (Rhizobium group) of added nitrogen (Hutton and Andrew,

Table 2

The effect of Mn on the mean dry matter production and root weight ratios of eleven bred lines of M. atropurpureum grown in the presence and absence of added nitrogen.

Cultural Series	D.M. (g/pot)		Root weight ratios	
	Mn (0.5 ppm)	Mn (40 ppm)	Mn (0.5 ppm)	Mn (40 ppm)
Nitrogen	15.9	8.0 (50% reduction)	0.15	0.19 (26% increase)
Rhizobium	10.4	6.3 (39% reduction)	0.27	0.33 (22% increase)

Hutton (Personal communication)

personal communication). A summary of some of the results from that experiment (Table 2) shows that the relative effect of manganese on the Rhizobium group was less than on the nitrogen group. This may be explained by the fact that plants of the nitrogen group produced more dry matter than those of the Rhizobium group. The effect of excess manganese was greater on the plant tops than on the roots especially in plants of the "nitrogen" group (Table 2). In the Rhizobium group the percentage weight of nodule tissue in the root system was reduced (32%) by manganese treatment but the reduction was minor (7%) when results were expressed as a percentage of the total plant. Excess manganese did not decrease the nitrogen concentration in the plant tops, in fact a slight increase occurred (Table 3).

These results show that manganese excess (40 ppm Mn) used under the experimental conditions employed had no marked effect on the Rhizobium-symbiosis of M. atropurpureum. Furthermore, manganese treatment in this and other experiments (4), did not adversely effect the morphology or dry matter production of the roots. Unlike aluminium, manganese excess over the range of 5 to 40 ppm Mn, caused no reduction in the calcium or phosphorus concentrations in plant tops. Manganese concentrations in plant tops increased; concentrations in roots exceeded those in the tops but the ratio of manganese in roots to that in the tops varied between species. Ratios for S. humilis, a tolerant species, and M. atropurpureum, a sensitive species, help to explain why species differ. In S. humilis the ratios were higher than those of M. atropurpureum at all treatments (4). This suggests that the

Table 3

The effect of Mn on the mean relative percentages of nodule tissue*, and nitrogen in the plant tops of eleven bred lines of M. atropurpureum (Rhizobium cultural series).

Index	Mn (0.5 ppm)	Mn (40 ppm)
Nodule tissue as a percentage of root + nodule	12.6	8.6 (32% reduction)
Nodule tissue as a percentage of total plant	3.1	2.9 (7% reduction)
Nitrogen in plant tops (% D.Wt)	3.32	3.49 (5% increase)

*Dry weight

Hutton (Personal communication).

relative tolerance of species such as S. humilis may be due to retention in the root system of a large proportion of the manganese absorbed. A similar explanation has been invoked in a study of differences between strains of M. sativa (24). Molybdenum and phosphorus deficiencies, two of the factors in soil acidity, should be considered in relation to the effect of manganese excess in plants. Many of the tropical pasture legumes respond in dry matter production following the addition of molybdenum to acid-neutral soils, and species differ e.g. the response of G. wightii > D. intortum > M. atropurpureum > S. humilis > L. bainesii (43). Reports of correction of manganese toxicity by substrate additions of molybdenum have been recorded (65,106). An investigation involving the tropical legumes, M. lathyroides, G. wightii, M. atropurpureum, S. humilis and C. pubescens, did not show any beneficial effect of added molybdenum on the degree of manganese toxicity (98). These authors suggested that beneficial effects of molybdenum in soils with excess manganese were easily attributed to an influence on nitrogen metabolism of the legume-Rhizobium symbiosis, and a consequent dilution of the manganese excess resulting from increased plant vigour.

High phosphate application has been shown to reduce manganese concentrations in plants (17,22). Other workers have results to show an increase in manganese concentrations and degree of toxicity

(1,64,67,88). Studies on this aspect in tropical pasture legumes have shown that phosphate increased manganese concentrations in the tops of G. wightii, M. atropurpureum, S. humilis and C. pubescens grown in water culture at constant pH (99). The increases were not accompanied by change in manganese transportation index, suggesting that the effect of phosphate was due to increased uptake. Acidic phosphate fertilisers, mono-ammonium phosphate > mono-calcium phosphate, aggravate manganese toxicity (88,99). In these cases it was considered that the effect of the acidic phosphate fertiliser was due to change of soil pH. However, the acidic effect should be added to that of phosphate on manganese uptake.

This review indicates that manganese toxicity is a simple factor in relation to soil acidity problems. Furthermore, toxicity does not impair initiation of nodulation or affect the efficiency of the legume-Rhizobium symbiosis, neither does it have an adverse effect on the uptake of calcium or phosphate. However, certain physical treatments of soil and some chemical ameliorants such as acidic-phosphate compounds, can aggravate manganese toxicity. Chemical analysis of soil and of plant tops offers a useful tool for assessment of the manganese status of the soil-plant system. Critical manganese toxicity concentrations for eight tropical legumes and M. sativa are available (4).

Phosphorus

Phosphorus deficiency is common in acid soils, however there is considerable evidence that liming of an acid soil partially overcomes the deficiency. As pointed out earlier, the addition of lime to soil, alters soil pH, increases available calcium and reduces aluminium in solution; all are factors that have some interaction with phosphorus. Reciprocally, the addition of phosphate to soil influences pH, aluminium, and the respective cation carrier of the phosphate, e.g. mono-calcium phosphate or mono and di-ammonium phosphate. It is not surprising therefore that phosphorus is very important in improving legume performance in acid soils. The gross effect is contained in the effect on plant dry matter production and on the nitrogen concentration of the legume tops. The dry matter response of a legume to addition of phosphorus is accompanied by a commensurate increase in the nitrogen concentration of the plant tops (9,32,60,79,84,92). The two latter publications were concerned with tropical legumes. Few data exist on detailed experiments that attempt to explain the above relationship.

Among the many benefits of applying phosphate to a deficient soil-plant system, stimulation of root growth is considered to be one of the prime factors. Certainly, the root dry matter is increased by phosphate, but results of several investigations show that this effect is less than that on plant tops i.e. phosphate addition increases the root weight ratios of responsive plants. However, this does not invoke the thesis that phosphate application brings about new root development and thus new sites for nodule initiation. The latter aspect was studied on S. humilis (32). It was found that phosphorus had a beneficial effect on the initiation of nodules; they were first detected at day 11 in high P plants, but not until day 14 in low P plants. Nodule numbers, volumes, and dry weights were also enhanced by increasing phosphate supply. Nodule relative growth rates were also increased, e.g. from 0.3 g/g/day at low P treatment to 0.7 g/g/day at high P treatment over the period day 23 to 26. Commensurate with observed and measured improvement in nodulation, the assimilation of nitrogen by the whole plant was increased from 17 mg/g nodule dry weight/day at low P treatment to 53 at high P treatment. Similar investigations were made on G. wightii (94) however in this experiment there was no improvement in the quantity of nitrogen fixed per unit weight of nodules. Other investigations (25) showed that species differ in this regard, e.g. data from M. atropurpureum showed an increase in nitrogen concentration whereas those for G. wightii and S. gracilis showed a decrease. In judging apparent interactions such as above, one must also consider the soil, e.g. it is known that the addition of phosphate to a deficient soil can stimulate soil microbial activity and so increase the available soil nitrogen. Furthermore this type of experimentation does not permit easy interpretation as to the prime effect of phosphate in relation to legume growth e.g. it may be argued that increasing phosphorus supply to a legume may permit more plant top growth, increase the supply of photosynthates, and in turn increase nodulation and efficiency of symbiosis. The reduction in root weight ratios in plants at high P treatment compared to those at low P treatment, i.e. greatly impaired top growth at low P, lends support to the latter thesis. This suggestion strengthens the argument of the dependency of nodule efficiency on a good photosynthate supply.

Inspection of the relationship between phosphorus and nitrogen concentrations in legume tops, shows that the best correlation applies to the dry matter response portion of the response curve, (8). From those data the slopes of the lines applicable to the correlations for the various species allows some assessment of the overall efficiency of nitrogen production in relation to phosphate supply, e.g. S. humilis is less efficient in this regard than M. atropurpureum (Figure 8). Nitrogen plateaus shown in Figure (8) indicate that legumes also differ in their capacities to attain high nitrogen concentrations. Recourse to the data of Andrew and Robins (8) also shows that the effect of phosphate on nitrogen concentrations continues beyond the phosphorus

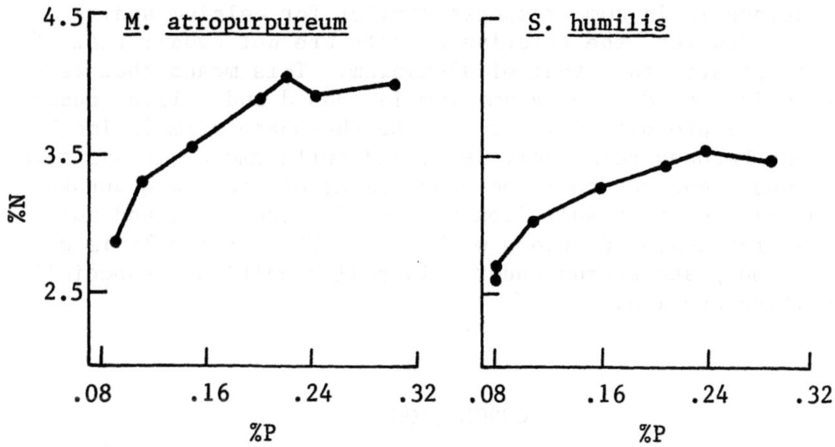

Figure 8. Effect of phosphorus on the concentration of phosphorus and nitrogen in the tops of two tropical legumes.

concentrations corresponding to maximum dry matter production.

The response of legumes to phosphate additions in acid soils is of course dominated by the species. In this regard it is well recognised that plant species differ; S. humilis and L. bainesii are considered very efficient species compared to a number of other legumes (8). S. gracilis has been shown to be less responsive than C. pubescens and M. atropurpureum (47). Differences also occur within genera and species (46). Differences in response of legumes to added phosphate may be attributed to many factors (5). Critical phosphorus concentrations in plant tops (8), and phosphate uptake mechanisms are of prime importance. The latter has particular relevance to acid soils particularly with regard to pH, calcium and aluminium.

Investigations using excised roots (78) and the application of kinetics to the data (34) have assisted in the understanding of uptake mechanisms. In such systems it has been shown that the absorption of phosphate by roots was greater at pH 4.0 than at pH 5.0, and in turn greater than at pH 6.0. These experiments were accomplished using barley; however, similar relationships hold for the roots of legumes (Andrew unpublished). On the other hand reduction of soil pH reduces the availability of phosphate to the plant, and this effect dominates that of the plant uptake mechanism, although there may be a degree of mitigation imposed by the reaction of the species to pH change.

The "Viets Effect" (101) imposed on phosphate uptake by polyvalent cations is in some respects similar for calcium and aluminium. However, the relative effects are not equal; that of calcium is greater than that of aluminium. This means that as soil acidity is increased, i.e. aluminium increased and calcium reduced, assuming equal proportional change, the phosphate stimulation by the cations becomes relatively less, and furthermore, it has also been discussed earlier that the enhancement of phosphate uptake due to aluminium may result from a mass flow phenomena and not from an energy assisted uptake mechanism. The net result is a low plant phosphate status under acid soil conditions, especially for sensitive species.

CONCLUSION

The understanding of the problem of legumes in acid soils requires a thorough knowledge of the soil and its chemical status, and also full mineral characterisation and requirements of the legume in question and the interactions involved.

Acid soils invariably have high concentrations of the hydrogen ion and free aluminium, associated with low concentrations of calcium and phosphate. In addition certain acid soils, depending on the parent rocks and physical status, may contain excess concentrations of soluble manganese. Lime application, the most efficient method of ameliorating acid soils, is beneficial to both the plant and the soil nutrition parameters mentioned above. However, one must guard against further induced deficiencies such as magnesium, and many of the micro plant nutrients e.g. copper, zinc, and boron.

In the nutrition of the legume in acid soils, it is evident that many compounded effects occur, and a series of chain events takes place. For example, in the case of aluminium toxicity such a sequence could be as follows: (a) an increase in soil hydrogen ion concentration is accompanied by an increase of aluminium in solution, and reductions in calcium and phosphorus in the soil solution, (b) the increase of both the hydrogen ion and aluminium damage the plant root system and reduce the uptake and translocation of calcium and phosphate by the plant, (c) low calcium in the plant further reduces the uptake of phosphate and this is accentuated by increased fixation of phosphate in the soil, (d) the combined effect of these factors inhibits or minimises the initiation of nodulation, and creates low efficiency of the legume - *Rhizobium* symbiosis, and (e) the low nitrogen status of the plant further reduces plant photosynthates and the possibility of further nodulation and growth. Manganese toxicity is usually accompanied by conditions of aluminium toxicity especially at low soil pH, but if the pH is approximately 5.5 to 6.0 and other conditions dictate manganese excess, then a simple state of manganese toxicity exists.

In the amelioration of acid soils accurate diagnosis is essential e.g. pH, aluminium, manganese, calcium, phosphate and molybdenum. Attempts to maximise N production by legumes will be dependent on each essential plant nutrient being at optimum supply and the absence of adverse factors. The most efficient method of diagnosis is the field experiment but this may not be possible in all instances. Valuable assistance can be obtained from soil and plant chemical analyses, and information from these, particularly the plant analysis approach assist in explaining plant-nutrient interactions. In the practical use of acid soils, the selection of plants, and indeed the breeding of plants with improved tolerance to excess hydrogen, aluminium, and manganese constitutes a major approach.

REFERENCES

1. Albrecht, W.A., and Smith, N.C. (1941) Bull. Torrey Botan. Club. 68,372.
2. Andrew, C.S. (1973) Lab. Pract. 23,20.
3. Andrew, C.S. (1976) Aust. J. Agric. Res. 27,611.
4. Andrew, C.S., and Hegarty, M.P. (1969) Aust. J. Agric. Res. 20,687.
5. Andrew, C.S., and Johansen, C. (1976) Plant relations in pastures, (Ed. Wilson, J.R.)(C.S.I.R.O. Melbourne)(in press).
6. Andrew, C.S., and Johnson, A.D. (1976) Aust. J. Agric. Res. 27,611.
7. Andrew, C.S., and Norris, D.O. (1961) Aust. J. Agric. Res. 12,40.
8. Andrew, C.S., and Robins, M.F. (1969a) Aust. J. Agric. Res. 20,665.
9. Andrew, C.S., and Robins, M.F. (1969b) Aust. J. Agric. Res. 20,675.
10. Andrew, C.S., and Vanden Berg, P.J. (1973) Aust. J. Agric. Res. 24,341.
11. Andrew, C.S., Johnson, A.D., and Sandland, R.L. (1973) Aust. J. Agric. Res. 24,325.
12. Arnon, D.I., and Johnson, C.M. (1942) Plant Physiol. 17,525.
13. Arnon, D.I., Fratzke, W.E., and Johnson, C.M. (1942) Plant Physiol. 17,515.
14. Asher, C.J., and Loneragan, J.F. (1967) Soil Sci. 103,225.
15. Banath, C.L., Greenwood, E.A.N., and Loneragan, J.F. (1966) Plant Physiol. 41,760.
16. Bear, F.E., and Wallace, A. (1950) Bull. N.J. Agric. Exp. Stn. No. 748.

17. Bolle-Jones, E.W. (1955) Plant Soil 6,45.
18. Carvalho, M.M.de., Frauza, G.E.de., Bahia, A.F.C., and Mozzei, D.L. (1971) Pesq. Agropec. bras. 6,285.
19. Clarkson, D.T. (1965) Ann. Bot. 29,309.
20. Clarkson, D.T. (1966a) J. Ecol. 54,167.
21. Clarkson, D.T. (1966b) Plant Physiol. 41,165.
22. Conroy, E., and Lamb, J.G.D. (1962) Nature (London) 194,500.
23. Damirgi, S.N., Frederick, L.R., and Anderson, I.C. (1967) Agron. J. 59,10.
24. Dessureaux, L., and Ouellette, G.J. (1958) Can. J. Soil Sci. 38,8.
25. Eira, P.A., Almeida, D.L.de., and Silva, W.C. (1972) Pesq. agropec. bras. Serie Agron. 7,185.
26. Epstein, E. (1972) Mineral Nutrition of Plants (Wiley, New York).
27. Epstein, E., and Leggett, J.E. (1954) Am. J. Bot. 41,785.
28. Foster, H.L. (1970) E. Afr. Agric. Forest. J. 36,58.
29. Foy, C.D., and Brown, J.C. (1964) Soil Sci. Soc. Am. Proc. 28.27.
30. Foy, C.D., Burns, G.R., and Brown, J.C. (1965) Soil Sci. Soc. Am. Proc. 29,64.
31. Freitas, L.M.M., and Pratt, P.F. (1969) Pesq. agropec. bras. 4,89.
32. Gates, C.T. (1974) Aust. J. Bot. 22,45.
33. Grassmanis, V.O., and Leeper, G.W. (1966) Plant Soil 25,41.
34. Hagen, C.E., and Hopkins, H.T. (1956) Plant Physiol. 30,193.
35. Helyar, K.R., and Anderson, A.J. (1971) Aust. J. Agric. Res. 22,707.
36. Henzell, E.F. (1962) Commonw. Bur. Pastures Field Crops Hurley, Berkshire Bull. 46,161.
37. Hewitt, E.J. (1949) Rep. agric. hort. Res. Stn. Univ. Bristol, 1948, 58-65.
38. Hewitt, E.J. (1952) Trans. 2nd and 4th Comm. int. Soc. Soil Sci. Dublin, 1952 vol 1, pp. 107-18.
39. Humphreys, L.R. (1967) J. Aust. Inst. Agric. Sci. 33,3.
40. Jackson, W.A. (1967) Soil Acidity and Liming (Am. Soc. Agron., Madison, Wis.)
41. Jensen, H.L. (1943) Proc. Linn. Soc. N.S.W. 68,207.
42. Jensen, H.L. (1944) Proc. Linn. Soc. N.S.W. 69,29.
43. Johansen, C., and Kerridge, P.C. (1977) Aust. J. Exp. Agric. Anim. Husb. (in press).
44. Johnson, R.E., and Jackson, W.A. (1964) Soil Sci. Soc. Am. Proc. 28,381.
45. Jones, L.H. (1961) Plant Soil 13,297.
46. Jones, R.K. (1974) Aust. J. Agric. Res. 25,847.
47. Jones, M.B., and Freitas, L.M.M.de. (1970) Pesq. Agropec. bras. 5,91.
48. Lee, M.T., and Wilson, G.L. (1972)Aust. J. Agric. Res. 23,257.
49. Leeper, G.W., and Swaby, R.J. (1940) Soil Sci. 49,163.

50. Lie, T.A. (1974) The Biology of Nitrogen Fixation (North Holland Publ. Co., Amsterdam), A. Quispel, ed.
51. Löhnis, M.P. (1951) Lotsya. 3,63.
52. Löhnis, M.P. (1952) Plant Soil 3,193.
53. Loneragan, J.F. (1973) Use of Isotopes for study of fertilizer utilisation in crops. (Int. Atomic Energy Agency, Vienna.)
54. Loneragan, J.F., and Dowling, E.J. (1958) Aust. J. Agric. Res. 9,464.
55. Lowther, W.L., and Loneragan, J.R. (1970) Proc. 11th Inter. Grass. Cong. 446.
56. MacLean, A.A., and Chiasson, T.C. (1966) Can. J. Soil Sci. 46,147.
57. MacLeod, L.B., and Jackson, L.P. (1956) Can. J. Soil Sci. 45,221.
58. MacLeod, L.B., and Jackson, L.P. (1967a) Agron. J. 59,359.
59. MacLeod, L.B., and Jackson, L.P. (1967b) Can. J. Soil Sci. 47,203.
60. McLachlan, K.D., and Norman, B.W. (1961) J. Aust. Inst. Agric. Sci. 27,244.
61. McLean, F.T., and Gilbert, B.E. (1927) Soil Aci. 24,163.
62. McNeur, A.J. (1954) N.Z.J. Sci. Technol. A. 36,167.
63. McIlvane, T.C., and Pohlman, W.E. (1949) Bull. W.Va. Agric. Exp. Stn. No. 337.
64. Messing, J.H.L. (1965) Plant Soil 23,1.
65. Millikan, C.R. (1947) J. Aust. Inst. Agric. Sci. 13,180.
66. Morris, H.D. (1948) Soil Sci. Soc. Am. Proc. 13,362.
67. Morris, H.D., and Pierre, W.H. (1947) Soil Sci. Soc. Am. Proc. 12,382.
68. Morris, H.D., and Pierre, W.H. (1949) Agron. J. 41,107.
69. Munns, D.N. (1965a) Aust. J. Agric. Res. 16,733.
70. Munns, D.N. (1965b) Aust. J. Agric. Res. 16,743.
71. Munns, D.N. (1968) Plant Soil 28, 129.
72. Munns, D.N. (1970) Plant Soil 32,90.
73. Munns, D.N. (1976) Soil Sci. Soc. Amer. J. (in press).
74. Munns, D.N. (1977) Exploiting the Legume-Rhizobium Symbiosis in Tropical Agriculture (College of Trop. Agric. Misc. Publ. 145).
75. Munns, D.N., and Fox, R.L. (1976) Plant Soil (in press).
76. Munns, D.N., Fox, R.L., and Koch, B.L. (1977) Plant Soil (in press).
77. Neenan, M. (1960) Plant Soil 12,324.
78. Nissen, P. (1973) Physiol. Plant.28,304.
79. Norman, M.J.T. (1959) C.S.I.R.O. Aust. Div. Ld. Res. reg. Surv. Tech. Pap. No. 5.
80. Norris, D.O. (1956) Empire J. Exp. Agric. 24,247.
81. Norris, D.O. (1959) Aust. J. Agric. Res. 10,651.
82. O'Toole, M.A., and Masterson, C.L. (1968) Irish J. Agric. Res. 7,129.

83. Ouellette, G.J., and Dessureaux, L. (1958) Can. J. Plant Sci. 38,206.
84. Parsons, J.L., and Davies, R.R. (1960) Agron. J. 52,441.
85. Peech, M., and Bradfield, R. (1948) Soil Sci. 65,35.
86. Ragland, J.L., and Coleman, N.T. (1962) Soil Sci. Soc. Am. Proc. 26,88.
87. Randall, P.J., and Vose, P.B. (1963) Plant Physiol. 38,403.
88. Randall, P.J., Schulte, E.E., and Corey, R.B. (1975) Agron. J. 67,705.
89. Robson, A.D., and Loneragan, J.F. (1970) Aust. J. Agric. Res. 21, 223.
90. Rorison, I.H. (1958) Nutrition of the Legumes (Butterworths Sci. Publ.,London).
91. Schmehl, W.R., Peech, M., and Bradfield, R. (1950) Soil Sci. 70,393.
92. Singh, A. (1958) Proc. Natl. Acad. Sci. India B. 28,81.
93. Snaydon, R.W. (1962) J. Ecol. 50,439.
94. Souto, S.M. and Johanna Döbereiner. (1969) Pesq. agropec. bras. 4,129.
95. Spurway, C.H. (1941) Bull. Mich. Agric. Exp. Stn. No. 306.
96. Sutton, C.H., and Hallsworth, E.G. (1958) Plant Soil 9,305.
97. Thoday, D., and Evans, H. (1931) Protoplasma 14,64.
98. Truong, N.V., Andrew, C.S., and Wilson, G.L. (1971a) Plant Soil 34,547.
99. Truong, N.V., Wilson, G.L., and Andrew, C.S. (1971b) Plant Soil 34,309.
100. Van Stevenick, R.F.M. (1965) Physiol. Plant. 18,54.
101. Viets, F.G. (1944) Plant Physiol. 19,406.
102. Vincent, J.M. (1965) Soil Nitrogen (Amer. Soc. Agron., Madison, Wis.).
103. Vincent, J.M. (1974) The Biology of Nitrogen Fixation. (North Holland Publ. Co.,Amsterdam), A. Quispel (ed.).
104. Vlamis, J. (1953) Soil Sci. 75,383.
105. Vose, P.B., and Randall, P.G. (1962) Nature (London) 196,85.
106. Warrington, K. (1951) Ann. Appl. Biol. 38,624.
107. Wright, K.E. (1943) Plant Physiol. 18,708.

MICRONUTRIENT REQUIREMENTS OF LEGUME-*Rhizobium* SYMBIOSIS IN THE TROPICS

Avilio A. Franco

Programa Fixacão Biológica de Nitrogenio
Convenio CNPq - EMBRAPA - UFRRJ, Km 47
Seropedica 23460, Rio de Janeiro, Brazil

Micronutrient deficiencies may constitute a major fertility problem in tropical soils. Low crop yield and lack of response to major nutrients and failure of the perennial legumes to persist in the pasture often are associated with micronutrient deficiencies. Tropical regions under high rainfall conditions normally present soils highly weathered, with low pH. In these conditions some elements (e.g., Zn, Mn, B and Fe) may be deficient because they were washed out of the soil while some even when present may be unavailable to the plants (e.g., Mo), because they are adsorbed to the soil particles. Responses of the legumes to micronutrients have been recorded extensively in tropical areas of Australia, Africa, Hawaii and Brazil (2, 5, 17, 22, 29, 41, 48, 54, 60, & 61).

This review emphasizes the effect of the elements on the host (e.g., Zn, Cu, Mn and Cl), on the *Rhizobium* (e.g., Co) and on both (e.g., Fe, Mo and B). Each element is discussed separately with emphasis on Mo and on the use of Fritted Trace Elements (FTE).

ZINC

Zinc deficiency primarily limits host plant growth (47) and several reports have shown legume yield responses to zinc fertilization (7, 30, 38 & 50). Jones *et al.* (41) obtained yield decrease of *Glycine wightii* in two out of 3 soils tested when Zn was withheld. In one of the responsive soils, *Stylosanthes guyanensis*, *Centrosem pubescens* and *Macroptilium atropurpureum* were also tested. Only with stylo was yield decreased by withholding Zn. Although no specific role of Zn is known to occur in nodulation and nitrogen fixation, results with *Desmodium intortum*, *G. wightii* and *M. atropurpureum* in water culture but in the absence of combined nitrogen, showed that Zn had an effect on nodulation but no significant effect on the percent nitrogen in the plant tops (5). Deficiencies of Zn have been observed when high levels of phosphorus are present but not otherwise, indicating the necessity to add this element to the soil when high doses of phosphorus are used (6).

MANGANESE

Manganese is a component of several host enzymes (24) with an important role in the Krebs cycle but without a specific role in the symbiosis. In some conditions manganese is deficient and in others it is toxic (22 & 31). In plant tops Mn should be 50 ppm for normal growth (26). However this level is influenced by other elements, e.g., in the presence of calcium, soybeans contained more Mn but showed less toxicity effects than in the absence of calcium (53). It was also shown that a Mn-tolerant soybean cultivar absorbed more calcium and less Mn than the sensitive cultivars (14). Similar results were obtained with P. vulgaris (31), and it was found that nodulation was more affected by Mn toxicity than plant growth.

Liming the soil from pH 5.3 to 6.3 decreased the yield of S. guyanensis with coincident reduction of Mn in the tops of the plant (40). In another soil it was observed that lime had a deleterious effect on growth of S. guyanensis. The addition of Mn and B ameliorated the deleterious effect of lime. However at pH 6.2 nodulation and nitrogen fixation performed better than at a lower pH (58). G. wightii differed considerably in nodulation and growth when tested in soil high in Mn (57). Phaseolus vulgaris is very sensitive to Mn toxicity (31).

Application of 20-30 kg/ha of manganese sulphate could be used to supply Mn to deficient soils (48). On the other hand the toxicity could be eliminated by liming the soil or adding organic matter to the soil (25). When toxicity is not too high, or the plant used is more tolerant, pelleting the seed with lime could improve establishment of the legume (33).

BORON

Nodule formation is prevented by boron deficiency; when nodules develop they are small and lack vascular strands and bacteriods (13). However non-legume plants also show similar meristematic failure when boron is deficient (45). This suggest a similar requirement of the plant and nodule for boron. In Australia, responses have been obtained experimentally as a result of liming (5). In Brazil several soils have shown responses to B (22, 30, 54 & 55). Boron deficiency could be corrected by applying 20 kg/ha of borax (48).

IRON

Iron is a constituent of leghemoglobin, which is important in supplying oxygen at high rates and low concentration for nodule function (11). Fe is a component of the iron-sulphur cluster, present in both the Fe-protein and Mo-Fe protein components of nitrogenase, and in the bacterial ferrodoxin which may function as a reductant of nitrogenase (27). However iron deficiency produces symptoms which are unrelated to nitrogen deficiency, and these are not eliminated by applying nitrogen. The host plant also needs iron in high quantities.

Functioning nodules need high concentrations of iron, but the absolute amounts may normally be a small fraction of the total assimilated by the plant. Oxisols and utisols normally contain high levels of iron oxides, and responses to Fe addition should not be expected. However, De-Polli et al. (22), in a pot experiment testing a complete factorial (2^5) design with an ultisol soil, found response to Fe. To avoid Fe immobilization in the soil, spraying of the leaves could be used (45). In tropical soils only a few cases of response to iron has been observed, in some cases with iron applied as a chelate. In this case a control using the chelate without iron should be used to avoid complicating effects of chelate in the soil.

Lie and Brotonegoro (43) studied the effect of Fe-EDTA on nodulation and found inhibitory effects like that of calcium deficiency or soil acidity. The action was only confined to 2 - 4 days after inoculation and if the roots were in direct contact with the chelate.

Chaney et al. (18) found that the Fe-EDTA is dissociated in the outer cells of the roots and the ferric ion was reduced to the ferrous state prior to absorption. Based on this, Lie (42) postulated that due to the accumulation of chelates at the roots where ion uptake occurs, calcium may be chelated from the cells: thus local calcium deficiency may occur.

COPPER

Copper-deficient plants bear nodules that incorporate carbon slowly to amino acids and protein, have fewer bacteriods, more starch, less cytochrome C oxidase and are smaller (16). Copper is a component of several enzymes such as ascorbic acid oxidase, phenolase and cytochrome oxidases. However these have not been shown to have any specific role in symbiosis. Hallsworth (36) studied the role of Cu in respect to nodulation and growth of *Trifolium subterraneum*, and showed that copper-deficient plants had poor root growth and small nodules, and that nodules were present mainly on the secondary roots, whereas with added Cu, the nodules developed in dense clusters close to the main root. Presence of Cu has also been shown to increase nodule abundance of *T. repens* growing in peat soils (51). Reports showing deficiency of copper are quite common. Andrew and Thorne (8) compared five temperate and five tropical pasture legumes in soil and water culture. Large differences between species were obtained. Among the tropical species, *Desmodium uncinatum* was the least responsive and *S. guyanensis* was the most responsive. *Lotononis bainesii* has also been shown to be relatively insensitive to Cu (5).

COBALT

Carefully conducted experiments with purified components in water culture systems have shown a specific requirement for Co for N_2 fixation by nodulated soybean, alfalfa and subterranean clover (1, 20 & 37). Cobalt deficient plants were N-deficient. Plants supplied with combined N, did not respond to added cobalt. Although cobalt deficiency has been occasionally reported in the field, there have been no reported cases of Co-deficiency in tropical areas. The supply of Co in pastures in usually more important for nutrition of cattle than for the legume symbiosis. Cobalt is a component of the coenzymes which are involved with conversion of propionyl CoA to succinyl-CoA and for nucleotide reductase activity in the rhizobia (27).

CHLORINE

Chlorine acts together with enzymes in photosynthetic release of oxygen in photosystem II (12). Thus this element has an indirect role in symbiosis by limiting energy. However no report showing the effect of chlorine on a legume host or symbiosis under field conditions has been found. One reason could be the capacity of the plants of extract chlorine from the air, especially near the oceans.

Molybdenum

Molybdenum is a constituent of a moiety common to nitrogenase and nitrate reductase (27). Thus the consensus that Mo is only necessary for plants dependent on N_2 is not true; plants dependent on NO_3^- need Mo as well. Extensive work carried out in Australia has shown Mo deficiency in several soils (3). In Brazil, Mo deficiency also has been shown in several soils (22, 54 & 55). Several experiments failed to show a response to Mo because of other limiting factors. Ruschel and Eira (55) obtained no Mo benefit in soybean until manganese toxicity was eliminated. It is known that plants well supplied with Mo produce fewer nodules but have higher efficiency in N_2 fixation (33 & 49).

The amount of Mo required is so small that plants grown from seeds with a high content of Mo should be self sufficient. An experiment in soils with different adsorption capacity and with soybean seeds containing different amounts of Mo showed that 19 μg of Mo/g of seed was equivalent to one foliar application of Mo. 28 g Mo/ha applied to soybean plants grown from seeds produced higher yields in the three soils tested (35).

There is no good correlation between Mo chemically analysed in the soil and the plant response. The most used method to predict Mo deficiency is extraction of soil with oxalate at pH 3.5. Walsh *et al*. (59) found that 0.12 ppm of extractable Mo was the turning point between deficient and non-deficient soils. Griegg (34) found 0.14 ppm to be the turning point. However, using a large number of soils, Davis (19) found that at pH 5.0, the response level was 0.20 ppm and at pH it was 0.05 ppm of extractable Mo.

Peres, Franco & Nery (52) used a microbial test for the percentage increase of N_2ase (C_2H_2) activity of *Azotobacter paspali* growing in a defined medium with soils + and -Mo, and they showed a generalized Mo deficiency in soils from Rio de Janeiro State. The microbial test correlated well with the total increase in plant N of *C. pubescens* due to Mo fertilization: $r = 0.97$ in oxisols, and $r = 0.96$ in utisols but not with entisols. The latter contain a higher proportion of organic matter (Fig. 1). Bataglia *et al*. (9) used a chemical determination and found generalized deficiency levels in S. Paulo State soils. The same soils used by them were tested with the microbial test. Values obtained by the two methods were positively correlated ($r = 0.76$). Soils with no oxalate-extractable Mo gave increases up to 500% of N_2ase activity of *A. paspali*. No increase in nitrogenase activity was obtained in soils where the extractable Mo was 0.10 ppm or higher (Franco, A. A. unpublished results).

With the same microbial test, 137 soils from several sites from Brazil were tested (Table 1). Of these soils 43% showed increases from 0 to 200% of activity indicating no probable Mo deficiency, and 16% appeared to be very Mo deficient. An attempt was made to correlate soil chemical properties with the Mo response in the microbial test. None of the soil chemical characteristics correlated well with N_2ase activity. Some correlation was obtained with pH and Ca + Mg content of the soil.

Although deficiencies of Mo in the soil are general, there are differences between legume species in their response to fertilization with Mo (39). In a red-yellow podzolic soil, large increases in yields of siratro and centrosema were obtained by Mo fertilization, however high yields of *Stylosanthes* were obtained without Mo fertilization (23). In the same soil, *P. vulgaris*

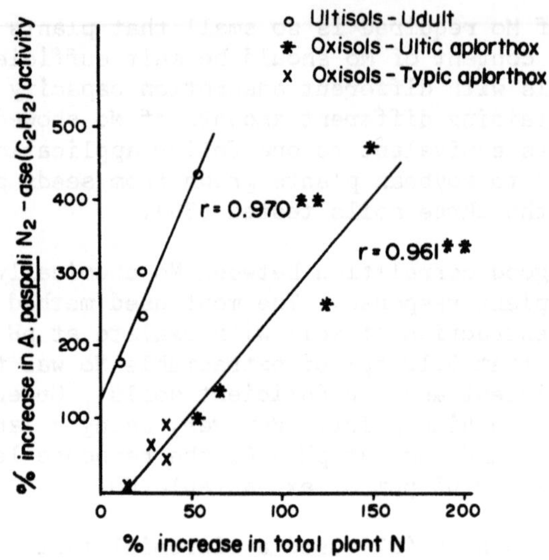

Fig. 1. Relation of increase in nitrogenase (as measured by C_2H_2 reduction) to increase in total plant nitrogen.

TABLE 1

Response of Azotobacter paspali to the addition of Mo in culture medium containing 100 mg of soil from different sites in Brazil.

% increase[a] N_2ase	Number of soils tested	% of soils
0-200	59	43
200-400	57	41
400-600	16	12
600	5	4

[a]Increase in relation to medium with 100 mg soil but without addition of Mo.

responded to Mo fertilization only after the soil was limed to a pH above 5.4. Six P. vulgaris cultivars behaved similarly in this soil. One of the cultivars was tested in 5 different soils. Four soils with a pH near 5 showed a very strong interaction of lime with

Mo fertilization. At a pH 5.0 to 5.2 there was minimum response to Mo addition, from pH 5.2 to 5.8 there was a good response to Mo, and at higher pH the Mo released from the soil was sufficient. The plants growing in a soil of pH 5.8 showed good response to Mo fertilization at the initial pH but the response disappeared after liming (32). A study made with 4 soils, with different initial adsorption capacity showed that between pH 5.4 and 5.6 all the soils lost the capacity to retain Mo (56). When Ca (as gypsum) was present in excess, plant growth was not increased when the pH was raised by addition of potassium carbonate, but plant-N increased. With no added gypsum, the plant yield and N-content increased almost linearly with pH (Laera and Franco, unpublished results).

Molybdate or molybdenum oxide at 100-200 g Mo/ha are indicated to correct a Mo deficiency. Rates up to ten fold higher than these may be needed, e.g. in copper-deficient soils consisting largely of organic matter (15), in manganese-deficient soils that are calcereous (10) or in Mo-deficient soils very high in colloidal iron oxides (3). The frequency of application of Mo to the soil depends on the rate of application, the crop removed and soil characteristics. Accumulated experience indicates that application at intervals of 3 to 4 years is sufficient (4). An experiment with *Desmodium intortum* for three years was carried out with increasing rates of applied Mo. In the first year 25 g Mo/ha was sufficient for maximum yield. After 3 years only the application of 200 g Mo/ha was sufficient to support maximum growth (4).

FRITTED TRACE ELEMENTS (FTE)

Fiskell & Winsor (28) obtained increases of 154% in N_2 fixed by clover in field experiments by applying FTE containing several micronutrients. These materials have been tested quite extensively under our conditions. The effect of FTE has been shown to more than double the yield of *C. pubescens*, *G. wightii*, *S. guyanensis* and *M. atropurpurem* in pot experiments (21). In a 4 year experiment in the same soil in the field, the same response was obtained with *M. atropurpureum* and *C. pubescens* but not with *S. guyanensis*. The latter showed a good growth without micronutrient fertilization (23). Studies with mycorrhiza have shown that stylo is well adapted to acid soils with low phosphorus content (46). Possibly the extraction of micronutrients from the soil is also mediated by the mycorrhiza that is abundant in this soil. De-Polli et al. (22) in greenhouse experiments showed differences between FTE formulas; the FTE containing Mn and Co performed better than without these elements (49). The addition of micronutrients favored the inoculated strain (50% increase in nodules produced) compared with only 20% in the pots without fertilization. They also found that the application of FTE directly in the soil was better than in pellets, that it was not lost in the percolated water and that concentrations up to 20 times the recommended rate were not toxic to the plant (Table 2).

TABLE 2

Effect of FTE on nodulation and total plant N of C. pubescens (49).

	Nodule dry weight (mg/pot)	Total plant N (mg/pot)
Control	448	139
40 kg BR-10/ha on soil (a)	583	198
200 kg BR-10/ha on soil (b)	457	222
Control	502	152
Percolated from (a)	485	157
Percolated from (b)	525	150
F values		
Levels of FTE	0.02	2.74
Application	0.27	17.10
Levels x Application	1.11	3.42
V.C. (%)	36	18

These results indicate that solubilization of FTE with consequent liberation of the elements may be a consequence of direct contact with the roots. This has special importance in tropical soils under high rainfall and with low exchange capacity. Attempts have been made to use FTE for soybeans and beans without any positive response. In one case in an acid soil the addition of FTE to beans gave depressed yields.

REFERENCES

1. Ahmed, S. and Evans, H. J. (1961) Proc. Nat. Acad. Sci. USA 47, 24.
2. Anderson, A. J. (1942) J. Aust. Inst. Agr. Sci. 8, 73.
3. Anderson, A. J. (1956) In: Inorganic Nitrogen Metabolism, W. D. McElroy and B. Glass, the Johns Hopkins Press, Baltimore 3.
4. Anderson, A. J. (1970) J. Aust. Inst. Agric. Sci. 36, 15.
5. Andrew, C. S. (1976) Exploiting the legume-Rhizobium symbiosis in tropical agriculture, J. M. Vincent, A. S. Whitney and J. Bose, Univ. of Hawaii, Miscell. Publ. 145, 274.
6. Andrew, C. S. (1977) personal communication.
7. Andrew, C. S. and Bryan, W. W. (1955) Austral. J. Agric. Res. 6, 625.
8. Andrew, C. S. and Thorne, P. M. (1962) Austral. J. Agric. Res. 13, 821.

9. Bataglia, O. C., Furlani, P. R. and Valadares, J. M. A. S. (1975) Ann. XV Congr. Bras. Cienc. Solo, Campinas, S. P., 107.
10. Batey, T. (1971) U.K. Min. Agr. Fish. Food Bull. 21, 137.
11. Bergersen, F. J. (1971) Ann. Rev. Plant Physiol. 22, 121.
12. Bové, J. M., Bové, C., Whatley, F. R. and Arnon, D. I. (1963) Z. Naturforsch 18b, 623.
13. Brenchley, W. F. and Thornton, H. G. (1925) Proc. Roy. Soc., London, Ser. B, 98, 373.
14. Cabeda, M. S. V. and Freire, J. R. (1968) Ann. IV BELAR, Univ. Fed. Rio Grande do Sul, Porto Alegre, 282.
15. Caldwell, T. H. (1971) U.K. Min. Agr. Fish. Food Bull. 21, 62.
16. Cartwright, B. and Hallsworth, E. G. (1970) Plant Soil 33, 685.
17. Carvalho, M. M. de, Franca, G. E. de, Bahia, F A. F. C. and Mozzer, O. L. (1971) Pesp. agropec. bras., Sér. Agron. 6, 285.
18. Chaney, R. L., Brown, J. C. and Tiffin, L. O. (1972) Plant Physiol. 50, 208.
19. Davis, E. B. (1952) Proc. 14th Conf. New Zealand Grassland Assn. 182.
20. Delwiche, C. C., Johnson, C. M. and Reisenauer, H. M. (1971) Plant Physiol. 36, 73.
21. De-Polli, H. and Döbereiner, J. (1974) Pesq. agropec. bras. Sér. Agron. 9, 93.
22. De-Polli, H., Suhet, A. R. and Franco, A. A. (1975) Ann. XV Congr. Bras. Cienc. Solo, Campinas, 151.
23. De-Polli, H., Carvalho, S. R., Lemos, P. F. and Franco, A. A. (1977) Intern. Symp. on Biol. Nitrogen Fixation in the Tropics, Brasilia, Brazil.
24. Dieckert, J. W. and Rozacky, R. (1969) Arch. Biochem. Biophys. 134, 473.
25. Döbereiner, J. and Alvahydo, R. (1963) IX Congr. Bras. Cienc. Solo, Fortaleza, Ceará.
26. Epstein, E. (1972) Mineral Nutrition of Plants, Wiley, New York. pp. 341.
27. Evans, H. J. and Russell, S. A. (1971) In: Chemistry and Biochemistry of Nitrogen Fixation, J. R. Postgate, Ed., Plenum, London, pp. 191-224.
28. Fiskell, J. G. A. and Winsor, H. W. (1958) Soil Agric. Res. Rep. Agric. Res. Sta. Gainesville, Florida, 1.
29. Fox, R. L. and Kang, B. T. (1976), In: Exploiting the Legume-Rhizobium Symbiosis in Tropical Agriculture, J. M. Vincent, A. S. Whitney and J. Bose, Univ. of Hawaii, Misc. public 145, 183.
30. França, G. E. de, Bahia Filho, A. F. C. and Carvalho, M. M. de (1973) Pesq. agropec. bras. Sér. Agron. 8, 197.
31. Franco, A. A. and Döbereiner, J. (1971) Pesq. agropec. bras. Sér. Agron. 6, 57.
32. Franco, A. A. and Day, J. M. (1975) V American Rhizobium Conference, Raleigh, North Carolina.

33. Franco, A. A., Maranhão, J. I. M. and Döbereiner, J. (1970) Ann. V. RELAR, Rio de Janeiro, 292.
34. Griegg, J. L. (1953) New Zealand J. Sci. Technol. A34, 405.
35. Gurley, W. H. and Giddens, J. (1969) Agron. J. 61, 7.
36. Hallsworth, E. G. (1958) In: Nutrition of the Legumes, E. G. Hallsworth, ed., Butterworths Scient. Publ., London, 183.
37. Hallsworth, E. G., Wilson, S. B. and Greenwood, E. A. N. (1960) Nature 187, 79.
38. Jain, G. L. (1972) Indian J. of Agronomy 17, 271.
39. Johnson, C. M., Pearson, G. A. and Stout, P. R. (1952) Plant Soil 4, 178.
40. Jones, M. B. and Freitas, L. M. M. (1969) Pesq. agropec. bras. 5, 91.
41. Jones, M. B. and Quagliato, J. and Freitas, L. M. M. (1970) Pesq. agropec. bras. 5, 209.
42. Lie, T. A. (1974) In: The Biology of Nitrogen Fixation, A. Quispel, ed., North-Holland Publishing Co., Amsterdam, 553.
43. Lie, T. A. and Brotonegoro, S. (1969) Plant Soil 30, 339.
44. Little, R. C. (1971) U.K. Min. Agr. Fish. Food Bull. 21, 45.
45. Martin, W. E. and Matocha, J. E. (1972) In: Soil Testing and Plant Analysis. Soil Sci. Soc. Am. Proc., Madison, 393.
46. Mosse, B. (1976) In: Exploiting the Legume-Rhizobium Symbiosis in Tropical Agriculture, J. M. Vincent, A. S. Whitney and J. Bose, eds., Univ. Hawaii, Miscell. Publ. 145, 275.
47. Mulder, E. G. (1948) Plant Soil 1, 179.
48. Munns, D. N. (1976) In: Exploiting the Legume-Rhizobium Symbiosis in Tropical Agriculture, J. M. Vincent, A. S. Whitney and J. Bose, eds., Univ. Hawaii, Miscell. Publ. 145, 211.
49. Nery, M., Peres, J. R. R. and Döbereiner, J. (1976) Ann. XV Congr. Bras. Cienc. Solo, Campinas, SP. 157.
50. North Carolina Univ. Annual Report. Agronomic-economic research on tropical soils, Annual Report for 1974. Soil Sc. Dept.- Raleigh, NC. pp. 230.
51. O'Toole, M. A. and Masterson, C. L. (1968) Irish J. Agr. Res. 7, 129.
52. Peres, J. R. R., Nery, M. and Franco, A. A. (1976) Ann. XV Congr. Bras. Cienc. Solo, Campinas, SP, pp. 163.
53. Ruschel, A. P. and Eira, P. A. da (1969) Pesq. agropec. bras. 4, 103.
54. Ruschel, A. P., Britto, D. P. P. de S. and Carvalho, L. F. (1969) Pesq. agropec. bras. 4, 29.
55. Ruschel, A. P., Rocha, A. C. de M. and Penteado, A. de F. (1970) Pesq. agropec. bras. 5, 49.
56. Siqueria, C. (1976) M.Sc. thesis, Univ. Rural do Rio de Janeiro, RJ., pp. 86.
57. Souto, S. M. and Döbereiner, J. (1970) Pesq. agropec. bras. 5, 365.
58. Vargas, M. A. and Döbereiner, J. (1974) Pesq. agropec. bras. Sér. Zootec. 9, 21.

59. Walsh, T., Neenan, M. and O'Moore, L. B. (1952) Eire Dept. Agr. J. 48, 32.
60. Werner, J. C. and Mattos, H. B. (1974) B. Industr. Anim., SP, 31, 313.
61. Werner, J. C. and Mattos, H. B. (1975) B. Indústr. Anim., SP, 32, 123.

ECOLOGY OF LEGUME-RHIZOBIUM SYMBIOSIS

Eli Sidney Lopes

Instituto Agronômico de Campinas, with fellowship from the CNPq
Caixa Postal 28, Campinas, SP - Brazil

INTRODUCTION

The legume family is one of the most numerous plant groups with the total number of species not precisely known. It comprises more than 12,000, and might reach up to 17,000 described species (40). One of the most important characteristics of Leguminosae is the high nitrogen fixing capacity of most species, when properly associated with the *Rhizobium* bacteria.

Although only 12% of the legume genera described by Taubert (84) are typical of temperate climates, the knowledge on almost all aspects of the plant-bacteria association stems from the work with cultivated species of that group, according to the summary history presented by Norris (64). From all the species grown in tropical and sub-tropical areas, soybean (*Glycine max* (L) Merr.) is the only one for which a great deal of information is available.

For better comprehension of the ecology of the association it is necessary to study the ecology of both partners, since they can live independently from each other, and be affected differently by environmental factors (27, 39, 90).

An ecological appreciation of the symbiosis would apply to the discussion of environmental effects on the nitrogen fixing dependent legumes, and also on the effects of the legumes on the environment. Both aspects are agronomically important.

The recognition of the nitrogen fixing potential, and the emphasis of study on tropical legumes is relatively recent. A great deal of systematic work has been carried out in the last decades, particularly in Australia. In that country new principles on pasture management were established, and followed, viewing the legume-rhizobia association as the best option for production of cheap protein forage (18). In Brazil the symbiotic fixation is also the principal mean of nitrogen nutrition of soybeans (59); several aspects of the symbiosis in this plant have been studied (24, 26, 32, 57, 58, 60, 75).

One of the major limitations for crop production and symbiotic efficiency in most soils of the tropics and subtropics is the high acidity and low fertility condition. The effects of acid soil nutritional factors, and the effects of micronutrients are dealt with in other chapters. Other aspects of legume nutrition and environmental factors have been fully discussed in recent papers (36, 49). The subject of this paper will be limited to the occurrence of the association in nature, and the implications of indigenous rhizobia on the establishment of symbiosis of some legume plants cultivated in tropical areas.

TAXONOMIC CONSIDERATIONS

A preview on the taxonomic position of the root nodule bacteria, and its host is desirable for the better understanding of the ecology of the association. For the host side the reader is referred to the articles by Tutin (88) and Heywood (40). The establishment to the taxonomic position of the root nodule bacteria into species is a difficult matter, since the isolation of the type species *Rhizobium leguminosarum* Frank-by Beijerinck, in 1888.

The genus *Rhizobium* comprises six recognized species (*R. leguminosarum, R. phaseoli, R. meliloti, R. lupini, R. trifolii* and *R. japonicum*), which together with species of the genera *Agrobacterium* and *Chromatium* are grouped in the family Rhizobiaceae. The main criteria which places the root nodule bacteria within the genus *Rhizobium* is its ability to nodulate legumes. The rank of species is mostly based on the preference of the bacteria on infecting a group of legumes, although it is widely accepted that speciation of rhizobia cannot be based exclusively on cross inoculation grouping (20, 27, 31, 37, 90).

Some complementary characteristics of *Rhizobium* spp useful in routine tests are the change in pH of the culture media, and the speed of growth. *R. leguminosarum*, *R. phaseoli* and *R. meliloti* are all fast growers which produce large circular colonies in 3-5 days on yeast extract-manitol-agar plates. They acidify the media with (pH 7,0 as starting point). Some strains, particularly of *R. phaseoli* can either alkalinize, or produce no apparent change in the culture media during growth. The slow growing *R. japonicum* and *R. lupini* are alkali producers and give small colonies (1-2 mm) after incubation for 5-6 days.

Among the legumes cultivated in the tropics and subtropics, particularly in Brazil, only soybeans and dry beans are associated with the defined species *R. japonicum* and *R. phaseoli*, respectively. Elkan (28) has observed that among 55 strains considered as *R. japonicum*, five were acid producers and gave a serum zone on litmus milk, a reaction which is not common of this species; those five strains had not been isolated from soybean, although they were able to nodulate this legume. The author considered that despite the heterogeneity of soybean nodule bacteria in regard to biochemical and genetic characteristics, it is to be considered as a species, namely *R. japonicum*. It is clear however that not all isolates from soybean are necessarily *R. japonicum*, since some root nodule bacteria isolated from some tropical legumes can nodulate soybean. It has been suggested (91) that the slow growing *R. lupini* has enough affinity to be considered as a symbiotype of *R. japonicum*. Neither the host relationships of the *P. vulgaris* root nodule bacterium, nor its genetics and biochemical characteristics are well studied. This bacterium is closely related to *R. leguminosarum*, but seems to deserve specific naming, according to its antigenic properties and others characteristics.

The root nodule bacteria associated with most tropical legumes, usually referred to as the cowpea type rhizobia, has not yet been elevated to species ranking. High cross-infection promiscuity exists and apparently much work is necessary before steps can be taken regarding speciation of the bacteria associated with this group of legume. The cowpea type *Rhizobium* has recently been isolated from *Trema canabina* (originally reported as *T. aspera*), a plant species of the Ulmaceae family (87), extending its promiscuity outside of the leguminous family.

The cowpea type rhizobia are slow growing and alkali-producing bacteria. However, not all symbionts isolated from tropical legumes share these properties. In fact some, like most bacteria isolated from nodules of *Sesbania* sp, are fast growers, strong acid producers and form large colonies. Most isolates from the Mimosoideae alkalinize-the media but are fast growers. Symbiotic specificity among tropical legumes have been reported in *Leucaena leucocephala;* this legume has established efficient symbiosis only with bacteria isolated from its nodules, or with isolates of four other species of Mimosoideae (86). Campelo and Döbereiner (12) have shown that effective or ineffective nodules on *Mimosa caesalpinifoliae* are formed only with bacteria isolated from the same species. On the other hand, as observed in *Leucaena,* the root bacteria of that legume do not associate efficiently with legumes of the Papilionoideae subfamily. On the basis of further cross-inoculation studies, Campelo and Döbereiner (13) have sugested that the Mimosoideae might constitute a cross inoculation group. Other examples of specificity on tropical legumes include *Centrosema pubescens* (7), *Latononis bainesii* (63). The bacteria from *Latononis* nodules have unusual characteristic of producing pink colonies in yeast-manitol-agar. Robinson *et al.* (72) have observed that there are two groups of *Stylozanthes guyanensis* in relation to symbiotic affinity. One group which included 30 of the 49 clones surveyed fixed nitrogen efficiently with 42% of the tested cowpea type strains. The other group, which included the remaining clones, associated efficiently with only 6% of the rhizobia. Specificity with the cultivar "fine stem stylo", and IRI-1022 of *S. guyanensis* had been reported previously (65, 81).

ECOLOGICAL MEANING OF THE SYMBIOTIC NITROGEN FIXATION

Naturally poor soils, or soils impoverished by cropping, when left on fallow under conditions favorable to biological activity will have a gradual increase in fertility, up to when a vegetation climax is reached (43, 67, 70). The nitrogen balance of a fallow ecosystem shows that the soil increase in this nutrient can be explained only on the basis of gains from the atmosphere, by biological fixation. Lysimeter studies carried out by Zinke (97) with individual forest plants indicate that there might be gains or losses of nitrogen, as compared to the original condition, depending on the plant species. Over a period of 13 years, the largest losses were observed in barren soil, and the largest gains (including soil, litter and above ground plant material) were observed with *Ceanothus crassifolius*, a symbiotic nitrogen fixing non-legume. The total nitrogen increase with this species averaged to 4.5 g/m^2/year. However, the greatest increase in soil nitrogen during the same period was observed with *Quercus dumosa* Nutt., which is not known to fix nitrogen.

In equilibrated natural ecosystems the nutrient balance indicates that the requirement for external nitrogen is minimum. This, and other nutrients circulate from the biomass to the soil, and upon mineralization are reabsorbed to the biomass. One could then expect that in equilibrated systems the biological nitrogen fixation is of minor importance. Legumes would then be deprived of nodules, or bear ineffective nodules. Surveying for nodulation in some natural ecosystem, such as the equatorial forest is difficult. Although no extensive surveys have been carried out, in such environment there are indications that some nodulating species are deprived of nodules (5). In contrast, legumes grown in disturbed agricultural lands, where nitrogen is limiting, are usually well nodulated. These observations were the basis put forward by Bonnier and Brakel (6) that the symbiotic nitrogen fixation by legumes is a process of adaptation to a disequilibrated nitrogen balance. The supply of nitrogen to the exosystem, via atmospheric fixation, would then be the ecological niche of rhizobia-legume association. The mechanism of regulation of the nitrogen fixation activity is supposedly the internal C/N ratio (6, 96). It is well known that the addition of combined nitrogen to the soil causes a reduction in nodule formation, and in nitrogen fixation. Reduction of nodulation can be also obtained by spraying the leaves with urea, or by decressing plant photosynthesis (80). However, the decrease of nodulation observed in pea by addition of combined nitrogen, or exposures of plants to low light regimes could not be overcome by spraying plants with a sugar solution. As pointed out by Schreven (79), not only internal C/N ratio, but nitrogen and carbohydrate levels in the surrounding media might also influence nodulation.

A regulatory action of nitrogen fixation by the C/N ratio has been pointed out in other nitrogen fixing organisms. As mentioned by Döbereiner (25), Hüser (42) has observed that when C/N is lower than 40, non-symbiotic nitrogen fixation does not occur in soil, probably due to the low competitive capacity of the nitrogen fixing microflora for energetic substrates.

NODULATION OF TROPICAL LEGUMES IN NATURE

Observations on the distribution of legumes, and of their nodulation status in nature are the first steps for complementing other information that is useful for preliminary estimates on the contribution of nitrogen fixation by spontaneous legumes. Surveying of tropical legumes deserves special attention not only because of the paucity of information but also due to the high promiscuity of the bacteria that associate with them. One has to consider also that the largest reserves of natural vegetation lie in the tropics.

Not many surveys take into account the frequency of legume occurrence and the appearance of nodulation. According to citations of Nye and Greenland (67), legumes constitute more than 50% of the forest trees in Colombia (43) and in British Guyana (19) rain forests; they are also the main component of humid forests in Nigeria (45) and the Central Congo Basin (4). However, in Puerto Rico legumes might represent not more than 3.0% of the total trees in drier forests (80). Evaluation of the contribution of nitrogen fixation in natural ecosystems is mostly speculative. Based on the organic matter increment on soils (assuming a constant C/N ratio of 12, and on total biomass, for a 10 year period of forest fallow, Nye and Greenland (67) estimated an annual increment of nitrogen in the order of 106 kg N/ha. Under high grass savanna fallow, the nitrogen gain was in the order of 30 kg N/ha. Using the acetylene reduction technique, in an equilibrated virgin grassland in Canada, Paul and others (68) observed that total nitrogen fixation is very low (aproximately 1 kg N/ha/annum). This is in agreement with the idea that in equilibrated ecosystems nitrogem fixation might be negligible. The above mentioned authors (68) indicated that high reserves of nitrogen were built in those soils, since in intensively cultivated fields in the same area nitrogen fixation and nitrogen requirements were also in the same order of that observed in virgin fields.

It should be recalled that only a small proportion of the total number of legume species have been checked for nodulation. According to Allen and Allen (1), 1285 legume species were observed in this respect until 1961. The great majority of species which belong to the Papilionoideae and Minosoideae bear nodules (93% and 87%, respectively). Nodules are frequently found among the Caesalpinoideae, which mostly comprise legume trees of tropical regions; only 23% of the observed species are known to be nodulated. The more recent reported surveys show the same trend regarding nodulation within subfamilies (14, 82). None of the Swartzioideae has been observed for nodulation until recently (90).

CULTIVATED LEGUMES

The contribution of symbiotic fixation in supplying cultivated plants with part of its nitrogen demand, depends on the efficiency of the association, and on the availability of soil nitrogen, when due attention is paid to environmental factors. Under certain conditions, such as in recently deforested soils where nitrogen is not expected to be limiting, the contribution from symbiosis might be nil; the soil meets the plant demands (43, 68, 70, 94). This however does not mean that legumes should not be cultivated in such areas. When pasture is the choice for land use, a mixture of grass and legumes is to be considered. Irrespective if soil nitrogen is being assimilated, or air nitrogen being fixed, legumes, by virtue of their constancy in nitrogen percentage, contribute to a more stable protein content in the forage throughout the year (39). Soybeans have been shown to be an adequate crop in areas where massive fertilization might pose a nitrogen pollution problem (44).

In certain soils, such as in those under "cerrado" vegetation in Brazil, where low fertility is a prominent condition (33), a different situation seems to occur. The poor nodulation of legumes observed in the first year of cropping seems to be related to low number of rhizobia in inoculant (Freire, J. R. J., personal communication), and not with high nitrogen in the soil. Although responses to nitrogen are erratic, most crops benefit from nitrogen dressing in the first year. Evolution of soil fertility and microbial activity as influenced by continuous cropping have been studied on those soils, but apparently there are changes in the dynamics of soil nitrogen transformations which favour mineralization with time. Some indirect evidence for this statment can be seen in results of experiments carried out by Ritchey and Naderman (71) for three consecutive years with corn; yields of the plots without nitrogen (PK added) increased gradually from the first to the third year. Apparently mineralization of the accumulated organic matter was somehow delayed.

Usually for full benefit from nitrogen fixation, seed inoculation with strains selected for high nitrogen fixing capacity, and others characteristics, is recommended. This is the case whenever rhizobia are absent from the soil, or when inefficient bacteria are known to be present. A pre-judgment for the need of inoculation is difficult. A great diversity of legumes is being cultivated in tropical areas and some of them, like soybean and dry beans have a specific rhizobia requirement. It is most probable that when a legume which associates with specific rhizobia is introduced in tropical areas, there is a need for inoculation. On the other hand, the great majority of tropical legumes, which associate with cowpea type bacteria will nodulate regardless of inoculation. There might be a case for concern regarding competition of the indigenous bacteria with strains selected for high nitrogen fixing capacity. Competition with indigenous rhizobia have been reported for legumes of temperate regions. Problems of competition for nodule site under field conditions, between *Rhizobium trifolii* associated with native *Trifolium polymorphum,* and inoculant strains selected for high efficiency on *T. subterraneum* have been reported (41, 47). The latter legume could not be established in the field with usual inoculation techniques, because the indigenous strains were more competitive than the inoculant strains. Although efficient on the native clover they were inefficient on subterranean clover. Such a problem does not seem to be common with legumes associated with the cowpea type rhizobia, which is a widespread organism. Evidence is accumulating that promiscuous legumes will establish efficient symbiosis without inoculation (23, 52, 64, 83). Evaluation of the autochthonous rhizobia population is an important matter in tropical areas. It has already been suggested that preliminary experiments should be carried out before further engagement on broad programs of strain selection are started (16). Guidelines on recognition of nodulation problems have been presented by Diatloff (21), and are useful in the process of reaching decisions.

Soybeans

Soybean *Glycine max* (L.) Merr. is considered to form a mono-specific cross inoculation group, requiring *R. japonicum* for the establishment of N_2 fixing associations (31). Strain specificity at cultivar level has been observed, but wide spectrum efficient strains can be selected. Non-nodulating isolines of soybeans have been bred (91) and are useful for differentiating N_2 fixation from soil nitrogen uptake, when they are grown comparatively with the nodulating partnership (92).

The observation that *R. japonicum* can nodulate some

tropical legumes which usually associate with cowpea type rhizobia is of great ecological interest, for those legumes can harbour the bacteria and help then in their survival and dispersal in soil. Approximately 75% of *R. japonicum* strains tested by Elkan (28) nodulated *Vigna sinensis*. Siratro *(Macroptilium atropurpureus)* can also be nodulated by the soybean bacteria (Stamer, E., personal communication). One *R. japonicum* strain tested by Galli (34) in cross incoulation studies did not form nodules on cowpea nor on twelve other tropical legumes. Bonnier (5) has observed "effective" nodules on soybean in non-fertilized plots which had not been previously cropped to that legume; however, inoculation with selected strains increased seed production up to 210% in comparison with non-inoculated treatments.

No *Rhizobium* capable of nodulating soybean cv. Santa Rosa was found in six soil samples collected in different areas of two important great groups of soils in São Paulo State, not previously cropped to soybean (56). One of them induced very few ineffective nodules. On the other hand, when the samples were collected from nearby areas, previously cropped to soybean, effective nodulation was observed. Although *R. japonicum* does not seem to be indigenous to those soils, the bacteria introduced with seed inoculation survive until next crop. Similar results were reported by Elkinset al. in USA (29); in soils never cropped to soybean, nodulation on Clark cultivar was not expressed, but in soils which had soybean in the past good nodulation was effected despite the fact that the rhizobia population was considered to be low (0.22 - 49 x 10^{10} bacteria/g of soil).

A hypothesis of "adaptation" of cowpea type rhizobia toward effective nodulation and efficient nitrogen fixation in soybean was sugested by Bonnier and Brakel (6). They postulate that with successive cropping with soybean, an adapted population will develop after three to four years. This hypothesis remains to be checked, with careful control of seed-carried rhizobia, and other means of accidental contamination. Increase in natural or naturalized soybean nodule bacteria under field conditions, as a function of continuous cropping was observed by Caldwell (11). In soil free of *R. japonicum*, 98-100% of the nodules could be attributed to the inoculant strains, using serological methods of checking; the remaining nodules were caused by native "c1" serogroup bacteria. In Brazil Kolling et al. (46) quoted by Freire (32) reported different results, with three soybean cultivars. In the first year the nodulation not related to the inoculant strains was in the order of 2-11% of the nodules, depending upon liming and cultivar. In the second year, the nodulation caused by soil strains tended to decrease, averaging 1-6% of total nodules.

There is much evidence showing that soybeans benefit from seed inoculation in the first year of cropping (32, 56, 93). Failures of response to inoculation are usually related to low quality inoculant, poor inoculation technique, or good soil nitrogen supply. Although there are reports indicating competition between natural or naturalized rhizobia and inoculant strain in this legume, usually the soil strains are efficient nitrogen fixers (38, 94). In one experiment carried out in Ribeirão Preto, SP., an inefficient strain caused abundant nodulation in the non-inoculated plots, but the inoculant strains were more competitive than the soil ones, as indicated by increase in seed production on plots inoculated by usual procedures (56).

Dry Beans

Like soybeans, dry beans *(Phaseolus vulgaris* L.) require a specific *Rhizobium (R. phaseoli)* for efficient symbiosis. This legume, together with *Phaseolus angustifolius* Roxb., and *P. coccineus* L. (Syn. *P. multiflorus* Willd) composes a small cross inoculation group (31, 62, 90). Most of the published information on cross inoculation studies involving the above mentioned legumes do not take into account the efficiency of N_2 fixation. It is known that cowpea type rhizobia might nodulate *P. vulgaris*, and some strains of the fast growing bacteria isolated from beans nodulate some of the promiscuous tropical legumes (95)

Burton et al. (9) indicate that dry beans are usually nodulated without inoculation, in commercial crops in the midwest of USA. Efficient and inefficient strains of root nodule bacteria could be isolated from one plant, in an experiment where 10 cultivars were inoculated with soil samples collected in different regions. In an evaluation of 85 isolated strains they observed that 14% were efficient and 69% were moderately efficient.

Dry beans cultivated in Egypt are usually non nodulated or ineffectively nodulated under field conditions; Mohamoud et al. (57) relate this fact to the low density of efficient rhizobia in the soil, since they were able to isolate efficient strains from plants grown in sand cultures, inoculated with soil.

Abundant natural nodulation in plants sampled from commercial crops, in Rio Grande do Sul, Brazil, have been observed by Pons et al. (69). Lack of response to inoculation, and nitrogen, probably related to the presence of indigenous rhizobia have been also observed in São Paulo (50,51).

Norris et al. (66) have observed considerable increase in percent of nodulated plants, under field conditions due to seed inoculation, in soils apparently free of *R. phaseoli;* however nodules per plant was low in general, and no increase in bean production was observed. In pot experiments carried out by Ruschel (74) there was an increase in dry matter production, but a further significant increase was obtained by the application of combined nitrogen. Using ^{15}N-Urea, Neptune and Muraoka (61) got no response to nitrogen applied at planting or flowering, at a rate up to 120 kg N/ha, in a field experiment. Soil plus fixed nitrogen amounted to 81.1 kg/ha, as compared to 3.4 kg/ha from the fertilizer, in the treatment with 30 kg N/ha.

There are some reports indicating clearly that beans benefit from inoculation (3, 8). Brakel (8) emphazises the importance of strain selection. Problems of competion of selected strains with soil strains have been reported in pot experiments, and under field conditions in a Latosol, in São Paulo (76, 77). Saito and Ruschel (77) observed no difference in total nodulation between control and inoculated treatments; the most inffective strain was responsble for 76% of nodules, and caused an average increase of total plant nitrogen in the order of 48%. However, in another experiment Saito and Ruschel (78) observed that natural nodulation, and total nitrogen can be significatively increased, by fertilization with PK plus micronutrients, in two out of four soils.

In Australia Diatloff (22) observed that beans might benefit from inoculation, but a further increase in seed production can be obtained from nitrogen application.

The comparative studies which have been undertaken at CIAT, Colombia with regard to the symbiotic efficiency of a number of cultivars have shown great differences among them (Graham, P., Personnal communication) Franco and Day (30) presented the view that dry beans seem to have specific requirements for the absorption and/or metabolism of molybdenun, which might interfere with the expression of maximum nitrogen fixing capacity of this legume. Foliar spray of molybdenun increased total nitrogen of nodulated dry beans, which was a response similar to application of combined nitrogen (73).

Symbiotic behaviour of cultivars, specific nutritional requirements, indigenous rhizobia in soils, are some aspects not yet fully understood that might be evoked to explain the conflicting results which have been observed with this legume.

Peanuts

Although peanut (*Arachis hypogaea* L.) associates with the cowpea type root nodule bacteria, it does not seem to be a legume so promiscuous as many others from tropical areas. In experiments reported by Burton (10) only two out of ten strains isolated from tropical legumes established efficient symbiosis with peanuts. Lopes et al. (54) observed that only three out of ten *Rhizobium* strains nodulated peanuts grown in Leonard jars. Crushed nodule suspensions prepared after sterilizing the surface of fresh nodules collected from 51 species of legumes indigenous to India, were able to induce nodulation in peanut (35); based on this observation Gaur et al. (35) concluded for high promiscuity on peanuts. Dadarwal et al. (15) have shown that rhizobia isolated from five spontaneus species of the genus *Arachis* (*A. duranensis*, *A. prostrata*, *A. villosa*, *A. glaberata*, and *A. marginata*), nodulated a cultivar of *A. hypogaea*. The percentage of nitrogen in plants inoculated with strains isolated from *A. duranensis* was greater than in plants inoculated with isolates of other species. By immunodiffusion technique it was possible to separate the 44 *Arachis* spp rhizobia into 33 different strain serotypes.

Sporadic observation of root systems of peanuts under farming conditions in São Paulo State shows that the great majority of plants are well nodulated. In an experiment under field conditions there was no response to inoculation by conventional procedures, and by lime pelleting; natural nodulation was abundant, and the plants did not respond to nitrogen application (53). Arora and Saini (2) observed, in India, an increase in oil and protein in the seed due to inoculation, but seed production was not greater either in presence or absence of phosphorus. Natural nodulation of peanuts grown in the Coastal regions of Gambia was not affected by phosphorus and potassium in experiments where seed production was favored by these fertilizers and also by nitrogen (85). In pot experiments inoculation with a selected strain (CB447) gave no increase in yield (85). There has been no report regarding peanut cultivar specificity, as for some other tropical legumes (7, 63, 65, 81). Observations on nodulation of 21 cultivars being tested in São Paulo State showed 100% naturally nodulated plants (55).

Sturkie and Buchanan (83) stated that usually there is no response to inoculation of peanuts in USA, and apparently the soils are supplied with effective root nodule bacteria. In most of the experiments reported, natural nodulation was observed, but in some cases there was a further yield gain from nitrogen additions. The control of nodules formed by the inoculant strains is not usually taken into due consideration. Such a control, as pointed out by Date (17), is indispensable for proper evaluation of the success of the inoculant strains in competing with native ones; such evaluation has recently been carried out in field experiments, using either serological techniques (23) or antibiotic resistant strains (48).

Diatloff and Langford (23) have shown, in Queensland, Australia, that the recovery of the CB 756 inoculating strain was in the order of 41 and 66% of nodules, in previously cultivated fields, and in virgin soils, respectively. The inoculation technique had some influence in recovery. Response to inoculation was erratic, and differences in yield could not be attributed to the strain used as inoculum. Nodule formation was higher (83.6 nodules/plant) in the previously cultivated field, than in areas never cropped to peanuts (22.3 nodules/plant).

In South Africa, Van Der Merwe et al. (89) employed antibiotic resistant strains to check the success of the inoculum strains. Natural nodulation was in general abundant, and the nodule formation by selected strains (CB 756, or a local isolate nº 6N) in some field experiments was usually higher than 50%. No increase in yield was attained with seed inoculation, indicating as observed by Diatloff and Langford (23) that the indigenous soil *Rhizobium* population established efficient nitrogen fixation symbiosis. Nodule formation by the inoculant strain varied with soil type, strain used and with the planting year.

FINAL CONSIDERATIONS

The final aim of the research efforts on legume-rhizobia associations is the economy of nitrogen fertilizers in farming practice. Results of such efforts have been most rewarding, since nitrogen fixation can supply most of the nitrogen demand of cultivated legumes.

It has normally been assumed that seed inoculation is necessary for full benefit from symbiotically fixed nitrogen. This is the case when inefficient rhizobia, or no rhizobia are present in the soil. Soybeans, which require specific rhizobia usually respond to seed inoculation when first cropped in tropical areas. The prompt survival of introduced *Rhizobium japonicum* in soils suggests that strain recommendation for inoculation purpose should be carefully controlled, and that the symbiotic characteristics of legumes should be taken into account in breeding programs.

Advantages from seed inoculation of dry beans and peanuts are not clearly established. The practice of inoculating seeds of those legumes is not usual. Natural nodulation is commonly observed in peanut plants, and nitrogen fixation by indigenous rhizobia seems to contribute significantly to meet the plant demand. More conflicting results are reported in the evaluation of the need for inoculation, and nitrogen fixing potential of dry beans. Responses to nitrogen fertilizers are not unusual. Host specificity, diversity of cultivars, and nutritional requirements not yet fully investigated are aspects that might explain some of the controversial results which have been reported with dry beans.

Investigations on the legume-rhizobia symbiosis in natural communities of tropical regions have been mostly restricted to observations on species nodulation. The quick landscape changes which are under way in these regions suggest that an effort should be made to stimulate research on legume symbiosis in natural communities of tropical areas. Collecting legumes and rhizobia are an aspect to be considered. Both are important natural resources which can be lost before being evaluated.

Acknowledgements - The author thanks the organizers of the Symposium for the invitation and for the funds to participate. The author also acknowledges the help of Dr. Alvaro Santos Costa in correcting the manuscript, and of Mrs. Dione Silva Lopes in typing it.

REFERENCES

1. Allen E. K., and Allen, O. N. (1961)*Recent Advan. Botany*
2. Arora, S. K., Saini, J. S., Gandhi, R. C., and Sandhu, R.S. (1970)*Oléagineux* 25,279
3. Basurco, J. C. P., Olivero, E. L. G., and Schiel, E. (19)*Rev. Invest. agropec., INTA, Ser. 2*, 6,373
4. Beirnaert, P. W. (1941)*I.N.E.A.C. Sér. Tech.* 6, Bruxelles, p.86
5. Bonnier, C. (1957)*I.N.E.A.C. Sér. Scient.*, 72, p.66
6. Bonnier, C., and Brakel, J. (1969)*Legumineuse-Rhizobium, Lutte Biologique contre la faim* J. Ducolot Gembloux, p.148
7. Bowen, G. D., and Kennedy, M. M. (1961)*Queensland J. Agr. Sci.* 18,161
8. Brakel, J. (1966)*Bull. Res. Agron. (Gembloux)* 1,525
9. Burton, J. C., Allen, O.N., and Berger, K.C. (1952) *Soil Sci. Soc. Am. Proc.* 16,167
10. Burton, J. C. (1952)*Soil Sci. Soc. Am. Proc.* 16,356
11. Caldwell, B. E. (1969) *Agron. J.* 61,813
12. Campelo, A. B., and Döbereiner, J. (1969)*Pesq. agropec. bras.* 4,67
13. Campelo, A. B., and Döbereiner, J. (1970)*Pesq. agropec. bras.* 5,333
14. Corby, H. D. L. (1971)*Plant and Soil, Special Vol.* pp. 305
15. Dadarwal, K. R., Singh, C. S., and Subba Rao, N.S. (1974)*Plant and Soil* 40,535
16. Date, R. A. (1974) *Proc. Indian natn. Acad. Sci.,Part B* 40,687
17. Date, R. A. (1975) in *Symbiotic Nitrogen Fixation in Plants* (Nutman, P.S.,ed.) IBP, vol. 7, pp. 137-150, Cambridge Univ. Press
18. Davies, J. G. (1965) *Proc. 9th Intern. Grassland Congress*, São Paulo, 1, 217
19. Davis, T. A. W., and Richards, P. W. (1934)*J. Ecol.* 21,350
20. De Ley, J., and Rassel, A. (1965)*J. Gen. Microbiol.* 41,85
21. Diatloff, A. (1972)*Tropical grasslands* 6,70
22. Diatloff, A. (1975)*5th Aust. Leg. Nod. Conf., Rhizobium Newsletter (Supplement)*, 27
23. Diatloff, A., and Langford, S. (1975)*Queensland J. Agric. Anim. Sci.* 32,95
24. Döbereiner, J. (1966)*Pesq. agropec. bras.* 1,357
25. Döbereiner, J. (1974)*The Biology of Nitrogen Fixation* (Quispel, A., ed.) Frontiers of Biology, Vol. 33,86 North Holland Publ. Co, Amsterdan
26. Döbereiner, J., Franco, A., and Guzman, I. (1970)*Pesq.*

agropec. bras. Sér. Agron. 5,155
27. Domergues, Y., and Mangenot, F. (1970)*Écologie Microbienne du Sol*, Massonet, Paris, p.796
28. Elkan, G. H. (1971)*Plant and Soil, Special Vol.* pp.85
29. Elkins, D. M., Hamilton, G., Chan, C. K. Y., Briskovich, M. A., and Vandervent, J. M. (1976) *Agron. Journal* 68,513
30. Franco, A. A., and Day, J. (1975)*5th American Rhizobium Conference*, Raleigh
31. Fred, E. B., Baldwin, I. L., and McCoy, E. (1932)*Root Nodule Bacteria and Leguminous Plants* Univ. Wisconsin, Madison p.343
32. Freire, J. R. J. (1976)Inoculation of Soybeans. *Workshop on "Exploiting the Legume-Rhizobium Symbiosis in Tropical Agriculture"*, Hawaii, mimeographed, p.33
33. Freitas, F. G., and Silveira, C. O. (1976) *IV Simposio sobre o Cerrado- Bases para Utilização Agropecuária* Brasília, mimeographed
34. Galli, F. (1958) *Rev. Agric. (Piracicaba)* 33,139
35. Gaur, Y. D., Sen, A. N., and Subba Rao, N. S. (1974) *Zbl. Bakt. Abt. II* 129,369
36. Gibson, A. H. (1971)*Plant and Soil, Special Vol.* pp. 139
37. Graham, P. H. (1964)*J. Gen. Microbiol.* 35,511
38. Ham, G. E., Caldwell, V. B., and Johnson, H. W. (1971) *Agron. Journal* 63, 301
39. Haydock, K. P., and Norris, D. O. (1967)*Aust. J. Agric. Sci.* 29,426
40. Heywood, V. H. (1971) in *Chemotaxonomy of the Leguminosae* (Harborne, J. B., Boulter, D., and Turner, B. L., eds.) Academic Press, London, p.1
41. Holland, A. A. (1970)*Plant and Soil* 32,293
42. Hüser, R. (1965)*Plant and Soil* 23,236
43. Jeny, H. (1950)*Soil Science* 69,63
44. Johnson, J. W., Welsh, L. F., and Kurtz, L. T. (1975) *J. Environ. Qual.* 4,303
45. Keay, R. W. (1953)*An Outline of Nigerian Vegetation* 2nd. ed., Government Printer, Lagos, Nigeria
46. Kolling, J., Freire, J. R. J., Hilgert, E. R., and Mendes, N. (1976)*III Reunião Conjunta de Pesquisa da Soja RS/SC*, Porto Alegre, p. 129 (sum.)
47. Kornelius, E., Freire, J. R., and Barreto, I. (1972) *Agron. sulgiograndense* 8,95
48. Law, I. J., and Strijdom, B. W. (1974) *Phytophylactica* 6,221
49. Lie, T. A. (1974)*The Biology of Nitrogen Fixation* (Quispel, A., ed.) Frontiers of Biology, Vol. 33,265

North Holland Publ. Co., Amsterdan
50. Lopes, E.S. (1974)*Circular nº 33* Inst. Agronômico pp. 16, Campinas, S.P., Brasil
51. Lopes, E. S., Deuber, R., Forster, R., Gargantini, H. and Bulisani, E. S. (1971)*Bragantia* 30,109
52. Lopes, E. S., Lovadini, L. A. C., Gargantini, H. and Igue, T. (1972)*Bragantia* 31,235
53. Lopes, E. S., Tella, R., Rocha, J. L.V., and Igue, T. (1972)*Bragantia* 31,XXVII
54. Lopes, E. S., Lovadini, L. A. C., Miyasaka, S., Igue, T., and Giardini, A. R. (1974)*Bragantia* 33,CV
55. Lopes, E. S., Savy, A., Oliveira, M. L. C., Giardini, A. R., and Pompeo, A. S. (1976)*Bragantia* 35,XI
56. Lopes, E. S., Giardini, A. R., e Kiihl, A. S. (1976) *Bragantia* 35,389
57. Mahamoud, S. A. Z., Taha, S. M., and Salem, S. H. (1971)*Zbl. Bakt. Abt. II* 126,33
58. Mascarenhas, H. A. A., Miyasaka, S., Freire,E. S., Di Sordi, G. and Lopes, E. S. (1967)*Bragantia* 26,143
59.Mascarenhas, H. A. A., Miyasaka, S. Braga, N., Miranda, M. A. C., and Tisselli Filho, O. (1977)*A Soja no Brasil Central* Fundação Cargill p. 85
60. Miyasaka, S. (1954)*Bragantia* 14,17
61. Neptune, A. M.L., and Muraoka, T. (1977) *Rev. Bras. Ciencia do Solo* (No prelo)
62. Norris, D. O. (1956) *Emp. J. Exp. Agric.* 24,247
63. Norris, D. O. (1958) *Aust. J. Agric. Res.* 9,629
64. Norris, D. O. (1965) *Proc. 9th Int. Grassl. Congr.* São Paulo, pp.1087
65. Norris, D. O. (1967)*Trop. Grasslands* 1,107
66. Norris, D.O., Lopes, E. S., and Weber, D. F. (1970) *Pesq. agropec. bras.* 5,129
67. Nye, P.H., and Greenland, D. J. (1965)*The Soil under Shifting Cultivation* Fârnham Royal, Comm. Agric. Bur. (Tech. Comm. nº 51) , p.156
68. Paul, E. A., Myrs, J. K., and Rice, W. A. (1971)*Plant and Soil, Special Vol.* pp.495
69. Pons, A. L., Goepfert, C. F., Kornelius, E. and Altmayer, M. B. (1976)*Agron. Sulgiograndense* 12,129
70. Richardson, H. L. (1938) *J. Agric. Sci.* 28,73
71. Ritchey, K. D., and Naderman, G. C. (1975)*Agronomic-Economic Research on Tropical Soils.* Annual Report Soil Science Dept., North Carolina State Univ. Raleigh, USA, pp.16
72. Robinson, P. J., and Date, R. A., and Megarrity, R. G. (1976)*Aust. J. Agric. Res.* 27,381
73. Robitaille, H. A. (1975) *Ann. Report Bean Impr. Coop.* 18,65

74. Ruschel, A. P., and Ruschel, R. (1975) *Pesq. agropec. bras., Sér. Agron.* 10,11
75. Ruschel, A., Ruschel, R., Almeida, D.L., and Suhet, A. R. (1974)*Pesq. agropec. bras., Sér. Agron.* 9,125
76. Saito, S. M. T., and Ruschel, A. P. (1976) *VIII RELAR, Colombia,* mimeographed, p.12
77. Saito, S. M. T, and Ruschel, A. P. (1976)*VIII RELAR,* Cali, Colombia, Mimeographed, p.15
78. Saito, S. M. T., and Ruschel, A. P. (1976)*VIII RELAR,* Cali, Colombia, mimeographed, p.11
79. Schreven,, D. A. Van (1959)*Plant and Soil* 11,93
80. Smith, R, M., Samuel, G., and Cernuda, C. F. (1951) *Soil Sci.* 72,409
81. Souto, S. M., Côser, A. C., and Döbereiner, J. (1972) *Pesq. agropec. bras., Sér- Zootec.* 7,1
82. Souza, D. I. A. (1966)*Trop. Agric., Trin.* 43,265
83. Sturkie, D. G., and Buchanan, G. A. (1973)in *Peanuts-Culture and uses* Stone Printing Co., Roanoke, Virginia, USA, pp.299
84. Taubert, P. (1894) *Leguminosae* In Die Naturlichen Pflanzenfamilien (Engler and Prantl) 3,3
85. Thornton, I. (1963) *Oléagineux* 18,871
86. Trinick, M. J. (1963)*Expl. Agric.* 4,234
87. Trinick, M. J. (1973) *Nature* 244,459
88. Tutin, T. G. (1958)in *Nutrition of the Legumes* (Hallsworth, E. G., ed.) Butterworths Sci. Publ. pp.3
89. Van der Merwe, Strijdom, B. W., and Uys, C. J. (1974) *Phytophylactica* 6,295
90. Vincent, J. M. *The Biology of Nitrogen Fixation* (Quispel, A, ed.) Frontiers of Biology, Vol. 33,265 North Holland Publ. Co., Amsterdam
91. Weber, C. R. (1966) *Agron. J.* 58,43
92. Weber, C. R. (1966) *Agron. J.* 58,46
93. Weber, D. F., Caldwell, B. E., Sloger, C., and Vest, H. G. (1971) *Plant and Soil, Special Vol.* pp.293
94. Williams, W. A. (1967) *Trop. Agric., Trin.* 44,103
95. Wilson, J. K. *Soil Science* 58,61
96. Wilson, P. W. (1940) *The Biochemistry of Nitrogen Fixation* Univ. Wisconsin Press., Madison
97. Zinke, P. J. (1969) *Biology and Ecology of Nitrogen Fixation- Proc. of a Conference.* Natl. Ac. Sci. Washington, D. C, pp.40

NITROGENASE SYSTEMS

R. H. Burris, T. Ljones and D. W. Emerich

Department of Biochemistry and Center for Studies of N_2 Fixation

College of Agricultural and Life Sciences
University of Wisconsin-Madison
Madison, Wisconsin 53706, U.S.A.

Nitrogenase is the term applied to the enzymatic system capable of fixing N_2. It is not a simple enzyme but actually consists of two protein components, a MoFe protein and an Fe protein. Both of these proteins are required for activity, and to date no one has demonstrated any catalytic activity of the individual proteins. In addition, the system must be supplied with MgATP and a strong reductant to reduce N_2. The physiological reductant is usually ferredoxin, but flavodoxin or azotoflavin can function

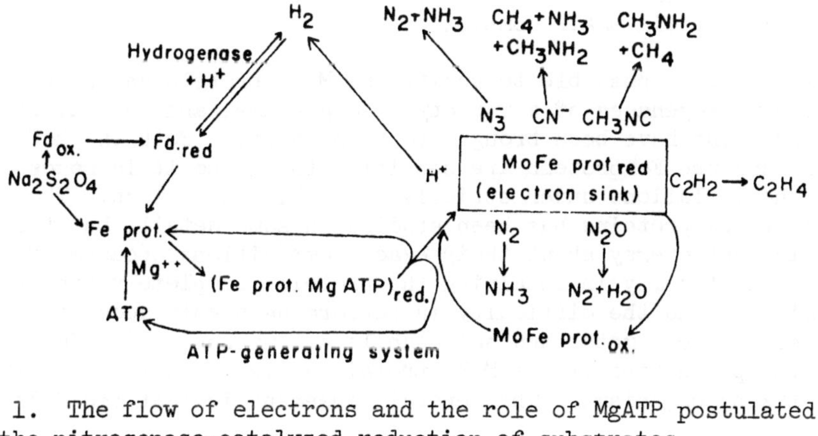

Fig. 1. The flow of electrons and the role of MgATP postulated for the nitrogenase-catalyzed reduction of substrates.

in some organisms, and in the laboratory one can substitute $Na_2S_2O_4$ for natural reductants to drive the reaction. It is apparent that the system operates at a very negative potential to reduce the N_2 molecule.

To understand the general reaction mechanism, let us consult Fig. 1 (32). Later we will discuss the evidence supporting the working hypothesis shown in this figure, but for the moment the figure will indicate the concept of how the nitrogenase system operates. The Fe protein is reduced by ferredoxin or dithionite, and the Fe protein is capable of binding MgATP. When the MgATP is bound to the Fe protein, the potential of the protein is lowered to somewhat below -400 mv (35). This confers upon the Fe protein the unique capacity to reduce the MoFe protein. Although $Na_2S_2O_4$ has a very low potential, it is not capable of reducing the MoFe protein directly. In fact, the only means that has been used successfully to reduce the MoFe protein artificially has been electrochemical reduction. The electrochemically reduced MoFe protein has shown no catalytic activity. The implication of these observations is that two proteins are necessary, because there is no physiological means for reducing the MoFe protein other than by way of the Fe protein-MgATP complex.

Apparently the MoFe protein serves to bind the various substrates of nitrogenase. The MoFe protein having been reduced by electrons from the Fe protein-MgATP complex then is in a position to pass these electrons to the substrate to reduce the substrate. Nitrogenase is a versatile enzyme, and in addition to hydrolyzing ATP and reducing N_2, it also is capable of reducing N_2O, N_3^-, CN^-, CH_3NC, C_2H_2, cyclopropene, and H^+. Apparently all of these substrates bind to the MoFe protein and compete for electrons from the MoFe protein. The capability of accepting electrons from the reduced MoFe protein varies with the substrate bound. The reduced substrates are released from the MoFe protein, and the oxidized MoFe protein then is in a state to accept more electrons from the reduced Fe protein-MgATP complex.

It has been possible to purify the MoFe protein and Fe protein from the nitrogenases of a variety of organisms, and several of these proteins have been brought to homogeneity. Both the MoFe protein and the Fe protein are sensitive to O_2, so it is necessary to do the isolations under strictly anaerobic conditions. The nature of the proteins has been studied in some detail, but there still is controversy about their exact compositions because of the inherent difficulty in obtaining the proteins completely free from contaminants and the difficulty in performing highly accurate analyses of their metal and acid-labile sulfur contents. Table I lists the properties of the MoFe protein as isolated from a number of N_2-fixing organisms. The general consensus is that the MoFe

TABLE I

Properties of the MoFe proteins of various nitrogenases.

Organism / Property	*C. pasteurianum*	*A. vinelandii*	*A. chroococcum*	*K. pneumoniae*	*C. vinosum*	*R. japonicum*	*R. lupini*
Molecular weight	210,000	216,000	227,000	218,000		200,000	194,000
Subunit composition	2 types: 60,000 51,000	1 type: 56,000	1 type: 60,000	2 types: 59,600 51,300	2 types:	1 type: 50,000	
Metal and sulfide content (mole/mole protein)							
Mo	1.95	1.54	1.9	2.01	1.3	1.3	1
Fe	22-24	24	23	32.5	17	29	18-20
$S^{=}$	22-24	20	20	18	14	26	-
EPR spectra, *g*	4.29 3.77 2.01	4.3 3.65 2.01	4.29 3.65 2.013	4.3 3.73 2.015	4.3 3.68 2.01	-	-
Specific activities, nanomoles substrate reduced/(min x mg Mo-Fe protein)							
N_2	620	316					
H_2		1488					
C_2H_2	2500	1638	2400	2150	1800	900	704
References	6,11,26,32	6,14,19,32	33	7,22	6,10,32	12,32	32

protein has from 24 to 32 Fe per molecule and that it has two Mo per molecule. These are the metal contents for a unit of molecular weight of approximately 220,000. This 220,000 unit can be broken into subunits. The MoFe protein from most organisms has two subunits with a molecular weight of about 60,000 and two with a molecular of about 50,000. There are indications that the MoFe protein from some organisms gives four indistinguishable subunits. This point is controversial and requires further clarification (13,34). The number of acid-labile sulfurs is approximately equivalent to the number of Fe atoms in the molecule.

As indicated in Table 2, the Fe protein is smaller than the MoFe protein; it has a molecular weight of about 56,000 to 67,000, depending upon the organism from which it is isolated. Convincing evidence indicates that it contains 4Fe and 4 acid-labile S per molecule. The 4Fe and 4S appear to be organized in a single active center that has properties very similar to those of a 4Fe-

TABLE 2

Properties of the Fe proteins of various nitrogenases.

	C. pasteurianum	*A. vinelandii*	*A. chroococcum*	*K. pneumoniae*
Molecular weight	56,000	64,000	64,000	66,800
Subunit composition	1 type 27,500	1 type 33,000	1 type 30,800	1 type 34,600
Metal & sulfide composition (moles/mole protein)				
Fe	4.05	3.45	4	4
S	4.0	2.85	3.9	3.85
EPR spectra	2.04 1.94 1.88	2.05 1.94 1.88	2.05 1.94 1.87	2.05 1.94 1.86
Specific activity, nanomoles substrate reduced/(min x mg Fe protein)				
N_2	755	470		275
H_2	-	-		1050
C_2H_2	3,100	1,815	2,500	980
References	3,26,32	14,19,32	33	7,22

4S center found in the ferredoxins. This appears to constitute the catalytically active center of the Fe protein, and such a structure explains the low potential of the Fe protein by analogy with the low potential of the ferredoxins. The structure of the ferredoxins has been established by x-ray crystallography (20), and the 4Fe and 4S are oriented in a cubical array with Fe and S alternating at the corners of the cube. It is probable that the O_2-inactivation of the Fe protein involves disruption of the integrity of the 4Fe-4S center. The Fe protein can be dissociated into two equivalent subunits. The 4Fe-4S center may be bound between the two subunits rather than to a single subunit, because the two isolated subunits appear equivalent. As in the case with the MoFe protein, it has been impossible to reassociate the Fe protein to yield the catalytically active unit after it has been dissociated by any means. The oxidation-reduction of the Fe protein is accompanied by a change in its absorption spectrum, and this change can be used to monitor the reduction state of the Fe protein (15).

In Fig. 1 we have presented a working hypothesis for the mechanism of biological N_2 fixation. What is the experimental basis for such a hypothesis? It is based primarily upon observations of changes in the electron paramagnetic resonance (EPR) spectra of the nitrogenase components under various conditions (17,34). As indicated in Fig. 2, the Fe protein of nitrogenase has a characteristic EPR spectrum; when MgATP is added to the Fe protein, there is a change in the spectrum as shown in the figure (pH 8.0). It can be demonstrated that MgATP binds specifically to the Fe protein, and that this binding is not influenced by the MoFe protein (28). It also can be shown that when this binding occurs, the potential of the Fe protein is lowered by about 100 mv (35), and this makes it possible for the Fe protein to reduce the MoFe protein, as indicated earlier. A molecule of Fe protein can bind two molecules of MgATP but only one molecule of MgADP with substantial affinity. The reactions of nitrogenase are inhibited by ADP when it binds to one of the ATP-binding sites. ADP does not bind effectively to the second ATP site, but it seems to increase the affinity of MgATP for its second site (28).

Figure 3 shows the response of the nitrogenase proteins in the absence of MgATP. You will note that when the MoFe protein is added to the Fe protein that the EPR spectrum of the mixture is approximately additive (17). That is, in the absence of MgATP the Fe protein does not influence the oxidation-reduction state of the MoFe protein. In contrast, when the two proteins are mixed in the presence of MgATP, the characteristic EPR signal of the MoFe protein is markedly suppressed, as shown in Fig. 4. The MoFe

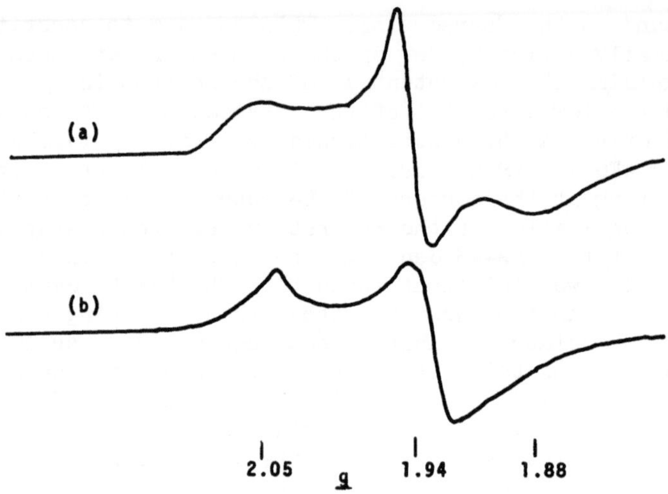

Fig. 2. EPR of the Fe protein of C. pasteurianum nitrogenase without (a) and with (b) MgATP. For experimental details see (17).

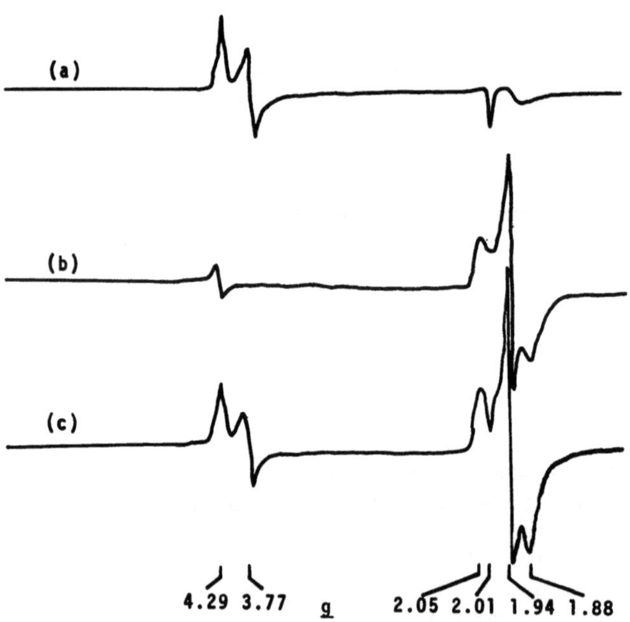

Fig. 3. EPR of the components of C. pasteurianum nitrogenase in the absence of MgATP. (a) 10 mg MoFe protein/ml; (b) 10 mg Fe protein/ml/ (c) 10 mg MoFe protein plus 10 mg Fe protein/ml. For experimental details see (17).

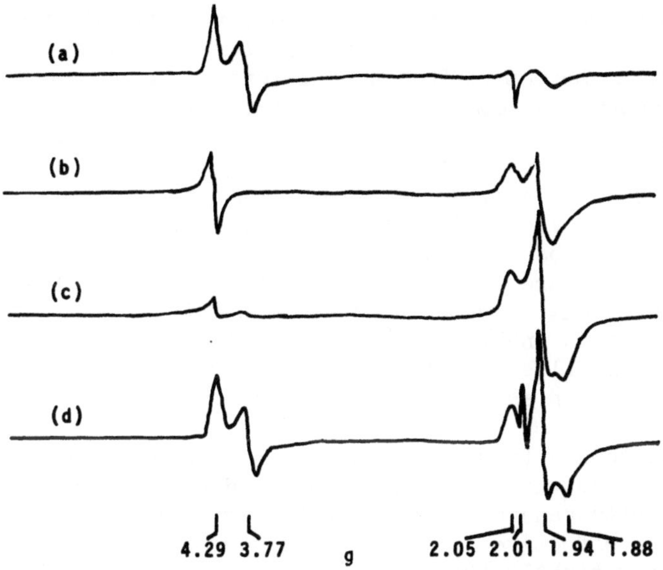

Fig. 4. EPR of the components of C. pasteurianum nitrogenase in the presence of MgATP. Each sample contained an ATP-generating system. (a) 10 mg of MoFe protein/ml plus 5 mM $Na_2S_2O_4$; (b) 10 mg Fe protein/ml plus 5 mM $Na_2S_2O_4$; (c) The two proteins, 10 mg of each/ml plus $Na_2S_2O_4$, frozen 45 sec after mixing; (d) same as (c) except that 0.5 mM $Na_2S_2O_4$ was initially present, together with 40 μM methylviologen, and incubation after mixing and before freezing was for 90 sec; after 90 sec the bluish cast of reduced viologen no longer was visible and the MoFe protein likewise had become reoxidized to its original partially reduced state. For further experimental details see (17).

protein can exist in three oxidation-reduction states. As it is isolated from the organisms, it is in a partially reduced state. In the presence of reduced Fe protein plus MgATP it is converted to the completely reduced state, and in this condition its EPR signal near g=3.77 is strongly suppressed. Apparently in the steady state the MoFe protein is about 95% reduced. If one allows the system to exhaust its supply of dithionite serving as reductant, the MoFe protein then returns to its partially reduced state and the g=3.77 signal is restored (Fig. 4d). It also is possible to titrate the MoFe protein with ferricyanide to a completely oxidized state, and under this condition the g=3.77 signal also is suppressed (27). It generally is agreed that the physiologically active change is from the partially reduced to the fully reduced state, and that the completely oxidized state probably is not physiologically active.

Most investigators accept the concept that the MoFe protein serves to bind the substrate and to transfer electrons to the substrate for its reduction. It can be demonstrated that the various nitrogenase substrates compete for electrons from the reducing pool, and hence it is assumed that they all extract electrons from a common pool. The interactions among the various substrates have been studied in some detail (18), but there will not be adequate time to review this material. It is interesting, however, that the various substrates are not completely equivalent. Although CN^-, CH_3NC and N_3^- are mutually competitive, none of these is competitive with N_2. N_2O and H_2 are competitive inhibitors of N_2. CO is a very potent inhibitor of N_2 reduction, and it inhibits the reduction of all substrates with the exception of H^+. In some fashion H^+ reduction escapes the CO block of electron transport.

It is customary to speak about the nitrogenase complex. The assumption is made that because an MoFe and an Fe protein are required for nitrogenase activity that they must form a complex in order to be effective. This however does not address the question of how tightly the components in the complex are bound or whether or not the complex associates and dissociates with each turn of the catalytic cycle. Interpretations usually are made in terms of a complex that remains associated during the electron transfer to substrate. In actual fact, there is little experimental evidence that compels one to take the viewpoint that the components of the complex are tightly associated and remain associated during catalytic activity. Let us assume that the complex associates and dissociates at each turn of the catalytic cycle. Is there any evidence that we can muster to support such an assumption?

First, it is apparent that the complex between the MoFe protein and the Fe protein is not a tightly binding complex. One can place a mixture of the two components on a column of DEAE cellulose and demonstrate that the two components are eluted separately with a change in salt concentration of the eluting fluid; no special effort is required to dissociate them. The individual components also will separate on a column of Sephadex, and the separations on DEAE and on Sephadex are routine steps in the purification of the individual proteins. Examinations of the effect of dilution on the catalytic activity of a mixture of MoFe and Fe proteins reveals a typical dilution effect (24,25). Apparently the dilution effect is evident because the complex dissociates as it is diluted to lower and lower concentrations of the components.

Recently Emerich (9) has demonstrated that the MoFe protein from *A. vinelandii* and the Fe protein from *C. pasteurianum* form a tightly-binding complex in contrast to the usual loose-binding complex of homologous proteins. This tightly-associated complex is not catalytically active. It is apparent that the mere binding between nitrogenase components is not sufficient to insure catalytic activity. In fact, when the two components are very tightly bound, as in this particular mixture, they lose their catalytic activity.

This in turn suggests that it probably is necessary for the MoFe protein and Fe protein complex to dissociate at each turn of the cycle. One can visualize a mechanism in which the Fe protein is reduced and complexed with MgATP independent of the MoFe protein. Following this, it could bind to the oxidized form of the MoFe protein, and in the complexed state it could transfer its electrons for the reduction of the MoFe protein. The complex then could dissociate either before or after the electrons were transferred to the bound substrates. There is no experimental evidence which convincingly establishes whether the dissociation occurs before or after the substrate is reduced. Substrate reduction may occur after electron transfer between proteins and after the dissociation of the oxidized Fe protein from the reduced MoFe protein. It seems logical that the site of electron transfer between the two proteins is opened to expose an adjacent site for transfer of electrons to the substrate after the transitory complex between the proteins has been formed and then dissociated.

Another point of some disagreement arises concerning the composition of the complex between the MoFe protein and the Fe protein. Evidence has been presented by Eady (5) and by Thorneley et al. (25) to suggest that one molecule of MoFe protein binds one molecule of Fe protein. Our data derived from titration of the MoFe protein against the Fe protein from C. pasteurianum or vice versa, suggested some years ago that one molecule of MoFe protein bound two molecules of Fe protein (29).

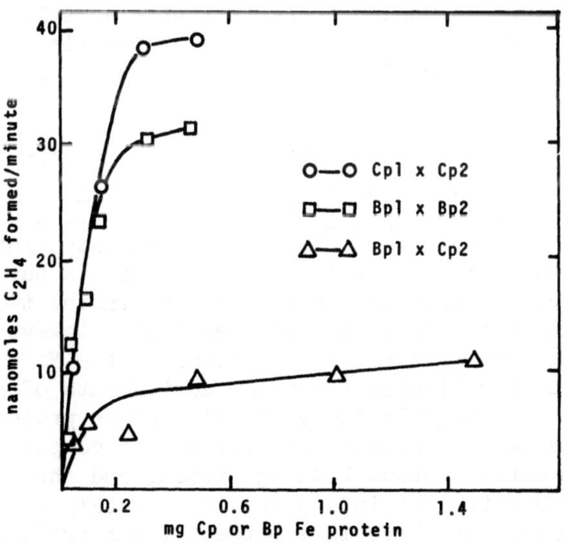

Fig. 5. The change in nitrogenase activity (C_2H_2 reduction) as increasing amounts of Fe protein are added to a constant amount of MoFe protein. Cp1 = the MoFe protein and Cp2 = the Fe protein from Clostridium pasteurianum; Bp = Bacillus polymyxa.

Titrations of increasing amounts of Fe protein against a constant amount of MoFe protein, characteristically show a linear increase in rates of nitrogenase activity (after the initial regions may show a dilution effect) followed by a tapering off to an essentially steady state with further additions of the Fe protein. The reverse titrations, in which increasing MoFe protein is added to constant Fe protein, show the same original response, but after rising to a maximum, further increases in MoFe protein decrease the rate of nitrogenase reducing activity; ATP hydrolysis remains vigorous. Fig. 5 (8) shows titration curves developed by adding C. pasteurianum Fe protein (Cp2) to constant C. pasteurianum MoFe protein (Cp1) and for a comparable titration with Bacillus polymyxa (Bp). When the nonhomologous titration was made (Bp1 x Cp2), an active nitrogenase was produced, but its activity was relatively low.

As indicated the Fe protein from C. pasteurianum can strongly inhibit catalysis by A. vinelandii nitrogenase. Fig. 6 (8) illustrates that as C. pasteurianum Fe protein (Cp2) is added to A. vinelandii nitrogenase the rate of acetylene reduction decreases; the decrease in rate of creatine hydrolysis is roughly parallel until 0.6 μM equivalents of Cp2 is reached. The fact that the decreases in rates are about parallel implies that the ATP/2e$^-$ ratio is nearly constant; the ratio is near 4. At this point formation of the tight complex between the A. vinelandii MoFe protein and the C. pasteurianum Fe protein is nearly complete and only a small residual catalytic activity of the A. vinelandii nitrogenase remains. Although there is little acetylene reduction, MgATP hydrolysis continues and the ATP/2e$^-$ ratio rises from 4 to 24 when the concentration of Cp2 is doubled beyond the inflection point of the curve.

The studies of Emerich (9) with the catalytically-inactive tightly-binding complex of the MoFe protein from A. vinelandii and the Fe protein from C. pasteurianum have provided another way to examine the binding between the two protein components. Analysis of Emerich's data from several types of experiments have suggested that one MoFe protein molecule is capable of binding two molecules of Fe protein as we had indicated earlier (29). However, other data from our laboratory suggest that although the MoFe protein has the potential for binding two molar equivalents of Fe protein, it is necessary to bind only one equivalent of Fe protein to achieve approximately maximum catalytic activity of the complex. So perhaps both groups are correct about binding ratios, and the difference between them is a matter of interpretation. Catalysis requires a one to one complex, but there is a potential for producing a two to one complex without increasing the catalytic activity appreciably.

How does formation of the tight-binding complex between Av1 and Cp2 affect the binding of MgATP? MgATP is bound to the catalytically

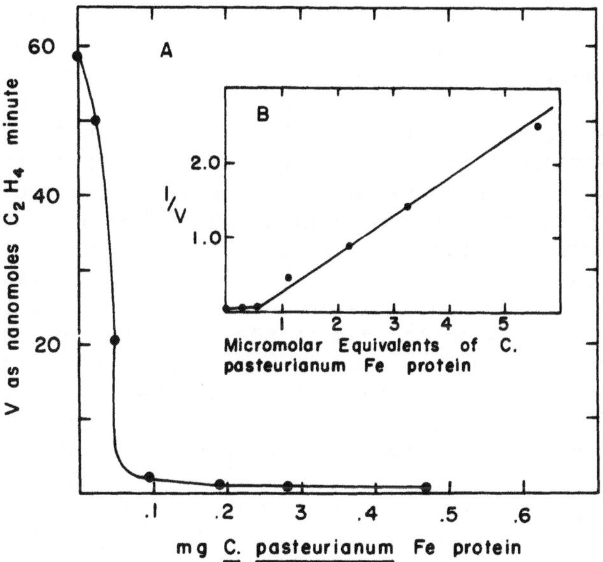

Fig. 6. A. Activity of A. vinelandii nitrogenase in the presence of increasing concentrations of C. pasteurianum Fe protein; the rate of ethylene production was measured.

B. Reciprocal velocity of A. vinelandii enzymatic activity versus concentration of C. pasteurianum Fe protein acting as an inhibitor. For further details see (9).

inactive complex (8) and analysis of the binding shows that about 4 MgATP are bound per unit of the complex. Thus, each of the Fe proteins in the complex still can bind its quota of 2 MgATP. This furnishes evidence that ATP is not hydrolyzed when MgATP is bound initially or when the Fe protein-MgATP complex joins the MoFe protein. The data are compatible with the concept that hydrolysis of MgATP accompanies electron transfer to the substrate.

Emerich (8) has made an extensive study of the interactions between the Fe proteins and MoFe proteins from a variety of N_2-fixing organisms. There have been several studies before this (1,2,4,16,23,25) in which the capacity of a nitrogenase component from one organism to form an active nitrogenase with a component from another organism has been tested, and it has been demonstrated that there is considerable homology among the nitrogenase systems from different organisms. In Emerich's current study he has examined the interactions of the nitrogenase components from eight different organisms. Table III shows the activities exhibited in these cross reactions. The diagonal region represents the homologous crosses between components, and each of these yields an active nitrogenase as anticipated. Note that about 80% of the nonhomologous crosses among these organisms were catalytically active for reduction of acetylene, H^+ or N_2. A distinguishing feature of this study is the determination of the percentage activity of the nonhomologous crosses relative to the activity of the homologous crosses. Some of the nonhomologous crosses were 100% as active as the homologous crosses. On the other hand, some showed activity as low as 1% that of the homologous crosses. It should be emphasized that the individual components had zero activity, so there was no contaminating component that could be held responsible for the crosses giving low activity. When 0% activity was recorded, it always was with a cross that involved one of the proteins from C. pasteurianum. C. pasteurianum gave the lowest percentage of active crosses, whereas Klebsiella pneumoniae gave active crosses with components from all other organisms. When nitrogenase activity of crosses was low, the preparations

TABLE 3

Nitrogenase activity generated by crossing homologous and nonhomologus nitrogenase proteins.

Source of Fe Protein	Source of MoFe Protein: Av	Kp	Rr	Sl	Cv	Rj	Bp	Cp
Av	+	100%	67%	25%	30%	100%	61%	0%
Kp	93%	+	4%	40%	23%	100%	57%	8%
Rr	91%	98%	+	26%	77%	50%	47%	0%
Sl	60%	100%	100%	+	60%	5%	4%	0%
Cv	100%	27%	50%	5%	+	N.D.	1%	0%
Rj	88%	88%	85%	24%	28%	+	71%	0%
Bp	19%	66%	28%	18%	5%	100%	+	38%
Cp	0%	17%	0%	0%	0%	0%	34%	+

+ = homologous crosses. N.D. = not determined.

Data expressed as % activity of that observed in homologous crosses. Av = Azotobacter vinelandii, Kp = Klebsiella pneumoniae, Rr = Rhodospirillum rubrum, Sl = Spirillum lipoferum, Cv = Chromatium vinosum, Rj = Rhizobium japonicum (bacteroids), Bp = Bacillus polymyxa, Cp = Clostridium pasteurianum. For further details see (8) frequently hydrolyzed considerable amounts of MgATP. In other words, the ATP hydrolysis was poorly coupled to the N_2 fixing reaction; Smith, et al. (21) observed poor coupling with nitrogenase reconstituted from Klebsiella pneumoniae MoFe protein and C. pasteurianum Fe protein.

Table 3 shows an impressive homology among the nitrogenase components. Apparently they derive from the same evolutionary stem and have retained a great deal in common during evolutionary development. The major divergence was observed with the C. pasteurianum nitrogenase. All nitrogenases examined to date require two protein components, the MoFe protein and the Fe protein. All require the energy from MgATP and from a strong reductant. All reduce the variety of substrates described earlier. The physical-chemical properties of the MoFe proteins and Fe proteins that have been isolated and purified are highly similar.

What can we say about the nature of the catalytic site? As indicated, the Fe protein has a 4Fe 4 acid-labile S center. Walker and Mortenson (30,31) have probed this site with α,α'-dipyridyl. In its normal state as isolated, the Fe protein shows very little reactivity with α,α'-dipyridyl, but when MgATP is added, the Fe of the active site is exposed so that it can react with α,α'-dipyridyl. Ljones has found that bathophenanthroline disulfonate serves as an excellent chelating agent for probing the active site. It has higher water solubility, it reacts more rapidly with the Fe of the protein, and its higher extinction coefficient increases sensitivity almost threefold over α,α'-dipyridyl.

Bathophenanthroline disulfonate when added to the Fe protein will chelate the small amount of free Fe that is present and also Fe bound to inactivated Fe protein. Upon addition of MgATP, the 4Fe-4S site is exposed and the bathophenanthroline rapidly chelates the exposed Fe with a concomitant development of a characteristic color that can be followed spectrophotometrically (Fig. 7). In addition to verifying the observation that MgATP exposes the active site of the Fe protein, the probe also has furnished a means for examining the state of the Fe under other experimental conditions. For example, it is possible to demonstrate that when the Fe protein becomes oxidized, the iron of its active site is exposed for reaction with bathophenanthroline disulfonate much as if MgATP had been added to the reaction mixture. The chelator reaction with enzymically oxidized Fe protein does not respond to increasing levels of MgATP, so MgATP seems to have no effect on the accessibility of the Fe-S site in oxidized Fe protein, as opposed to reduced Fe protein. Another interesting observation is that cold-inactivated Fe protein has an exposed Fe-S center that is reactive with bathophenanthroline. During storage of Fe protein from C. pasteurianum at 0° C, the time course for disappearance of the ATP-

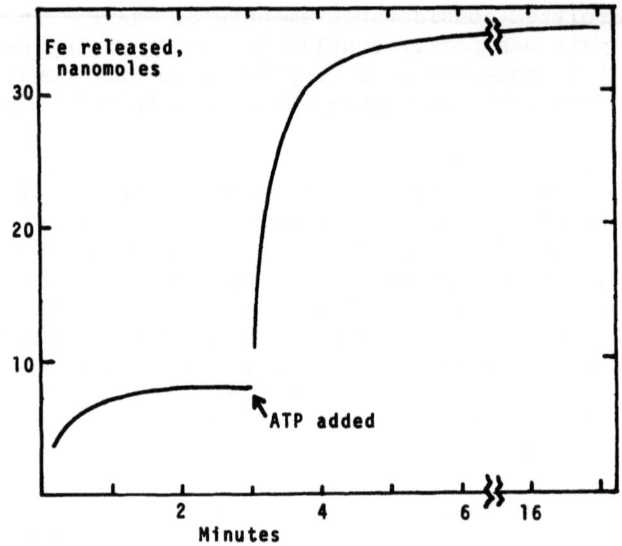

Fig. 7. Record from a strip chart recorder of the change in absorbance as bathophenanthroline disulfonate chelates the Fe from C. pasteurianum Fe protein when MgATP is added to expose the Fe-S center of the protein.

specific reaction with chelator is similar to previously reported data on the time course for loss of enzymic activity of the Fe protein during cold inactivation.

Formation of the inactive complex between the Fe protein from C. pasteurianum and the MoFe protein from A. vinelandii completely inhibits the reaction between the Fe protein and bathophenanthroline disulfonate in the presence of MgATP. Together with the observation that the Fe protein in the inactive complex still binds 2 molecules of MgATP per protein molecule, this indicates that in the protein-protein complex, the 4Fe-4S site in the Fe protein is covered by the MoFe protein, whereas the binding sites for MgATP are freely exposed. Thus, the Fe-S site on the Fe protein is probably located some distance away from the MgATP sites.

Considerable information has been accumulated during the past decade concerning the nature of nitrogenase. However, a number of important questions remain unanswered. To indicate a few: (a) at what point is the MgATP hydrolyzed in the electron transfer scheme? Apparently it is not hydrolyzed upon binding to the Fe protein, because by itself the Fe protein is incapable of hydrolyzing MgATP. Possibly the hydrolysis occurs when the reduced Fe protein-MgATP complex joins the MoFe protein. Or, the hydrolysis may occur at the time the electrons are transferred from the MoFe protein to the substrate; this concept is favored by our observations on tight-binding complexes. (b) Relative to the interactions among substrates, we are interested in the reason acetylene is a noncompetitive inhibitor of N_2 reduction whereas N_2 is a competitive inhibitor of acetylene reduction. (c) As indicated, CO blocks electron transfer to all of the substrates except H^+. What is the mechanism by which H^+ escapes the CO block in electron transfer? (d) We have discussed the evidence supporting the concept that the complex between the Fe protein and the MoFe protein is transitory and is dissociated at each turn of the catalytic cycle, but resolution of the problem will require support from independent lines of evidence. (e) What is the rate limiting partial reaction in the overall nitrogenase reaction? (f) How does MgATP lower the potential of the Fe protein? (g) Is the physiological ratio of MoFe protein/Fe protein 1:1 or 1:2? (h) What is the role of Mo in the catalyst? All of these are worthy problems that will serve as a challenge for future research.

REFERENCES

1. Biggins, D.R., Kelly, M., and Postgate, J.R. (1971) Eur. J. Biochem. 20, 140.

2. Dahlen, J.V., Parejko, R.A., and Wilson, P.W. (1969) J. Bacteriol. 98, 325.

3. Dalton, H., and Mortenson, L.E. (1972) Bacteriol. Rev. 36, 231.

4. Detroy, R.W., Witz, D.F., Parejko, R.A., and Wilson, P.W. (1968) Proc. Nat. Acad. Sci. U.S.A. 61, 537.

5. Eady, R.R. (1973) Biochem. J. 135, 531.

6. Eady, R.R., and Postgate, J.R. (1974) Nature 249, 805.

7. Eady, R.R., Smith, B.E., Cook, K.A., and Postgate, J.R. (1972) Biochem. J. 128, 655.

8. Emerich, D.W. (1977) Studies on Nitrogenase. I. Purification and Properties of Nitrogenase from Bacillus polymyxa. II. Interactions between heterologous nitrogenase components. Ph.D. thesis, Univ. of Wis.-Madison.

9. Emerich, D.W., and Burris, R.H. (1976) Proc. Nat. Acad. Sci. U.S.A. 73, 4369.

10. Evans, M.C.W., Telfer, A., and Smith, R.V. (1973) Biochim. Biophys. Acta 310, 344.

11. Huang, T.C., Zumft, W.G., and Mortenson, L.E. (1973) J. Bacteriol. 113, 884.

12. Israel, D.W., Howard, R.L., Evans, H.J., and Russell, S.A. (1974) J. Biol. Chem. 249, 500.

13. Kennedy, C., Eady, R.R., Kondorosi, E., and Rekosh, D.K. (1976) Biochem. J. 155, 383.

14. Kleiner, D., and Chen, C.H. (1974) Arch. Microbiol. 98, 93.

15. Ljones, T. (1973) Biochim. Biophys. Acta 321, 103.

16. Murphy, P.M., and Koch, B.L. (1971) Biochim. Biophys. Acta 253, 295.

17. Orme-Johnson, W.H., Hamilton, W.D., Ljones, T., Tso, M.-Y.W., Burris, R.H., Shah, V.K., and Brill, W.J. (1972) Proc. Nat. Acad. Sci. U.S.A. 69, 3142.

18. Rivera-Ortiz, J.M., and Burris, R.H. (1975) J. Bacteriol. 123, 537.

19. Shah, V.K., and Brill, W.J. (1973) Biochim. Biophys. Acta 305, 445.

20. Sieker, L.C., Adman, E., and Jensen, L.H. (1972) Nature 235, 40.

21. Smith, B.E., Thorneley, R.N.F., Eady, R.R., and Mortenson, L.E. (1976) Biochem. J. 157, 439.

22. Smith, B.E., Thorneley, R.N.F., Yates, M.G., Eady, R.R., and Postgate, J.R. (1976) in Proceedings of First International Symposium on Nitrogen Fixation (Newton, W.E., and Nyman, C.J., eds.) p. 150, Washington State University Press, Pullman

23. Smith, R.V., Telfer, A., and Evans, M.C.W. (1971) J. Bacteriol. 107, 574.

24. Thorneley, R.N.F. (1975) Biochem. J. 145, 391.

25. Thorneley, R.N.F., Eady, R.R., and Yates, M.G. (1975) Biochim. Biophys. Acta 403, 269.

26. Tso, M.-Y.W. (1974) Arch. Microbiol. 99, 71.

27. Tso, M.-Y.W. (1973) Purification and Properties of Nitrogenase Proteins from Clostridium pasteurianum. Ph.D. thesis, Univ. of Wis.-Madison.

28. Tso, M.-Y.W., and Burris, R.H. (1973) Biochim. Biophys. Acta 309, 263.

29. Vandecasteele, J.P., and Burris, R.H. (1970) J. Bacteriol. 101, 794.

30. Walker, G.A., and Mortenson, L.E. (1973) Biochem. Biophys. Res. Comm. 53, 904.

31. Walker, G.A., and Mortenson, L.E. (1974) Biochemistry 13, 2382.

32. Winter, H.C., and Burris, R.H. (1976) Ann. Rev. Biochem. 45, 409.

33. Yates, M.G. and Planqué, K. (1975) Eur. J. Biochem. 60, 467.

34. Zumft, W.G., Cretney, W.C., Huang, T.C., Mortenson, L.E., and Palmer, G. (1972) Biochem. Biophys. Res. Commun. 48, 1525.

35. Zumft, W.G., Mortenson, L.E., and Palmer, G. (1974) Eur. J. Biochem. 46, 525.

RELATIONSHIP BETWEEN HYDROGEN METABOLISM AND NITROGEN FIXATION IN LEGUMES

Harold J. Evans, Tomás Ruiz-Argüeso[1] and Sterling A. Russell
Department of Botany and Plant Pathology
Oregon State University
Corvallis, Oregon 97331 U.S.A.

INTRODUCTION

The urgent need for conservation of energy throughout the world has created a renewed interest in biological N_2 fixation. Increasing the extent and efficiency of biological N_2 fixation hopefully will lead to a decreased demand for natural gas that is now being used extensively in the manufacture of nitrogen fertilizers. Loss of energy through H_2 evolution from nodules of legumes has been known for twenty years, but realization that H_2 evolution from a great majority of legumes represents a substantial proportion of the energy supplied to the nitrogenase system was not generally appreciated until recently (25,11). Legume cultivar-rhizobium strain combinations that efficiently utilize energy for metabolic processes might be expected to show increases in N_2 fixation and plant growth. Several research and review papers concerning the role of H_2 metabolism in nodulated legumes have been prepared in the last two years (25,26,11). It is the purpose of this paper to summarize the status of our present understanding of the relationship between H_2 metabolism and N_2 fixation in nodulated symbionts.

Wilson et al. (31) reported that H_2 was a specific inhibitor of N_2 fixation in clover plants. Phelps and Wilson (21) observed that nodules formed on garden peas (Pisum sativum) by strain 311 of Rhizobium leguminosarum exhibited hydrogenase activity. Initial evidence that nodules produced H_2 via the N_2-fixing system was obtained by Hoch et al. (16,17) and this has been confirmed by others

[1]On leave from Departmento de Microbiologia, E.T.S. de Ingenieros Agronomos, Universidad Politecnica de Madrid.

including Bergersen (1) and Dart and Day (5). Initial demonstration of the capacity of cell-free nitrogenase from soybean nodules to catalyze ATP-dependent H_2 evolution was published in 1967 by Koch, Evans and Russell (19).

In a series of papers beginning in 1967 (6,7,8) Dixon has provided considerable insight into H_2 metabolism of pea root nodules. *Pisum sativum* cv. Meteor inoculated with *R. leguminosarum* strain 311 formed nodules that evolved little or no H_2, but took up H_2 when this gas was provided in a mixture over them. By use of deuterium Dixon (6) demonstrated the involvement of two separate systems, one of which evolved H_2 via the nitrogenase system and the other consumed H_2 in an O_2-dependent series of reactions that included a hydrogenase. Furthermore Dixon (7) showed that bacteroids from pea nodules possessed the capability for H_2 oxidation that could be coupled to ATP synthesis. In considering the physiological significance of the H_2 uptake mechanism in nodules three possibilities were suggested (8). (a) Oxidation of H_2 via the oxyhydrogen reaction utilizes excess O_2 in the nodule that may assist in the maintenance of the nitrogenase system in an environment where O_2 damage to the enzyme is minimal. (b) Since inhibition of nitrogenase by H_2 has been established, the hydrogenase system in nodules may serve as a mechanism to initiate the oxidation of H_2 produced by the nitrogenase system, thus preventing H_2 build-up and inhibition. (c) Since the evolution of H_2 by the nitrogenase system requires energy in the form of ATP and reductant, H_2 evolution from the nitrogenase system is wasteful, but oxidation of H_2 through a hydrogenase system could lead to ATP synthesis and therefore energy conservation. A mechanism for conservation of energy through H_2 recycling therefore might lead to an increase in the overall efficiency of the N_2-fixing process.

Since it was generally considered that nitrogenase-dependent H_2 evolution under *in vivo* conditions was negligible (4) our laboratory initiated research in 1976 to evaluate the influence of energy loss through H_2 evolution on the efficiency of the N_2-fixing process in nodulated legumes.

Nitrogenase Reactions

The nitrogenases that have been purified from different sources are similar in most respects (22). Purified preparations consist of a molybdenum-iron component with reported molecular weights ranging from 200,000 to 270,000 daltons and an iron component with molecular weights from 56,000 to 69,000 daltons. The larger component from different organisms contains about 2 gram atoms of Mo, 22 to 36 gram atoms of non-heme iron and 14 to 28 gram atoms of acid-labile sulfur per mole of protein. Approximately four gram atoms each of non-heme iron and acid-labile sulfide per mole have been observed in the smaller protein components from different sources. Both proteins are O_2 sensitive and essential for nitrogenase catalysis, but the ratio of the two in the functional nitrogenase complex has not been conclusively resolved. The stoichiometry of the nitrogenase reaction *per se* is indicated as follows (22):

$$N_2 + 6e^- + 6\ H^+ + 15\ ATP \longrightarrow 2NH_3 + 15\ MgADP + 15\ Pi$$

This expression describes the reduction of N_2 to two moles of NH_3 but does not reveal the experimental fact that purified nitrogenases catalyze the reduction of protons concomitantly with the reduction of N_2. The original experiments of Bulen *et al*. (3) showed that about 30 percent of the electron flow through nitrogenase from *Azotobacter vinelandii* was utilized in H_2 evolution. The results of Koch, Evans and Russell (19) who used a partially purified nitrogenase from soybean root nodules were consistent with Bulen's observations (Figure 1). The kinetic experiments of Rivera-Ortiz and Burris (23) suggested that an infinite concentration of N_2 would not completely inhibit H_2 evolution. In the absence of N_2 the entire electron flux through nitrogenase is expended in the reduction of protons to H_2 (3). The number of ATP molecules hydrolyzed per pair of electrons transported during the nitrogenase

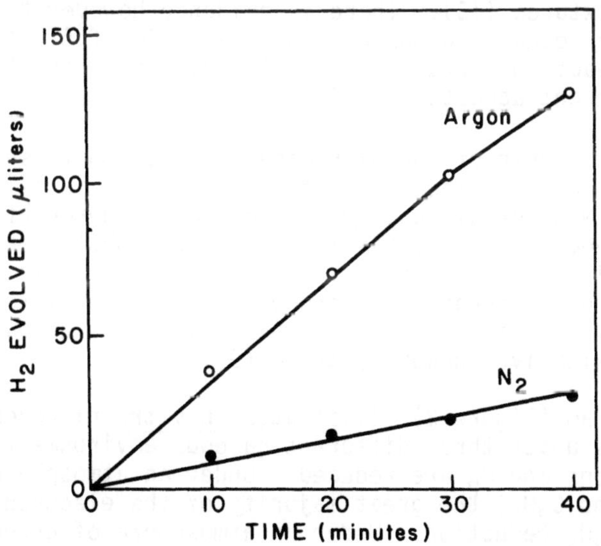

Figure 1. ATP-dependent H_2 evolution by a reaction mixture containing cell-free nitrogenase from soybean nodule bacteroids. The total equivalents of $Na_2S_2O_4$ utilized for H_2 evolution and N_2 fixation under N_2 was approximately equal to the equivalents of $Na_2S_2O_4$-dependent H_2 evolution under argon. Conditions were as described by Koch *et al*. (19).

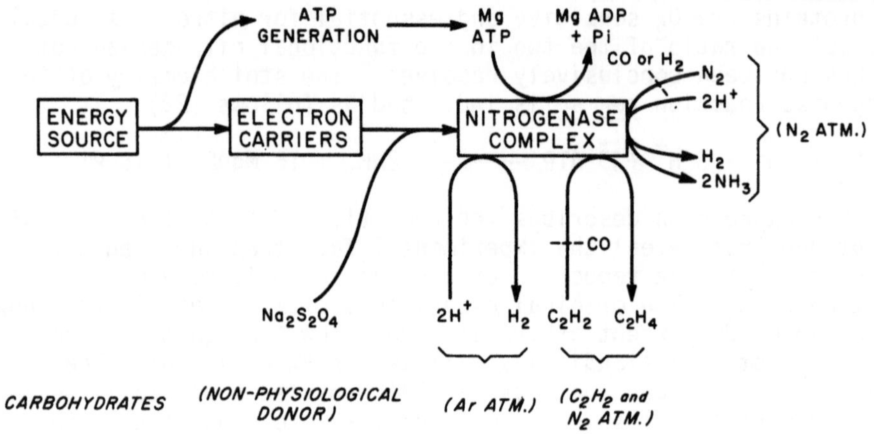

Figure 2. A scheme illustrating the energy requirements for nitrogenase-dependent reactions. (After Evans and Barber, 10).

reaction is relatively independent of the type of acceptor present in reaction mixtures (13). There is evidence however that the ratio of the two nitrogenase components and the ratio of ATP to ADP in nitrogenase reaction mixtures may influence the allocation of electrons to different acceptors (27).

The relationship of the nitrogenase system to those metabolic processes that provide the energy for nitrogenase is illustrated in Figure 2. Nitrogenase catalysis requires a source of reductant which is derived from the oxidation of carbon substrates. In aerobic organisms such as rhizobium bacteroids, details of the pathways of electron transport to nitrogenase are incompletely understood. The ATP required for nitrogenase catalysis undoubtedly is supplied by oxidative phosphorylation (9).

The diagram (Figure 2) illustrates in vitro nitrogenase catalyzed reactions under three different gaseous environments. Under N_2, both protons and N_2 are reduced. Under an atmosphere of N_2 and C_2H_2 (or Ar and C_2H_2) the great majority of the electron flow is utilized in C_2H_2 reduction. Under an atmosphere of argon the entire ATP-dependent electron flow through nitrogenase is utilized in the reduction of protons to H_2. This reaction, in contrast to the nitrogenase reactions in which N_2 or C_2H_2 are used as acceptors is insensitive to CO inhibition. The total electron flow through nitrogenase in an intact organism in an environment containing Ar and O_2 may be estimated by measurement of H_2 evolution, provided that the organism lacks a hydrogenase and therefore the capacity for recycling H_2. Also it may be measured by the rate of C_2H_2 reduction. For example, the evolution of H_2 from soybean nodules exposed to 0.1 atmosphere of C_2H_2 and 0.2 atmosphere of O_2 was not detectable by a relatively insensitive gas chromatographic procedure (25).

In considering the reduction of one mole of N_2 via the nitrogenase system, about 15 moles of ATP and three pairs of electrons are required. Since approximately one mole of H_2 (about 25% of the electron flux) is evolved from the nitrogenase system under ordinary atmospheric conditions one must include an additional five moles of ATP and one pair of electrons as energy expenditure associated with H_2 evolution. The total energy expenditure during the reduction of one mole of N_2 to ammonia and two protons to H_2 is about 20 moles of ATP and four pairs of electrons. Some symbionts form nodules that possess a mechanism for oxidation of H_2 and therefore the capability of conserving some of the energy that is lost through evolution of H_2 from the N_2 fixation reaction.

It is clear that the energy demanded for the biological N_2 fixation process in legumes is great. Experimental results (29,15, 14) provide strong evidence supporting the conclusion that the capability of a legume to supply energy to nodules in the form of photosynthate is a major factor limiting the rate of N_2 fixation. Any steps that can be taken to improve the efficiency of energy utilization by the nitrogenase system in nodules therefore are highly desirable.

Hydrogenase Reactions

In considering the role of H_2 metabolism in the N_2-fixing process it is necessary to emphasize three types of H_2 reactions. The first, is the ATP-dependent H_2 evolution reaction catalyzed by nitrogenase. As discussed already this reaction utilizes protons as an alternate acceptor in the nitrogenase reaction. The second reaction, is the "classical" reversible hydrogenase in organisms such as *Clostridium* species (12). This type of hydrogenase catalyzes a reaction between two protons and two electrons yielding molecular H_2. A third type of hydrogenase occurs in *Azotobacter* (18), nodules of *Pisum sativum* (6) and perhaps N_2-fixing blue-green algae (2,20). In these organisms hydrogenase catalyzes H_2 uptake but not H_2 evolution. This hydrogenase is more active in *Azotobacter* cells grown on N_2 than in cells cultured in a medium containing ammonia. This latter type of hydrogenase is synthesized in nodules formed by some but not all rhizobium strains.

Hydrogen Evolution from Nodules

Our laboratory has described methods for estimating energy loss from nodules through H_2 evolution (25,26) and relative efficiency of energy utilization during N_2 fixation. These estimates are based upon measurements of H_2 evolution in air and the total energy flux through the nodule nitrogenase system. Experiments with soybean nodules lacking a hydrogenase have shown that rates of acetylene reduction and rates of H_2 evolution under argon are approximately equivalent. When nodules of this type are examined, the total electron flux through the nitrogenase system may be estimated on the basis of either H_2 evolution under argon and O_2 or the rate of C_2H_2 reduction. When the hydrogenase status of nodules is unknown it is more reliable to use the rate of C_2H_2 reduction for the total electron flux estimate despite the fact that C_2H_2 may have unknown biochemical effects on the system. Relative efficiency estimates are made by subtracting the rate of nitrogenase-dependent H_2 evolution in air from the total electron flux through nitrogenase and then expression of this value as a fraction of the total electron flux through the nodule nitrogenase system. The relative efficiency estimate is nothing more than the decimal fraction of the total nitrogenase electron flux that is used in N_2 reduction. The estimate by this method involves several assumptions (11) and is not claimed to be rigorous, however, the method has provided a useful index for assessment of the relative efficiencies of nodules formed by different rhizobium strains.

The curves in Figure 3 illustrate typical H_2 evolution measurements by soybean nodules formed by a commercial inoculum. This tracing which was obtained by use of a H_2 gas electrode (30) demonstrates that H_2 is evolved in air and that the nitrogenase-dependent H_2 evolution requires O_2. Furthermore, the experiment reveals that the replacement of N_2 by argon in the gas mixture over the nodules results in a marked increase in the rate of H_2 evolution. This is caused by the removal of N_2 as a competitive acceptor. From the H_2 evolution measurements in air (Figure 3) and from the rate of C_2H_2 reduction by a parallel sample of nodules in this particular case, about 74% (relative efficiency of 0.74) of the total electron flux through nitrogenase was estimated to be utilized in N_2 reduction.

In an initial survey (25) a large number of nodulated symbionts were collected from the field and greenhouse, and estimates of energy loss through H_2 evolution were determined. Relative efficiencies of nodules from most of the legumes in the initial study ranged between 0.52 and 0.70. Cowpeas inoculated with Rhizobium strain 32H1, however exhibited no H_2 evolution in air and the relative efficiency value was near 1.0. Likewise, soybeans inoculated with USDA 110 produced nodules that evolved no measurable quantity of H_2 in air (26). Nodules from three of four species of non-legumes collected from native habitats in Oregon showed relative efficiencies

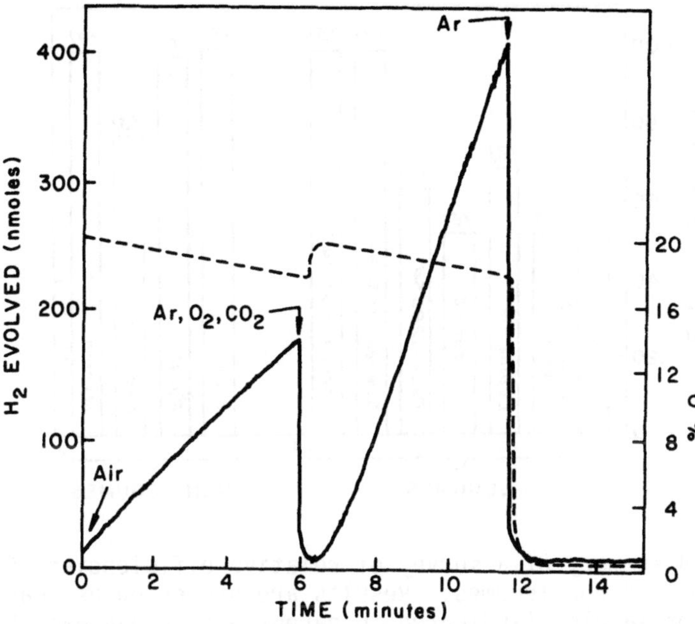

Figure 3. Continuous amperometric tracings of H_2 evolution and O_2 consumption by nodules of soybeans (*Glycine max* cv. Chippewa 64) inoculated with commercial inoculant. Plants were grown under bacteriologically controlled conditions in a greenhouse with 5200 Lux supplementary illumination and a 16-hour light period. The temperature was maintained near 27°C during the day and 21°C at night. Two days before the assays, the plants were transported to a growth chamber provided with a 21,500 Lux illumination during a 16-hour day period. The temperature regime was the same as that used in the greenhouse. The assay was initiated 3 hours after the beginning of the light period. Reactions were initiated in air and as shown by the arrows, the chamber was flushed for 30 seconds first with 79.96% Ar, 20% O_2, 0.04% CO_2 and then with 99.96% Ar, 0.04% CO_2. A sample of 341 mg fresh nodules from 31 day-old plants was used. Dashed line represents % O_2 in the reaction chamber. (After Evans *et al*., 11).

approaching 1.0. These data suggest that native nodulated species that have not been subjected to the usual agricultural practices including nitrogen fertilization may have greater efficiencies than legume-rhizobium combinations commonly used in agriculture.

Since the initial survey our laboratory has examined a series of different leguminous species and over 100 different strains of rhizobium to evaluate the extent of energy loss by H_2 evolution from the nodules formed by the various combinations. Six strains of

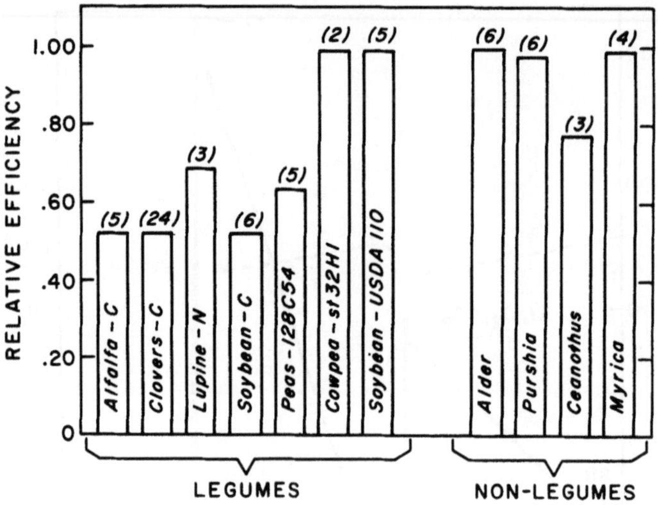

Figure 4. A summary of a survey of relative efficiencies of nodules from legumes and non-legumes. Results are presented as mean values of numbers of samples (indicated in parenthesis above each bar) of each species examined. (After Schubert and Evans, 25,26).

Rhizobium japonicum have been identified that form nodules that lose little or no H_2* in air and these show relative efficiency values of 0.95 or greater (Table 1). An examination of strains of cowpea rhizobia on the Whippoorwill cultivar of cowpeas has identified nine out of 13 strains tested which produce nodules that evolve only traces of H_2 and therefore are relatively efficient. Results obtained so far indicate that the rhizobium strain is the primary component of the symbiosis that determines the relative efficiency value.

Capacity to Recycle Hydrogen

Nodules that fail to evolve H_2 have been examined for a capacity to consume H_2 when this gas is placed over them (26). In all cases those nodules that do not evolve H_2 during the N_2 fixation process show a capacity for O_2-dependent H_2 uptake. Nodules from non-

*The R. japonicum strains that form nodules without H_2 evolution have been listed in a paper by Carter, K., Jennings, N., and Evans, H. presented at Western Division of The American Society of Plant Physiologists, San Francisco State University, June 12-16, 1977.

Summary of the Relative Efficiencies of Some Legumes Inoculated With Selected Rhizobium Strains

Legumes	Strains tested	Numbers of strains with relative efficiencies[1] ranging: < 0.75	0.75 - 0.95	> 0.95
White clover	11	11	0	0
Alfalfa	20	10	10	0
Austrian winter peas	15	10	5	0
Soybeans	31	23	2	6
Cowpeas	13	2	2	9

[1]Relative efficiency: $1 - \frac{H_2 \text{ evolved in air}}{C_2H_2 \text{ reduction}}$. H_2 evolution determined by use of an amperometric method and C_2H_2 reduction by gas chromatography (25).

legumes such as Alnus rubra, also are capable of catalyzing an O_2-dependent H_2 consumption (26).

Some legume nodules that evolve H_2 in air also contain a hydrogenase that is capable of recycling only part of the H_2 evolved from the nitrogenase system. Nodules from field samples of red clover (Trifolium pratense) exhibit net H_2 evolution in air, but treatment of these nodules with a gas mixture containing 66% O_2 preferentially destroyed the capacity for nitrogenase catalyzed H_2 evolution (Figure 5). When these O_2-treated red clover nodules were placed in an atmosphere containing O_2, N_2 and H_2, a capacity for consumption of H_2 through the oxyhydrogen reaction was exhibited (24). Furthermore it has been shown that bacteroid suspensions may be prepared from nodules that exhibit O_2-dependent H_2 consumption. Oxygen dependent H_2 uptake by a preparation of Rhizobium leguminosarum bacteroids isolated from field-grown Vicia sativa is shown in Figure 6. A series of experiments with peas inoculated with Rhizobium leguminosarum strains also has shown capacities for H_2 uptake by intact nodules and bacteroids (24). Further experiments are now in progress to characterize the hydrogenase that is present in the bacteroids from several legumes. Already it is clear that some nodules contain sufficient hydrogenase to catalyze the oxidation of the entire H_2 output from the nitrogenase system while others do not.

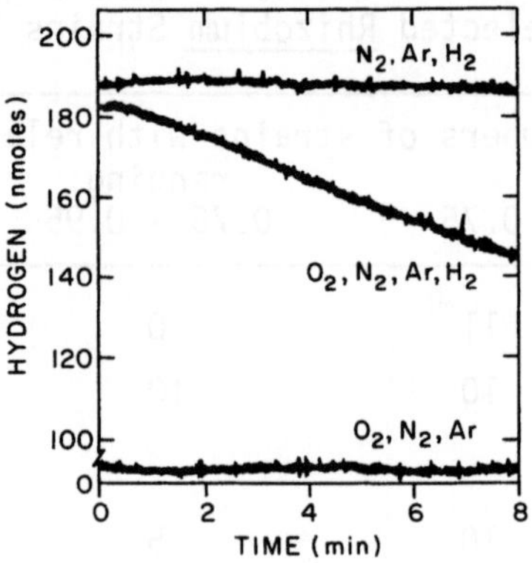

Figure 5. Tracings of continuous amperometric measurements of H_2 uptake by intact nodules (0.1 g) from red clover (Trifolium pratense) from the field. Concentrations of H_2 were recorded for periods of 8 minutes after successively flushing the chamber for about 30 seconds with: 66% O_2, 24% N_2, 10% Ar (lower tracing); 66% O_2, 24% N_2, 9.85% Ar, 0.15% H_2 (middle tracing); 24% N_2, 75.85% Ar, 0.15% H_2 (upper tracing).

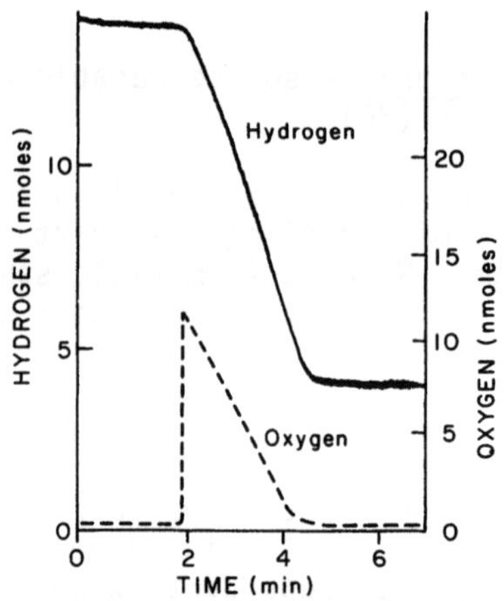

Figure 6. Continuous amperometric measurements of H_2 and O_2 uptake by bacteroids from nodules of vetch (Vicia sativa) from the field. The electrode chamber contained 2.8 ml of bacteroid suspension (104 µg of N/ml) in buffer (0.05 M potassium phosphate, 0.0025 M $MgCl_2$, pH 7). The procedure of Dixon (1958) was followed in the preparation of the bacteroids with the exception that anaerobic conditions were not maintained. H_2 and O_2 were added to the chamber as H_2- or O_2-saturated buffer solutions at the times indicated. The concentrations of H_2 or O_2 in the figure refer to amounts in the 2.8 ml reaction chamber.

The hydrogenases present in Azotobacter vinelandii and Anabaena cylindrica utilize the H_2 that is evolved by the ATP-dependent nitrogenase system. As a consequence H_2 evolution is not observed in these N_2-fixing organisms. In the presence of CO and C_2H_2 which inhibit hydrogenase activity it has been demonstrated that H_2 is produced by whole cells of both organisms (28,2).

The recycling of H_2 produced by the nitrogenase enzyme may be the natural function of the hydrogenase associated with aerobic N_2-fixing systems. ATP production resulting from the oxidation of this H_2 has been reported (18,7,20) and some of the energy expended during nitrogenase-dependent H_2 evolution therefore may be conserved through the H_2 recycling process.

Conclusions

The relationship of the N_2-fixing apparatus in bacteroids to other physiological processes in legume nodules is illustrated in Figure 7. Rhizobium bacteroids inside nodules are surrounded by

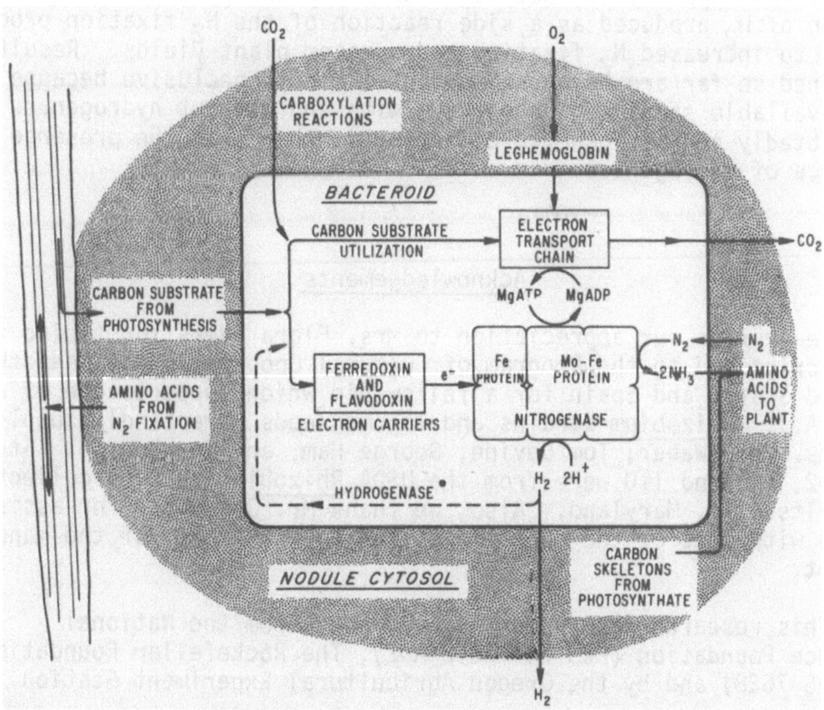

Figure 7. A diagram illustrating the interrelationship of reactions involved in the nitrogen-fixing process in a legume nodule. (After Evans and Barber, 10).

a membrane of plant origin. The bacteroids are supplied with O_2 through the aid of leghemoglobin which is present in the cytosol at a relatively high concentration. Carbon substrates are supplied through the vascular system of the plant and oxidation of them supplies the energy for synthesis of the ATP and the reductant that is needed to support the nitrogenase reaction. The initial product of N_2 fixation is ammonia which is formed in the bacteroids and excreted into the cytosol where glutamine, asparagine and several amino acids are synthesized. These are exported into the xylem and transferred to different parts of the plant. During the nitrogenase reaction, in most legumes, an average of about 30 percent of the electron flux through the nitrogenase system is expended in the reduction of protons to H_2. If this H_2 is not recycled the nodule system has expended about five ATP molecules and one pair of electrons for each mole of H_2 evolved. Some rhizobium strains have been identified by our research group that induce nodules that exhibit a capacity to oxidize H_2 and presumably utilize some of the energy derived from H_2 oxidation. It is expected that this would be available as a source of electrons or as a source of energy for synthesis of ATP for the nitrogenase or other reactions. Experiments are in progress to determine whether or not the capacity of nodules to conserve energy through the oxidation of H_2 produced as a side reaction of the N_2 fixation process leads to increased N_2 fixation and greater plant yields. Results obtained so far are encouraging, but are not conclusive because the available strains of rhizobium with and without hydrogenase undoubtedly contain genetic differences other than the presence or absence of hydrogenase.

Acknowledgements

We express our appreciation to Mrs. Flora Ivers for typing the manuscript and to the Program of Cultural Cooperation between the United States and Spain for a fellowship which supports one of us (T.R.A.). *Rhizobium* strains and legume seeds were kindly supplied by Drs. Dean Weber, Tom Devine, George Ham, and Joe Burton. Strains B6, 42, 143 and 110 were from the USDA *Rhizobium* culture collection at Beltsville, Maryland. Also, we thank Mr. Joe Hanus for assistance with some of the experiments and for criticism of the manuscript.

This research was supported by grants from the National Science Foundation (PCM 74-17812-A02), The Rockefeller Foundation (GA AS 7628) and by the Oregon Agricultural Experiment Station.

References

1. Bergersen, F.J. (1963) Aust. J. Biol. Sci. 16, 669.

2. Bothe, H., Tennigkeit, J., and Eisbrenner, G., and Yates, M. G. (1977) Planta 133,237.

3. Bulen, W.A., Burns, R.C., and LeComte, J.R. (1965) Proc. Nat'l. Acad. Sci., USA 53, 532.

4. Burns, R.C., and Hardy, R.W.F. (1975) Nitrogen Fixation in Bacteria and Higher Plants, Springer-Verlag, New York.

5. Dart, P.J., and Day, J.M. (1971) Plant and Soil, Special Volume, p. 167.

6. Dixon, R.O.D. (1967) Annals of Botany 31, 179.

7. Dixon, R.O.D. (1968) Arch. Mikrobiol. 62, 272.

8. Dixon, R.O.D. (1972) Arch. Mikrobiol. 85, 193.

9. Dixon, R.O.D. (1975) In "Nitrogen Fixation by Free-Living Microorganisms", pp. 421-435, (W.D.P. Stewart, Ed.), Cambridge University Press, Cambridge.

10. Evans, H.J., and Barber L.E. (1977) Science (in press).

11. Evans, H.J., Ruiz-Argüeso, T., Jennings, N., and Hanus, J. (1977) Conference on Genetic Engineering for Nitrogen Fixation, Brookhaven National Laboratory, New York, (A. Hollaender, Ed.) (in press).

12. Gray, C., and Gest, H. (1965) Science 148, 186.

13. Hadfield, L.K., and Bulen, W.A. (1969) Biochemistry 8, 5103.

14. Hardy, R.W.F. (1976) In "Proc. 1st Intern. Symp. on Nitrogen Fixation", pp. 693-717, W.E. (Newton and C.J. Nyman, Eds.), Washington State University Press, Pullman, Vol. 2.

15. Hardy, R.W.F., and Havelka, U.D. (1975) Science 188, 633.

16. Hoch, G.E., Little, H.N., and Burris, R.H. (1957) Nature 179, 430.

17. Hoch, G.E., Schneider, K.C., and Burris, R.H. (1960) Biochim. Biophys. Acta 37, 273.

18. Hyndman, L.A., Burris, R.H., and Wilson, P.W. (1953) J. Bact. 65, 522.

19. Koch, B., Evans, H.J., and Russell, S.A. (1967) Proc. Nat'l. Acad. Sci., USA 58, 1343.

20. Peterson, R.B. (1976) Ph.D. Thesis, Dept. of Biochemistry, University of Wisconsin, Madison, Wisconsin.

21. Phelps, A.S., and Wilson, P.W. (1941) Proc. Soc. Exp. Biol. and Med. 47, 473.

22. Postgate, J.R. (1975) In "Genetic Manipulations with Plant Material", pp. 107-122, (L. Ledoux, Ed.), NATO Advanced Study Series A, Vol. 3, Plenum Press.

23. Rivera-Ortiz, J.M., and Burris, R.H. (1975) J. Bact. 123, 537.

24. Ruiz-Argüeso, T., Hanus, J., and Evans, H.J. (1977) Manuscript in preparation, Oregon State University.

25. Schubert, K.R., and Evans, H.J. (1976) Proc. Nat'l. Acad. Sci., USA 73, 1207.

26. Schubert, K.R., and Evans, H.J. (1976) In "Proc. of II International Symp. on Nitrogen Fixation", Interdisciplinary Discussions, (C. Rodriquez-Barrueco and W.E. Newton, Eds.), Salamanca, Spain, September 13-17.

27. Silverstein, R., and Bulen, W.A. (1970) Biochemistry 9, 3809.

28. Smith, L.A., Hill, S., and Yates, M.G. (1976) Nature 262, 209.

29. Streeter, J.G. (1974) J. of Exp. Bot. 25, 189.

30. Wang, R., Healey, F.P., and Myers, J. (1971) Plant Physiol. 48, 108.

31. Wilson, P.W., Umbreit, W.W., and Lee, S.B. (1938) Biochem. J. 32, 2084.

AMMONIA ASSIMILATION IN N_2-FIXING SYSTEMS

D. B. Scott

Programa Fixacão Biológica de Nitrogênio
Convênio CNPq-EMBRAPA-UFRRJ, Km 47
Seropédica 23460, Rio de Janeiro, Brazil

INTRODUCTION

There is increasing evidence that the assimilation of ammonia by N_2 fixing organisms proceeds by way of the dual enzyme system of glutamine synthetase (GS) (EC. 6.3.1.2 - Equation 1) and glutamate synthase (GOGAT) (EC. 2.6.1.53 - Equation 2) (34, 12). Glutamate synthase was originally isolated by Tempst *et*

$$\text{L-glutamate} + NH_4^+ + \text{ATP} \longrightarrow \text{L-glutamine} + \text{ADP} + \text{Pi} \quad (1)$$

$$\text{L-glutamine} + \text{2-oxoglutarate} + \text{NAD(P)H} \longrightarrow 2 \text{ glutamate} + \text{NAD(P)}^+ \quad (2)$$

al., (66) from ammonia-limited chemostat cultures of *Klebsiella aerogenes*. This discovery provided an alternative pathway for the assimilation of ammonia other than the reductive amination of 2-oxoglutarate, catalysed by glutamate dehydrogenase (GDH) (EC. 1.4. 1.3., Equation 3). Although the GS/GOGAT pathway proceeds at the

$$\text{2-oxoglutarate} + NH_4^+ + \text{NAD(P)H} \longrightarrow \text{L-glutamate} + \text{NAD(P)}^+ \quad (3)$$

expense of an extra ATP, the greater affinity of GS and GOGAT for their substrates compared to glutamate dehydrogenase would suggest that the GS/GOGAT system is the major route for ammonia assimilation in microorganisms under conditions of limiting ammonia (33, 67, 7, 12). This pathway has also been demonstrated in higher plants (17, 13, 26).

The purpose of this review is to describe the pathways of ammonia assimilation of N_2-fixing organisms, both in the free-living situation and in symbiotic association with plants, as well as the regulatory mechanisms involved during N_2-fixation.

THE ASSIMILATION OF AMMONIA IN N_2-FIXING ORGANISMS

1. N_2-fixing Bacteria.

The assimilation of ammonia in microorganisms proceeds via glutamate dehydrogenase or the GS/GOGAT pathway depending on the organism and on the ammonia concentration of the medium (67, 7). Under conditions of ammonia limitation, assimilation occurs by way of the GS/GOGAT pathway (7, 12).

The GS/GOGAT pathway has been found to operate in a number of N_2-fixing bacteria viz: *Clostridium pasteurianum* (34, 9), *Klebsiella pneumoniae*, *Azotobacter vinelandii*; *Chromatium* sp., *Rhizobium japonicum* *Chlorobium thiosulfatophilum*, *Rhodospirillum rubrum* (34), *Azotobacter chroococcum* (14) and several species of *Rhizobium* (6).

The uptake of ammonia in *K. pneumoniae* is well documented (34, 22). Cultures grown on N_2 have high nitrogenase activity, high levels of GOGAT and unadenylylated GS, and low levels of GDH (34, 22). Addition of NH_4^+ to chemostat cultures of *K. pneumoniae* results in the induction of glutamate dehydrogenase and an increase in the cellular glutamine concentration, implying regulation by GS (22). There is an increase in the apparent adenylylation (56) of glutamine synthetase and this correlates with the repression of nitrogenase synthesis and the induction of glutamate dehydrogenase (22). Recent investigations using mutants of *K. pneumoniae* which have been altered in inthe structural and regulatory genes of the enzymes catalyzing the primary steps of ammonia assimilation, suggest that GS has a major role in the regulation of N_2-fixation and ammonia assimilation (70, 64, 51). The regulation of nitrogen assimilation by GS recently has been reviewed by Magasanik (30). In *Salmonella typhimurium*, non-adenylylated GS was shown to act as a positive effector at the level of transcription of the histidine utilization genes (*hut* genes) (71). The adenylylation of GS was shown to be regulated by the levels of 2oxoglutarate and glutamine. A low ratio of 2-oxoglutarate to glutamine stimulates the adenylylation of GS, whereas conditions of ammonia limitation lead to de-adenylylation of GS (16). Furthermore, adenylylated GS has been shown to repress the synthesis of further GS protein (16). The regulation of nitrogenase synthesis in K. *pneumoniae* appears to occur by the same mechanism (70, 64) although the direct demonstration of the activation of transcription of *nif* by non-adenylylated GS *in vitro* has not as yet been achieved.

Further support for the functioning of the GS/GOGAT pathway during N_2-fixation is the greater affinity of glutamine synthetase (e.g., $KmNH_4^+$ = 0.33 mM in *K. pneumoniae* and $KmNH_4^+$ = 0.30 mM in *A. vinelandii*) and glutamate synthase (e.g., Km glutamate = 1.0 mM, *K. pneumoniae*) for their substrates compared to glutamate dehydrogenase (e.g. $KmNH_4^+$ = 12 mM in K. pneumoniae) (22). In chemostat cultures of A. chroococcum, an increased concentration of ammonia caused a proportionate repression of nitrogenase activity (14). Nitrate also caused repression, but aspartate, glutamate and glutamine had no effect and were not metabolized. GOGAT and GDH were present in almost equal amounts in N_2-fixing cultures and showed no substantial changes from cultures grown on NH_4^+ or NO_3^-. This implies that control in this organism must have been on GS, although this still remains to be demonstrated. In certain strains of C. pasteurianum GDH has not been demonstrated (9). Glutamate synthase in unaffected by ammonia concentration in this organism but high ammonia causes repression of GS (9).

In N_2-fixing cultures of K. pneumoniae and A. vinelandii, NH_4^+ normally represses N_2-fixation. However, in the presence of methionine sulfone (MSF) and methionine sulfoximine (MSX), [known inhibitors of GS and GOGAT but not GDH in K. aerogenes, (5)] N_2 fixation is derepressed (18). Furthermore, MSF causes a high proportion of the fixed NH_4^+ to be excreted in A. vinelandii. Excretion of NH_4^+ under N_2-fixing conditions in the presence of these inhibitors has also been observed in the blue-green alga, Anabaena cylindrica (60) and in the photosynthetic bacterium, Rhodospirillum rubrum (72). Mutants of K. pneumoniae deficient in GOGAT are also known to excrete ammonia (53). These studies emphasize the importance of the GS/GOGAT pathway for the assimilation of ammonia in free-living N_2-fixing bacteria.

A detailed study of the pathways for the assimilation of ammonia in rhizobia has been carried out by Brown and Dilworth (6). These authors demonstrated that chemostat cultures of R. leguminosarum, R. trifolii and R. japonicum, under conditions of ammonia or nitrate limitation, assimilated ammonia via the GS/GOGAT route. Under glucose-limiting conditions with an excess of inorganic N, neither GS nor GOGAT was detectable, and ammonia was assimilated via GDH. The coenzyme specificity for GOGAT varies with species. The slow-growing species of Rhizobium (e.g., R. lupini and R. japonicum) contain an NADH-dependent enzyme, whereas the fast-growing species (e.g., R. trifolii and R. leguminosarum) contain an NADPH-dependent enzyme. Ammonia or glutamate had little effect on the NADH-dependent GOGAT from R. japonicum but both N sources repressed the activity of the NADPH-dependent GOGAT from R. leguminosarum. Glutamine synthetase however, was repressed in all species by high ammonia concentrations (6). Bishop et al., (4) have recently demonstrated that NH_4^+ repression of GS in R. japonicum is accompanied by an increase in the adenylylation of this enzyme. Although these species of Rhizobium appear to have different control mechanisms for the assimilation of ammonia, they all have sufficient GS and GOGAT to account for the assimilation of ammonia under nitrogen-limiting conditions (6).

The recent demonstration that strains of rhizobia can synthesize nitrogenase in either solid (28, 39, 23), or liquid culture (20, 69) has now made it possible to examine the pathways for the assimilation of ammonia under N_2-fixing conditions.

In O_2-limited cultures of a cowpea Rhizobium sp. (strain CB 756) high levels of nitrogenase activity were obtained; the cultures were insensitive to an excess of NH_4^+ or glutamine (3).

Under these conditions relatively high levels of GS and GOGAT were found but GDH was not detectable. When the O_2-limitation was relieved, nitrogenase was repressed by NH_4^+ and there was a concomitant increase in the degree of adenylylation of glutamine synthetase. This elegant piece of work indicates that the mechanism for the control of nitrogenase synthesis in Rhizobium (CM 756) is similar to that found in Klebsiella.

The assimilation of ammonia in rhizobia appears to be tightly regulated by the nature of the organic N source present in the medium. Free-living cultures of R. japonicum and Rhizobium sp. (32 Hl) grown in the presence of glutamate have been shown to export most of fixed $^{15}N_2$ as ammonia, into the medium (36). Several strains of Rhizobium grown on histidine also exported NH_4^+ (up to 7.0 µmol/h/mg protein, R. leguminosarum). O'Gara and Shanmugan (36) suggest that the inability of rhizobia to assimilate ammonia in the presence of glutamate is due to repression of the ammonia assimilatory enzymes. Ammonia was found to repress GS and glutamate repressed GOGAT (36). In the presence of glutamate, R. trifolii is unable to utilize ammonia as a N source for growth. However, in the presence of aspartate, increased GOGAT activity was observed, and ammonia was assimilated for growth. In fact, an organic N source such as aspartate, leucine, or serine is essential for R. trifolii to utilize ammonia for growth (37). Glutamate synthase is essential for ammonia assimilation in this organism. Mutants of R. trifolii in which assimilation of NH_4^+ is not subject to repression by glutamate, no longer require an organic N source such as aspartate for growth in a medium containing inorganic N(NH_4^+ or NO_3^-) (37). These mutants contained elevated levels of GOGAT. Whether GOGAT is derepressed in these mutants still remains to be determined.

The possibility of these regulatory mechanisms operating in the bacteriods of leguminous nodules during N_2 fixation, resulting in the overproduction of NH_4^+ for plant growth, is an exciting possibility demanding further study.

2. Leguminous Nodules

The pathway for the assimilation of ammonia in leguminous nodules has received considerable attention in the last few years.

The demonstration of the GS/GOGAT enzyme system in the bacteriod fraction of nodules from a number of different legumes (34, 21, 15, 55) led these authors to conclude that the bacteriod was the site of ammonia assimilation during nitrogen fixation. However in all of these studies full data indicating levels of GS and GOGAT compatible with nitrogenase activity were not presented.

Recently, several groups have demonstrated that the levels of glutamine synthetase in the bacteriods were insufficient to account for the observed rates of N_2 (C_2H_2) fixation (6, 24, 45, 46). Brown and Dilworth (6) examined the levels of GS and GOGAT in both the bacteriod and plant fractions of a number of legumes. In all legumes examined the levels of GS found in the bacteriods were insufficient to account for the observed rates of N_2 (C_2H_2)-fixation, but the levels of the plant GS were more than sufficient to account for the primary assimilation of ammonia. The levels of GOGAT in the bacteriods were also very low, except in soybean and lupin, where there may be an inherent lack of regulation of this enzyme (12).

The inability of the bacteriod enzymes to account for ammonia assimilation during N_2-fixation has also been confirmed by the work of Kurz _et al_. (23). Although Brown and Dilworth (6) found high levels of GS in the plant fraction they were unable to demonstrate the presence of a plant GOGAT. However, Robertson _et al_. (46) demonstrated high levels of an NAD-dependent GOGAT in the plant fraction of lupin nodules. Furthermore, they found that both GS and GOGAT activities increased in the plant fraction, but not in the bacteriod fraction, over a time course of development which followed the induction of both leghemoglobin and nitrogenase (45, 46). This observation led these authors to propose that the major product exported from the bacteriod during N_2-fixation was ammonia, which was assimilated into glutamine and glutamate predominantly in the plant cytosol of the nodule. However, the major nitrogenous compound exported from the nodule to the leaves of legumes is in fact asparagine (41, 62, 46). It seemed likely therefore that asparagine would be synthesized in the plant cytosol of the nodule by a glutamine-dependent asparagine synthetase (EC 6.3.5.4.; Equation 4) (47, 63).

$$\text{L-asparate} + \text{L-glutamine} + \text{ATP} \longrightarrow \text{L-asparagine} + \text{L-glutamate} + \text{AMP} + \text{PPi} \qquad (4)$$

Asparagine synthetase was detected in the plant fraction of lupin nodules and was found to increase in activity with nodule development in a similar manner to that reported for the plant GS and the plant GOGAT (50). The role of asparagine as the major carrier of fixed N throughout nodule development was also demonstrated. Scott _et al_., (50) also found a plant asparaginase (3.5.1.1.; Equation 5) in very young (non-N_2fixing) lupin nodules.

$$\text{L-asparagine} + H_2O \longrightarrow \text{L-aspartate} + NH_4^+ \quad (5)$$

It has been proposed that this enzyme functions in supplying the developing nodule tissue and bacteriods with ammonia by hydrolysing asparagine transported from the cotyledons (50). This function is supported by the observation that activity is lost with the onset of N_2 fixation.

On the basis of the studies reported by Robertson et al., (45, 46) and Scott et al., (50) a scheme has been proposed (50) for the assimilation of ammonia by enzymes located in the plant fraction of leguminous nodules (Figure I). A similar scheme has also been proposed by Miflin and Lea (32) in a recent review. The first step in the assimilation of the ammonia is the synthesis of glutamine by the plant GS. The high level of plant GS reported in lupin (45) and other legumes (6), coupled with the low Km for NH_4^+ [20 μM for the pea leaf enzyme, (38)] would provide an efficient mechanism for removing NH_4^+ excreted from the bacteriod. The excretion of NH_4^+ from free-living cultures of N_2-fixing R. japonicum (36) and from bacteriods in vitro (3), would strongly suggest that NH_4^+ is also excreted in vivo. O'Gara and Shanmugam (36, 37) have suggested that the bacterial genes coding the ammonia assimilatory enzymes may be repressed by plant amino acids, thus resulting in the excretion of ammonia with the onset of N_2 fixation. Glutamine synthetase has recently been purified from the plant fraction of soybean nodules (29). The enzyme was found to constitute about 2% of the total soluble protein in the nodule cytosol. The presence of such large quantities of GS together with its high affinity for NH_4^+ would result in the rapid removal of NH_4^+ in the plant cytoplasm and may facilitate diffusion of NH_4^+ from the bacteriod. Activity of GS is regulated by feedback inhibition and by energy charge (29).

Following the incorporation of ammonia into glutamine in the plant fraction of the nodule tissue (Figure 1) the amide nitrogen of glutamine could then be transferred to oxaloacetate via glutamate using GOGAT and a transaminase. An aspartate amino transferase has been characterized from the plant fraction of soybean nodules (48). This enzyme has also been identified in lupin nodules and is also induced during nodule development (Farnden, personal communication). The conversion of oxaloacetate to aspartate by this enzyme would provide the link between carbohydrate and nitrogen metabolism in the nodule. The synthesis of asparagine by the glutamine-dependent asparagine synthetase could then be accomplished using this aspartate and a second amide from glutamine (Figure 1). The net result of these reactions would be the synthesis of asparagine from oxaloacetate and 2 moles of ammonia with an energy requirement from the cytoplasm of 1 mole of NADPH and 3 moles of ATP.

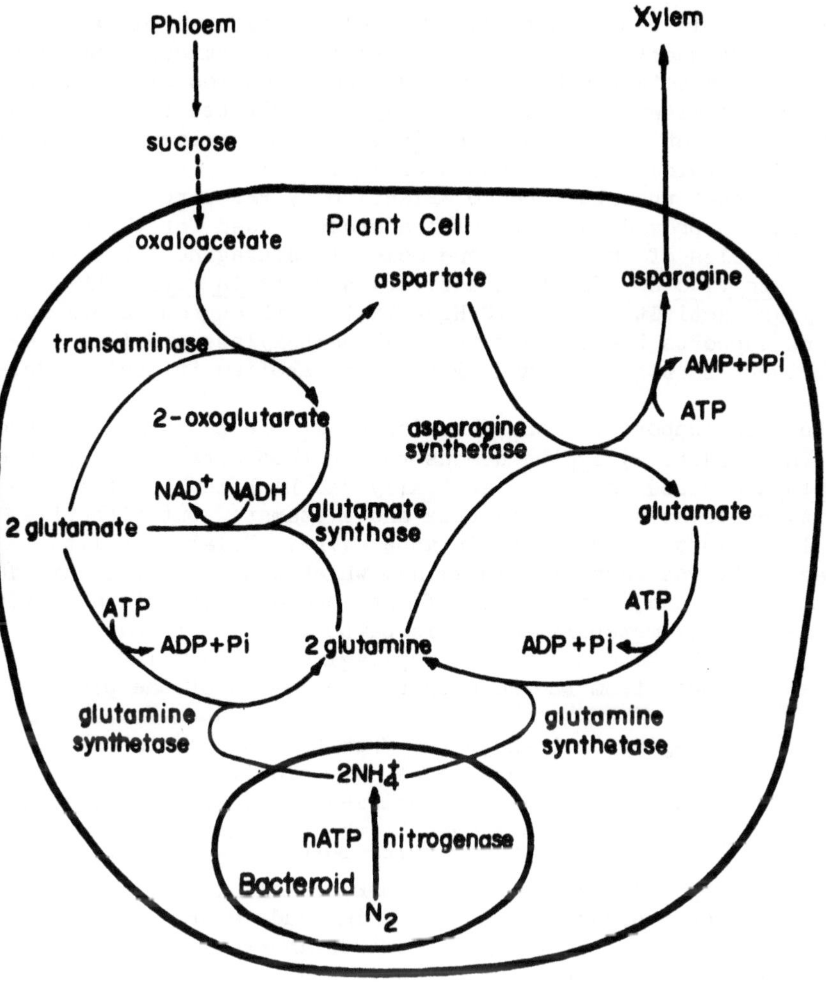

Figure 1. Pathway for the assimilation of ammonia in leguminous nodules [redrawn from Scott et al., (1976)].

The source of the oxaloacetate for this reaction may be provided by the dark fixation of CO_2, catalyzed by phosphoenol pyruvate (PEP) carbosylase (EC.4.1.1.31.; Equation 6). Lawrie and

$$\text{phosphoenol pyruvate} + CO_2 + H_2O \quad \text{oxaloacetate} + H_3PO_4 \qquad (6)$$

Wheeler (25) have shown that the activity of PEP carboxylase in the plant fraction of Vicia faba nodules was 50-fold higher than the activity of the corresponding enzyme in root extracts. Both malic enzyme and malate dehydrogenase levels were also higher in nodule extracts than in root extracts. The role of PEP carboxylase in providing oxaloacetate as a carbon acceptor for ammonia assimilation and amino acid synthesis in leguminous nodules has been pursued by Christeller *et al.*, (8). They demonstrated that both the *in vivo* CO_2 fixation rate and the *in vitro* PEP carboxylase activity increased (3-fold) with the onset of N_2-fixation. Furthermore, they calculated that the observed CO_2-fixation by PEP carboxylase is sufficient to provide all the oxaloacetate needed for the assimilation of 2 moles of ammonia to one mole of asparagine. The correlations of *in vivo* CO_2-fixation activity and the *in vitro* PEP carboxylase activity with $N_2(C_2H_2)$-fixation throughout nodule development supports the concept that PEP carboxylase provides the link between carbohydrate and nitrogen metabolism in lupin nodules.

Further support for the functioning of the pathway described above (see Figure 1) is the demonstration that detached nodules of *Vicia faba* exposed to $^{14}CO_2$, initially had high levels of radioactivity associated with aspartate and glutamate, but with time, accumulated most of the label in asparagine. Relatively little radioactivity was found in glutamine, which would support the idea that this compound, although the first formed in ammonia assimilation, is rapidly metabolized. This pattern of incorporation is consistent with the concept that asparagine is the main carrier of fixed nitrogen from the nodules to the leaves of the plant.

Once asparagine is transported from the nodules to the leaves of the plant it appears to be only slowly metabolized, while other amino acids are readily metabolized (40). The utilization of asparagine as a source of nitrogen to the plant appears to occur in situations where nitrogen is limiting (42). This concept is supported by the recent work of Atkins *et al.*, (1) who have shown high asparaginase activity in crude embryo extracts from *Lupinus albus* seeds. Furthermore, the demonstration of asparaginase activity in the plant fraction of lupin nodules before the onset of N_2-fixation (50) and in crude extracts from root tips (49), provides additional support for the idea that the mobilization of asparagine is important in a rapidly developing tissue, where there is a high requirement for nitrogen.

3. Symbiotic Associations Involving Blue-green Algae

The overproduction of NH_4^+ by bacteriods for plant growth is analogous to the excretion of NH_4^+ by several N_2-fixing blue-green algae in symbiotic association with pteridophytes, liverwort and fungi (lichens). A similar mechanism, as described for the regulation of ammonia assimilation in rhizobia, may also function in blue-green algae.

The functioning of the GS/GOGAT pathway for the assimilation of ammonia in blue-green algae has only recently been demonstrated. Glutamine synthetase was originally demonstrated in crude extracts of *Anabaena cylindrica* (10), and the high affinity of this enzyme for its substrates ($KmNH_4^+$ = 1 mM, Km glutamate = 2 mM) (11), provides an efficient mechanism for the assimilation of ammonia during N_2 fixation. The site for the assimilation of ammonia appears to be the heterocyst, as these cells have higher GS activity than the vegetative cells (11).

Low levels of GOGAT were also reported for *A. cylindrica* (10) but subsequent work showed that the activity was most likely due to the combined action of a glutaminase and GDH (19). However, the free amino acid pools of *A. cylindrica*, *A. flos-aquae* and *Westelliopsis prolifica* are largely aspartate, glutamine and glutamate, which is indicative of a functional GS/GOGAT system for the assimilation of ammonia (10).

The finding of a ferredoxin-dependent GOGAT (EC 2.6.1.53) in the chloroplasts of higher plants (26) led Lea and Miflin (27) to examine algae for a similar enzyme. High activities of a ferredoxin-dependent GOGAT were detected in intact cells of *A. cylindrica* and in extracts in which GOGAT was solubilized by ultra sonication (27). Tracer studies in which *A. cylindrica* cells were exposed to ^{13}N-labelled N_2 gas for very short intervals, demonstrated that the amide-N of glutamine was labelled first and the α-amino group of glutamate only secondarily, and thus confirms the functioning of the GS/GOGAT pathway in this alga (68). Further evidence for the incorporation of N_2 into amino acids by this pathway is discussed in a recent review (32).

In symbiotic associations with higher plants, blue-green algae excrete large quantities of fixed N_2 for plant growth. Algal clumps of *Anabaena azollae* isolated from the fronds of *Azolla* and incubated under N_2, release NH_4^+ into the medium (43). No ammonia was released for incubations carried out under Ar or N_2 with 2% CO. This work has recently been confirmed using $^{15}N_2$ (44). From these tracer studies it has been estimated that about 50% of the fixed-N was incorporated into the symbiont and the other 50% was excreted into the medium, principally as ammonia (44). Up to 50% of the total N in the intracellular pool of the intact association, grown in the absence of combined - N is also ammonia (35). Substantial levels of glutamine, glutamate and cystathionine were also found. Disks of the lichen *Peltigera canina* (incubated in the presence of digitonin to inhibit fungal metabolism) have been shown to excrete large quantities (50%) of fixed ^{15}N into the medium (61).

Half of this extracellular N was released as NH_4^+ and the other half as organic N products, probably amino acids. The symbiotic algae of the liverwort, Blasia pusilla, are also known to excrete newly fixed N as ammonia (59).

Ammonia completely inhibits nitrogenase activity in the free-living algae (Nostoc) but has less effect (40% inhibition) on the symbiont. Similarly, NO_3^- inhibited nitrogenase activity (50% inhibition) in free-living cultures of Nostoc but had no effect on the activity of lichen disks (61). Nitrogen fixation by the isolated symbiont from Azolla is completely insensitive to NH_4^+ under both aerobic and microaerophilic conditions (43); a state similar to the derepressed situation of the free-living algae, Anabaena cylindrica, in the presence of MSX (60). These studies suggest that the eucaryotic host produces some factor, or creates an environment, which results in the inhibition or repression of the ammonia assimilatory enzymes of the symbiotic algae (57).

In support of this concept, the levels of GS in Anabaena isolates from Azolla are 50-fold less than levels in free-living cultures of Anabaena cylindrica (44). Furthermore, the levels of GS in the host are high, suggesting that the plant provides a sink for the rapid removal of ammonia from the site of fixation. Rapid transfer of fixed-N to the host has been demonstrated in Gunnera (54), cycad root nodules and liverworts (see 58 and references therein), but the assimilatory enzymes involved have not been studied. In the P.canina system, GS is low in the symbiotic alga but high in free-living cultures of Nostoc (61). In this association, high NADP-dependent GDH was found in extracts of the thallus, and only low activities of GS, implying that GDH provides the major route for the assimilation of the fixed-N. This is consistent with the fact that GDH is the major route for ammonia assimilation in fungi (32). Further studies are necessary on the role of plant enzymes for the assimilation of ammonia during N_2-fixation in these associations, as well as information on the carbon exchange between host and symbiont. As yet there are no reports of GOGAT in the host tissue of the blue-green algae associations. With the exception of lichens, the GS/GOGAT pathway is probably the major route for the assimilation of ammonia in these associations.

REFERENCES

1. Atkins, C. A., Pate, J. S., and Sharkey, P. J. (1975), Plant Physiol. 56, 807.
2. Bergersen, F. J. and Turner, G. L. (1967), Biochim. Biophys. Acta 141, 507.
3. Bergersen, F. J. and Turner, G. L. (1976), Biochem. Biophys. Res. Commun. 73, 425.
4. Bishop, P. E., Guevara, J. G., Engelke, J. A., and Evans, H. J. (1976), Plant Physiol. 57, 542.

5. Brenchley, J. E. (1973), J. Bacteriol. 114, 666.
6. Brown, C. M. and Dilworth, M. J. (1975), J. Gen. Microbiol. 86, 39.
7. Brown, C. M., MacDonald-Brown, D. S. and Meers, J. R. (1974), Adv. Microbiol. Physiol. 11, 1.
8. Christeller, J. T., Laing, W. A., and Sutton, W. D. (1977), Plant Physiol. (in press).
9. Dainty, R. H, (1972), Biochem. J. 126, 1055.
10. Dharmawardene, M. W. N., Stewart, W. D. P. and Stanley, S. O. (1972), Planta 108, 133.
11. Dharmawardene, M. W. N., Haystead, N., and Stewart, W. D. P. (1973), Arch. Microbiol. 90, 281.
12. Dilworth, M. J. (1974), Annu. Rev. Plant Physiol. 25, 81.
13. Dougall, D. K. (1974), Biochim. Biophys. Res. Commun. 58, 639.
14. Drozd, J. W., Tubb, R. S, and Postgate, J. R. (1972), J. Gen. Microbiol. 73, 221.
15. Dunn, S. D. and Klucas, R. V. (1973), Can. J. Microbiol. 19, 1493.
16. Foor, F., Janssen, K. A., and Magasanik, B. (1975), Proc. Natl. Acad. Sci. 72, 4844.
17. Fowler, M. W., Jessup, W., and Sarkissian, G. S. (1974), FEBS. Lett. 46, 340.
18. Gordon, J. K. and Brill, W. J. (1974), Biochim. Biophys. Res. Commun. 59, 967.
19. Haystead, A., Dharmawardene, M. W. N., and Stewart, W. D. P. (1973), Plant Sci. Lett. 1, 439.
20. Keister, D. L. (1975), J. Bacteriol. 12 , 1265.
21. Kennedy, I. R. (1973), Proc. Aust. Biochem. Soc. 6, 33.
22. Kleiner, D. (1976), Arch. Microbiol. 111, 85.
23. Kurz, W. G. W. and LaRue, T. A. (1975), Nature 256, 407.
24. Kurz, W. G. W., Rokosh, D. A. and LaRue, T. A. (1975), Can. J. Microbiol. 21, 1009.
25. Lawrie, A. C. and Wheeler, C. T. (1975), New Phytol. 74, 437.
26. Lea, P. J. and Miflin, B. J. (1974), Nature 251, 614.
27. Lea, P. J. and Miflin, B. J. (1975), Biochem. Soc. Trans. 3, 381.
28. McComb, J. A., Elliott, J., and Dilworth, M. J. (1975), Nature 256, 409.
29. McParland, R. H., Guevara, J. G., Becker, R. R., and Evans, H. J. (1976), Biochem. J. 153, 597.
30. Magasanik, B. (1977), Trends in Biochem. Sci. 2, 9.
31. Meers, J. L., Tempest, D. W., and Brown, C. M. (1970), J. Gen. Microbiol. 64, 187.
32. Miflin, B. J. and Lea, P. J. (1976), Phytochem. 15, 873.
33. Miller, R. E. and Stadtman, E. R. (1972), J. Biol. Chem. 247, 7407.
34. Nagatani, H., Shimizu, M., and Valentine, R. C. (1971), Archives Mikrobiol. 79, 164.
35. Newton, J. W. and Cavins, J. F. (1976), Plant Physiol. 58, 798.

36. O'Gara, F. and Shanmugam, K. T. (1976), Biochim. Biophys. Acta 437, 313.
37. O'Gara, F. and Shanmugam, K. T. (1976), Biochim. Biophys. Acta 451, 342.
38. O'Neal, D. and Joy, K. W. (1974), Plant Physiol. 54, 773.
39. Pagan, J. D., Child, J. J., Scowcroft, W. R., and Gibson, A. H. (1975), Nature 256, 406.
40. Pate, J. S., Walker, J., and Wallace, W. (1965), Ann. Bot. 29, 475.
41. Pate, J. S., Gunning, B. E. S., and Briarty, L. G. (1969), Planta 85, 11.
42. Pate, J. S., Sharkey, P. J., and Lewis, O. A. M. (1975), Planta 122, 11.
43. Peters, G. A. (1976), In: Proc. Int. Symp. N_2-fixation (eds. W. E. Newton and C. J. Nyman) Washington State Univ. Press. p. 592.
44. Peters, G. A. (1977) Proc. Symp. Genetic Engineering for Nitrogen Fixation. Brookhaven National Laboratory, Brookhaven, New York, p. 231.
45. Robertson, J. G., Farnden, K. J. F., Warburton, M. P., and Banks, J. M. (1975), Aust. J. Plant Physiol. 2, 265.
46. Robertson, J. G., Warburton, M. P., and Farnden, K. J. F. (1975), FEBS Lett. 55, 33.
47. Rognes, S. E. (1970), FEBS Lett. 10, 62.
48. Ryan, E., Bodley, F., and Fottrell, P. F. (1972), Phytochem. 11, 957.
49. Scott, D. B. (1976), Ph.D. thesis, University of Otago, Dunedin, New Zealand.
50. Scott, D. B., Farnden, K. J. F., and Robertson, J. G. (1976), Nature 263, 703.
51. Shanmugam, K. T., Chan, I., and Morandi, C. (1975), Biochim. Biophys. Acta 408, 101.
52. Shanmugam, K. T. and Morandi, C. (1976), Biochim. Biophys. Acta 437, 322.
53. Shanmugam, K. T. and Valentine, R. C. (1975), Proc. Natl. Acad. Sci. USA 72, 136.
54. Silvester, W. B. and Smith, D. R. (1969), Nature 224, 1231.
55. Sloger, C. (1973), Plant Physiol. 51, (suppl.), 354.
56. Stadtman, E. R., Ginsburg, A., Ciardi, J. E., Yeh, J., Henning, S. B., and Shapiro, B. M. (1970), Adv. Enzyme Regul. 8, 99.
57. Stewart, W. D. P. (1977), Ambio. 6, 166.
58. Stewart,W. D. P. (1977), In: A Treatise on Dinitrogen Fixation" pp. 63-123 Eds. R. W. F. Hardy and W. S. Silver, John Wiley and Sons, New York.
59. Stewart, W. D. P. and Rodgers, G. A. (1977), New Phytol. 78, 459.
60. Stewart, W. D. P. and Rowell, P. (1975), Biochim. Biophys. Res. Commun. 65, 846.
61. Stewart, W. D. P. and Rowell, P. (1977), Nature 265, 371.

62. Streeter, J. G. (1972), Argon. J. 64, 311.
63. Streeter, J. G. (1973), Archives Biochem. Biophys. 157, 613.
64. Streicher, S. L., Shanmugan, K. T., Ausubel, F., Morandi, C., and Goldberg, R. B. (1974), J. Bacteriol. 120, 815.
65. Sutton, W. D., Jepsen, N. M., and Shaw, B. D. (1977), Plant Physiol. 59, 741.
66. Tempest, D. W., Meers, J. L., and Brown, C. M. (1970), Biochem. J. 117, 405.
67. Tempest, D. W., Meers, J. L., and Brown, C. M. (1973), In "The Enzymes of Glutamine Metabolism," S. Prusiner and E. R. Stadtman, eds., Academic Press, New York. pp. 167-182.
68. Thomas, J., Wolk, C. P., Shaffer, P. W., Austin, S. M., and Gabonsky, A. (1975), Biochem. Biophys. Res. Commun. 67, 501.
69. Tjepkema, J. D. and Evans, H. F. (1975), Biochem. Biophys. Res. Commun. 65, 625.
70. Tubb., R. S. (1974), Nature 251, 481.
71. Tyler, B., Deleo, A. B., and Magasanik, B. (1974), Proc. Natl. Acad. Sci. USA 71, 225.
72. Weare, N. M. and Shanmugam, K. T. (1976), Arch. Microbiol. 110, 207.

GENETICS AND REGULATION OF NITROGEN FIXATION

Winston J. Brill

Department of Bacteriology and Center
for Studies of Nitrogen Fixation
University of Wisconsin
Madison, Wisconsin 53706 U.S.A.

INTRODUCTION

The genetics and regulation of N_2 fixation has only recently begun to be understood. Breakthroughs from a variety of laboratories have yielded both techniques and ideas that will be useful for understanding genetic and regulatory mechanisms of microbial N_2 fixation. Furthermore, the potential for using this knowledge for practical applications of N_2 fixation is very great.

This chapter will discuss three N_2-fixing systems--*Klebsiella pneumoniae*, *Azotobacter vinelandii*, and various species of *Rhizobium*. The best understood system is *K. pneumoniae*, an organism that will grow on N_2 only anaerobically but will grow on fixed N both aerobically and anaerobically. *Azotobacter vinelandii*, a strict aerobe, is one of the fastest-growing N_2-fixing bacteria. *Rhizobium* normally only fixes N_2 in nodules on the roots of a particular host legume. Studies on the genetics and regulation of the latter system probably will have the most immediate practical use for agriculture.

KLEBSIELLA PNEUMONIAE

Regulation

All N_2-fixing systems seem to share the property of lowering N_2 fixation when sufficient available fixed N is present in the environment. When *K. pneumoniae* grows anaerobically with excess ammonium, no nitrogenase is synthesized (16). The mechanism by which nitrogenase synthesis is regulated in this organism is just beginning to be understood. It seems that the enzyme, glutamine synthetase, is directly responsible for initiating transcription of the nitrogenase (nif) genes (22,43,44). Glutamine synthetase plays a key role in glutamate synthesis when the organism is limited in growth by fixed N (22,27). High levels of glutamine synthetase are not required when Klebsiella grows on ammonium, because glutamate dehydrogenase supplies glutamate for protein synthesis and metabolism. Regulation by glutamine synthetase is quite complex because the enzyme is inhibited and activated by a modifying system that adenylylates and deadenylylates glutamine synthetase (reviewed in 22).

Klebsiella pneumoniae is unable to protect its nitrogenase from being O_2 inactivated, and therefore the organism does not synthesize nitrogenase in the presence of O_2, even when the cell is starved of a N source (40). Nothing is known about the mechanism of this regulation, but it may share common regulatory factors that are required for specific repression and derepression when a cell switches from an aerobic environment to an anaerobic one.

All active nitrogenases contain molybdenum, which is a part of a cofactor making up an active site (34). When a cell is starved of molybdenum, no active nitrogenase can be made. It seems that molybdenum also plays a regulatory role in the synthesis of nitrogenase in K. pneumoniae. Under conditions during which the organism is starved of both N and molybdenum, neither of the nitrogenase proteins is made (6).

Genetics

Klebsiella pneumoniae is the organism with which the most detailed genetic studies of nif have been performed. The reason for this is that K. pneumoniae is closely related to Escherichia coli. The order of genes in the chromosome of K. pneumoniae is very similar to the order of genes in E. coli (25). Another similarity (42) between the two organisms is that both can be infected by phage P1, which is a generalized transducing phage useful for genetic mapping by transduction. Another phage, Mu, also infects both species and has been useful because it integrates randomly in the chromosome and thus causes mutations (1,31). Phage Mu also in valuable for generating random deletions within specific genes of interest (1, 17,31).

All of the nif^- mutations in K. pneumoniae are cotransducible (39,42) with the genes specifying the enzymes of histidine biosynthesis (his genes). Several nif mutations have been mapped by transduction with respect to his (39).

There seem to be at least eight nif genes in the cluster, but they make up at least three operons, with DNA that is not required for N_2 fixation separating several of these operons (T. MacNeil, D. MacNeil, W. J. Brill, unpublished results).

Potential Applications

Mutant strains that are derepressed for nitrogenase synthesis and are blocked in their ability to utilize ammonium, excrete ammonium into the medium. This situation can be achieved both genetically (37) as well as by addition of the glutamine analog, methionine sulfoximime (15), to the medium. In both of these situations, high levels of glutamate or glutamine are required for the organism to grow, thus the practicality of using these conditions for ammonium production is minimized. If the amino acid requirement is removed, however, such strains may be able to enrich an anaerobic, N-deficient medium with fixed N.

After it was realized that all of the nif genes in *K. pneumoniae* are clustered in a small region of the chromosome (39,42), the nif genes were then genetically transferred to an F factor as well as to a drug-resistance transfer factor (9,10). After the nif genes were placed on these plasmids, it was possible to mobilize the plasmids via conjugation and therefore transfer nif genes to bacteria normally unable to fix N_2 (9,11). The first example of such a transfer was with *E. coli* (11). When nif was transferred to an aerobic plant pathogen, *Agrobacterium tumefaciens*, the pathogen still was unable to grow on N_2 because O_2 inactivated the nitrogenase that was synthesized (9). It seems, therefore, that any strain that is to be a recipient of nif genes must have some mechanism to prevent O_2 from inactivating nitrogenase.

What about the possibility that nif genes will be transferred to cereal plants, therefore allowing such plants to grow without a requirement for fertilizer N? Many barriers must be overcome if this situation is ever to be achieved in agriculture. An O_2 protection mechanism must be introduced, mechanisms for sequestering very high levels of iron and molybdenum must be available, and the very high ATP requirement also must be met without concomitant disruption of the normal life cycle of the cereal plant. It also would seem that selective pressures for N_2 fixation by cereals has been present for a long time and one might have expected that these plants should have acquired the ability to fix N_2 during their evolution if the mere integration of nif genes would have been sufficient for N_2 fixation.

AZOTOBACTER VINELANDII

Regulation

Excess ammonium also prevents the synthesis of nitrogenase in *A. vinelandii* (35, 41). When a N source, such as nitrate is used by the organism, nitrogenase synthesis is repressed about 50%. Nitrate must be reduced to ammonium for this repression to occur (38). It is not yet known whether glutamine synthetase plays a role in regulating nitrogenase synthesis in *A. vinelandii*. When methionine sulfoximine is added to *A. vinelandii* growing in the presence of excess ammonium, nitrogenase synthesis is derepressed (15). Mutant strains of *A. vinelandii* have been isolated that are derepressed for nitrogenase synthesis (14), but unlike the derepressed mutant strains of *K. pneumoniae*, these strains have no amino acid requirement for growth and have no obvious defects in enzymes involved with ammonium assimilation (J. K. Gordon, unpublished results).

One interesting regulatory mutant strain produces up to eight times the normal level of component II of nitrogenase but at the same time does not produce any component I (36). This provides evidence that both nitrogenase components are coded by different operons.

Genetics

Mutations in *A. vinelandii* have been mapped by transformation (29). Unlike the situation in *K. pneumoniae*, in which the *nif* genes are closely clustered, the *nif* genes in *A. vinelandii* seem to be scattered around the chromosome (3). Other techniques must be devised for more detailed analyses of *nif* genes in *A. vinelandii*.

Potential Applications

There are many examples of claims that *Azotobacter* inoculants have been useful in agriculture (for example, see 32). Such inoculants still are being produced in several countries, including the United States. The value of these inoculants is debatable, and any aid to crop plants has been presumed to be through production of plant-growth hormones rather than by N_2 fixation (7).

Azotobacter, however, is very easy to grow because it is a strict aerobe and is able to protect its nitrogenase through a very high respiratory activity (30). This bacterium has one of the fastest doubling times on N_2 of any N_2-fixing organism. No special growth requirements are necessary, and Azotobacter can use a wide variety of carbon sources including monosaccharides, disaccharides, alkanes, fatty acids, alcohols, etc. (32). Ammonium-excreting mutant strains of A. vinelandii have been isolated and are able to grow well on such substrates as paper-mill waste (J. K. Gordon and W. J. Brill, unpublished results). This laboratory is currently exploring means for increasing ammonium excretion by such mutant strains. Perhaps these strains can be used for enriching suitable environments with fixed N.

Azotobacter has been found in high numbers on the roots of a tropical grass, Paspalum (12). Perhaps a stable N_2-fixing association can be obtained between an ammonium-excreting Azotobacter and a carbohydrate-or fatty acid-excreting cereal plant.

RHIZOBIUM

Regulation

As found with the other N_2-fixing systems, the Rhizobium-legume symbiosis is disrupted when sufficient fixed N is in the environment. No nodules are formed on plants growing with excess available fixed N (13). The plant and Rhizobium each lives independently in this situation. When fixed N is limiting, however, all of the N can be supplied to the legume through N_2 fixation from the bacteria in the nodules. It seems that the regulation of this symbiosis occurs primarily at the level of the initial steps in the infection process. When certain Rhizobium strains fix N_2 asymbiotically, the addition of fixed N does not depress the nitrogenase activity to any great extent (19). There is a possibility that glutamine synthetase also plays a role in the regulation of nitrogenase synthesis because a revertant of a glutamine auxotroph has altered regulatory properties (20,21). It is not known whether the regulatory target in the infection process is in the bacterium or plant or both. This area requires much more research effort.

Genetics

Several groups are now in the process of mapping by conjugation, the chromosome of various Rhizobium species and preliminary genetic maps have been published (2,26). Conjugation experiments have indicated that the gene order on the chromosomes of R. leguminosarum R. trifolii, and R. phaseoli may be similar (18). So far, no genes required specifically for N_2 fixation or infection have been mapped, even though a mutation in the nitrogenase structural gene and mutations in the initial stages of infection have been described (23).

Transduction and transformation also have been used for mapping mutations in strains of Rhizobium. DNA from Rhizobium is capable of transforming Nif^- mutant strains of A. vinelandii to the Nif^+ phenotype (28). Interestingly, genes required for production of Rhizobium surface antigens required for infection also may be transformed to A. vinelandii (4).

Since N_2 fixation in these systems normally requires both the bacterium and plant, it is important to study the genetics of the plant with respect to N_2 fixation. The plant plays an important role because it supplies the ATP and electron source for the reaction and also protects the bacteria from adverse soil conditions and competing bacteria. Some plant genes have been described that seem to be important in the symbiosis (reviewed in 5). It is clear that our understanding of the genetics of N_2 fixation in this important system is in its infancy, but techniques and concepts gained from studying the simpler free-living N_2-fixing bacteria should aid in our understanding of this complex system.

Potential Applications

The Rhizobium-legume symbiosis already has many applications in agriculture, but there are many ideas that may possibly allow greater N input by legumes. For instance, it should be possible to breed for more effective plants and, at the same time, screen mutant strains of Rhizobium to optimize the N_2-fixing capacity of the system. Another application, that would help to further enrich the soil with N, is to find symbioses that are less responsive to the negative effect on N_2 fixation by fixed N.

A rapid technique for screening the N_2-fixing potential of the symbiosis is an effectiveness assay that is completed within two weeks after the seed is germinated (45). This assay has been useful for isolating mutant strains of R. japonicum that fix more N_2 for the plant than their wild-type parent strain (24).

It may be important to screen symbioses for their ability to recycle electrons lost through hydrogen evolution by nitrogenase (33), and thus conserve some of the energy expended by a N_2-fixing nodule. The importance of the ability to successfully compete with indigenous soil rhizobia should not be minimized. Nothing is really know about the properties required for competitiveness, but it may some day be possible to transfer or induce competitiveness into desirable strains of Rhizobium.

SUMMARY

At this point, there is very little known about the regulation and genetics of N_2 fixation in K. pneumoniae, A. vinelandii, and species of Rhizobium. However, the techniques of molecular biology are just beginning to be applied to nif genes and we should expect a great wealth of data within the next several years. Other organisms, including cyanobacteria (8) and photosynthetic bacteria (46), also are being examined with respect to the regulation and genetics of N_2 fixation. When we understand how nitrogenase is produced and regulated, it should be possible to modify genes or regulatory proteins so that new and more efficient uses can be made with N_2-fixing bacteria.

REFERENCES

1. Bachhuber, M., Brill, W. J. and Howe, M. M. (1976) J. J. Bacteriol. 218, 749.

2. Beringer, J. E. and Hopwood, D. A. (1976) Nature 264, 291.

3. Bishop, P. E. and Brill, W. J. (1977) J. Bacteriol. 130, 954.

4. Bishop, P. E., Dazzo, F. B., Appelbaum, E. R, Maier, R. J. and Brill, W. J. (1977) Science. In press.

5. Brill, W. J. (1975) Ann. Rev. Microbiol. 29, 109.

6. Brill, W. J., Steiner, A. L., and Shah, V. K. (1974) J. Bacteriol. 118, 986.

7. Brown, M. E. (1976) J. Appl. Bact. 40, 341.

8. Currier, T. C., Haury, J. F., and Wolk, C. P. (1977) J. Bacteriol. 129, 1556.

9. Dixon, R., Cannon, F., and Kondorosi, A. (1976) Nature 260, 268.

10. Dixon, R. A. and Postgate, J. R. (1971) Nature 234, 47.

11. Dixon, R. A. and Postgate, J. R. (1972) Nature 237, 102.

12. Döbereiner, J., Day, J. M., and Dart, P. J. (1972) J. Gen. Microbiol. 71, 103.

13. Fred, E. B., Baldwin, I. L., and McCoy, E. (1932) Root Nodule Bacteria and Leguminous Plants, University of Wisconsin Press, Madison.

14. Gordon, J. K. and Brill, W. J. (1972). Proc. Nat. Acad. Sci. U.S.A. 69, 3501.

15. Gordon, J. K. and Brill, W. J. (1974) Biochem. Biophys. Res. Commun. 59, 967.

16. Hamilton, R. B., and Wilson, P. W. (1955) Ann. Acad. Sci. Fenn. Ser. A. 60, 139.

17. Howe, M. M. and Bade, E. G. (1975) Science 190, 624.

18. Johnston, A. W. B., and Beringer, J. E. (1977) Nature 267, 611.

19. Keister, D. L. (1975) J. Bacteriol. 123, 1265.

20. Kondorosi, A., Svab, Z., Kiss, G. B. and Dixon, R. A. (1977) Molec. Gen. Genetic. 151, 221.

21. Ludwig, R. A. and Signer, E. R. (1977) Nature 267, 245.

22. Magasanik, B., Prival, M. J., Brenchley, J. E., Tyler, B. M., DeLeo, A. B., Streicher, S. L., Bender, R. A., and Paris, C. G. (1974). In "Current Topics in Cellular Regulation" (A.Ginsburg and E. R. Stadtman, eds.), Vol. 8, pp. 119-138. Academic Press, New York.

23. Maier, R. J. and Brill, W. J. (1976) J. Bacteriol. 127, 763.

24. Maier, R. J., and Brill, W. J. (1977) Abst. Ann. Meet. Amer. Soc. Microbiol. 77, 198.

25. Matsumoto, H. and Tazaki, T. (1971) Jap. J. Microbiol. 15, 11.

26. Meade, H. M. and Signer, E. R. (1977) Proc. Nat. Acad. Sci. U.S.A. 74, 2076.

27. Nagatani, H., Shimizu, M., and Valentine, R. C. (1971) Arch. Mikrobiol. 79, 164.

28. Page, W. J. (1977) Abst. Ann. Meet. Amer. Soc. Microbiol. 77, 199.

29. Page, W. J. and Sadoff, H. L. (1976) J. Bacteriol. 125, 1080.

30. Phillips, D. A. and Johnson, M. J. (1961) J. Biochem. Microbiol. Tech. Eng. 111, 277.

31. Rao, R. N. (1976) J. Bacteriol. 218, 356.

32. Rubenchik, L. I. (1960) Azotobacter and its Use in Agriculture. Translated from Russian. Israel Program for Scientific Translations, Jerusalem.

33. Schubert, K. R. and Evans, H. J. (1976) Proc. Nat. Acad. Sci. U.S.A. 73, 1207.

34. Shah, V. K. and Brill, W. J. (1977) Proc. Nat. Acad. Sci. U.S.A. In press.

35. Shah, V. K., Davis, L. C., and Brill, W. J. (1972) Biochim. Biophys. Acta 256, 498.

36. Shah, V. K., Davis, L. C., Stieghorst, M., and Brill, W. J. (1974) J. Bacteriol. 117, 917.

37. Shanmugam, K. T., and Valentine, R. C. (1975) Proc. Nat. Acad. Sci. U.S.A. 72, 136.

38. Sorger, G. J. (1969) J. Bacteriol. 98, 56.

39. St. John, R. T., Johnston, H. M., Seidman, C., Garfinkel, D., Gordon, J. K., Shah, V. K., and Brill, W. J. (1975) J. Bacteriol. 121, 759.

40. St. John, R. T., Shah, V. K., and Brill, W. J. (1974) J. Bacteriol. 119, 226.

41. Strandberg, G. W., and Wilson, P. W. (1968) Can. J. Microbiol. 14, 25.

42. Streicher, S. L., Gurney, E. G., and Valentine, R. C. (1971) Proc. Nat. Acad. Sci. U.S.A. 68, 1174.

43. Streicher, S. L., Shanmugam, K. T., Ausubel, F., Morandi, C., and Goldberg, R. B. (1974) J. Bacteriol. 120, 815.

44. Tubb, R. S. (1974) Nature 251, 481.

45. Wacek, T. J., and Brill, W. J. (1976) Crop Science 16, 519.

46. Wall, J. D., Weaver, P. F., and Gest, H. (1975) Nature 258, 630.

LEGHAEMOGLOBIN, OXYGEN SUPPLY AND NITROGEN FIXATION : STUDIES WITH SOYBEAN NODULES

F.J. Bergersen

Division of Plant Industry, CSIRO, P.O. Box 1600

Canberra, 2601, Australia

SUMMARY

This paper reviews some aspects of the occurrence, properties and role of leghaemoglobins in N_2-fixation. These myoglobin-like haemoproteins are invariably present in N_2-fixing root nodules of legumes where they undergo reversible oxygenation in response to changes in ambient oxygen concentration and bacteroid respiration. Blocking their O_2-binding functions with carbon monoxide depresses nitrogenase activity of intact soybean nodules and depresses respiration of nodule slices. Leghaemoglobin in soybean nodules is present at concentrations of about 1 mM and *in vivo* is partially oxygenated. Experiments in which purified oxyleghaemoglobin has been added to N_2-fixing soybean bacteroids, allowed the study of the relationships between the concentration of free, dissolved O_2 (monitored by leghaemoglobin oxygenation), nitrogenase activity and bacteroid ATP concentration. Partially oxygenated leghaemoglobin enhances bacteroid respiration in a range of concentrations of free O_2, in which production of ATP is more efficient than at higher concentration of free O_2, where leghaemoglobin is fully oxygenated. Consequently, in the presence of partially oxygenated leghaemoglobin, nitrogenase activity is greater. Distinct components of bacteroid respiration can be recognised in terms of substrate supplied, sensitivity to inhibitors and free oxygen concentration. These components vary in the efficiency with which they support nitrogenase activity.

In 1940, Allison *et al.*[1] concluded that the respiration of soybean nodules was restricted at normal atmospheric concentrations of oxygen. About the same time Kubo [28] discovered that the pink colour of nodules of soybean and other legumes, was due to a soluble, haemoglobin-like pigment. He concluded that these haemoglobins were connected with oxygen consumption in the nodules. Keilin and Wang [27] and Keilin and Smith [26] confirmed the oxygenation cycle of soybean leghaemoglobin and showed that the only spectroscopically detectable reaction in the nodule was

$$\text{ferrous leghaemoglobin (Lb)} + O_2 \rightleftarrows \text{ferrous oxyleghaemoglobin } (LbO_2)$$

Nevertheless, some experiments by Smith [34] seemed to indicate that nodule respiration was not affected by carbon monoxide, which blocks O_2-binding by Lb. Consequently, for many years, alternative roles for Lb were sought, since it was clear that, in pea nodules [41] and in nodules of many other legumes, the content of Lb was correlated with nitrogen fixing capacity. All of these alternative roles have now been eliminated and it is clear that the reversible oxygenation of leghaemoglobin is central to its function in legume nodules (reviewed by Appleby, ref. 4).

In this paper, previously published developments in understanding the role of leghaemoglobin in nodules will be briefly described and some more recent results outlined. (For other viewpoints about the role of Lb, the reader should consult Appleby *et al*. [7] and Wittenberg [43]).

PROPERTIES OF LEGHAEMOGLOBINS

Leghaemoglobins are monomeric haemoproteins of molecular weight about 15-17000 daltons, containing protohaem IX as the prosthetic group. Nodules of some species of legumes contain several distinct components of leghaemoglobin, differing in iso-electric point and amino-acid composition (see review by Appleby, ref. 4). The ratio of the components may change with time in soybean nodules[30] and different roles have been proposed [22] but there is no experimental evidence for this. The amino acid sequences of leghaemoglobins from soybean [21], broad bean [33], kidney bean [29] and preliminary 3-dimensional X-ray structures of lupin leghaemoglobin [38], show some striking similarities to mammalian myoglobin. Soybean leghaemoglobin has a much higher affinity for O_2 ($K' = 0.037\text{-}0.073 \times 10^{-6}$ M O_2, refs. 25,44) than myoglobin ($K' = 1 \times 10^{-6}$ M O_2). Studies of the molecular structure of soybean leghaemoglobin, including properties of proton-dependent reactions with various ligands, are helping to develop understanding of the reasons for the high O_2-affinity (e.g. Appleby *et al*., ref. 6). These studies follow from similarities between the optical spectra and kinetics of formation of oxy- and carboxyleghaemoglobin and those of oxy- and carboxyhaemoglobin which suggest that O_2 and CO compete for the vacant 6th coordination position of ferrous haem in both [4].

Leghaemoglobins are readily extracted from nodules crushed in dilute buffers of pH about 6.4-7.4. The fraction precipitating between 0.6 and 0.8 $(NH_4)_2SO_4$ saturation is used for further purification. In our laboratory we routinely obtain preparations of unfractionated oxyleghaemoglobin of high quality and 85-95% purity (based on haem:protein ratios) with the following procedure. After dialysis, $(NH_4)_2SO_4$-fractionated material is chromatographed on a 90 x 5 cm column of Sephadex G75 equilibrated with 25 mM tris-HCl buffer, pH 7.4, at 4°C. For greatest stability it may be advantageous to convert the leghaemoglobin completely to the ferric form, with an excess of ferricyanide, before chromatography. After concentration of the pooled main band fractions to about 5 mM haem, a portion of the ferric leghaemoglobin is treated with an excess of anaerobically prepared $Na_2S_2O_4$ and then rapidly chromatographed on a small column of G10 Sephadex, equilibrated with air - saturated 25 mM tris-HCl buffer, pH 7.4. Aliquots of these preparations, concentrated to about 2 mM over an Amicon Diaflo UM 10 membrane, can be stored indefinitely in liquid N_2, for subsequent use in experiments with bacteroid suspensions. The optical spectrum of such a preparation is shown in Fig. 1. Preparation of leghaemoglobin components is conveniently done by chromatography of the G75 preparation on DE52 cellulose (Whatman), using a Na-acetate gradient at pH 5.2 [4].

Leghaemoglobins are characteristic of the host legumes from which they are isolated [19,15]. Verma et al.[40] showed that apo-leghaemoglobin was synthesized in an experimental wheat embryo system containing 9S mRNA associated with 80S-type plant ribosomes from soybean nodules. There is evidence that the haem moiety is synthesized by the bacteroid component of nodules [16,23], but the site of assembly of the functional leghaemoglobin is uncertain.

INTRACELLULAR LOCATION OF LEGHAEMOGLOBINS

Leghaemoglobins are freely-soluble proteins wholly located within the bacteroid-containing central tissue cells of nodules [34]. They are present in high concentration for a cell component (15-25 mg/ml - ref. 10) and are not readily immobilized by the fixation procedures normally used for the preparation of specimens for optical or electron microscopy. Consequently, the intracellular location has been controversial. Dilworth and Kidby [20] used electron microscopy and autoradiography to determine the localization of ^{59}Fe in thin sections of serradella nodules from plants supplied with ^{59}Fe in the nutrient solution in which they were grown. In a carefully-reasoned analysis of their results, the authors concluded that leghaemoglobin-Fe was located between the bacteroid surface and the enclosing membrane envelope. There was insufficient ^{59}Fe in the host cytoplasm to account for the leghaemoglobin, known by analysis to be present. In contrast, Dart [17] and Dart and Chandler [18] concluded that leghaemoglobin in several

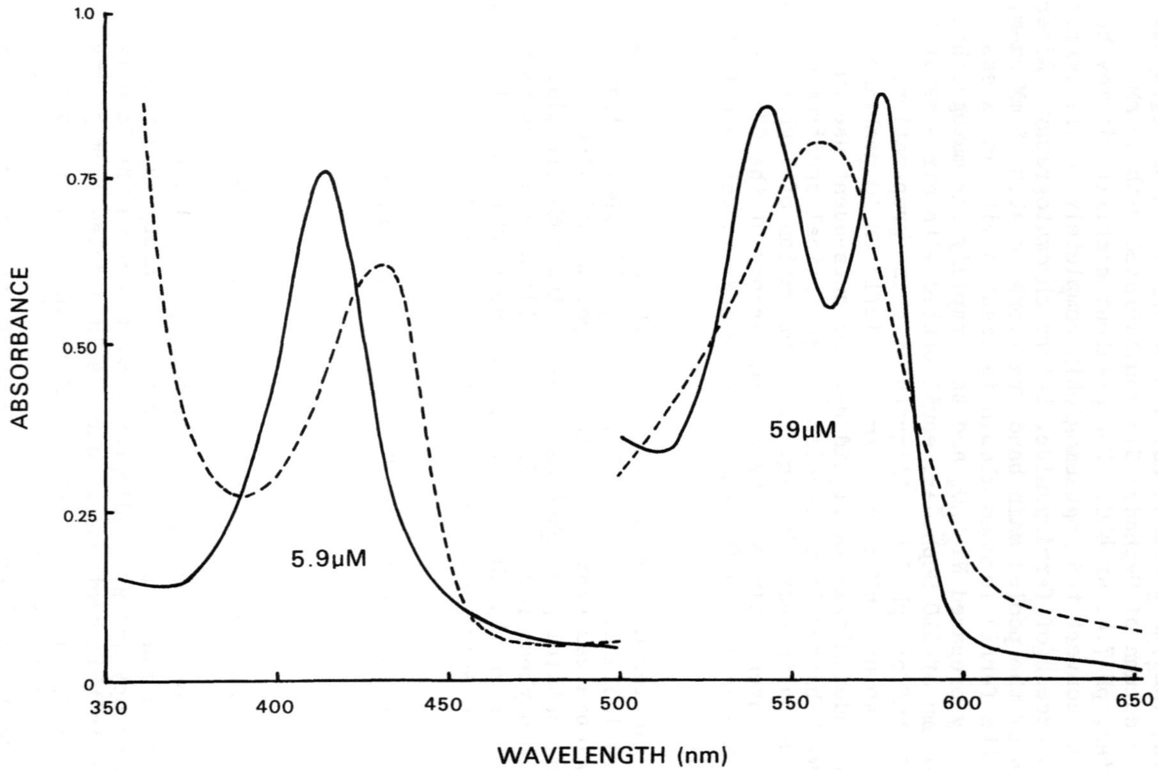

Fig. 1. Absorption spectrum of a sample of soybean leghaemoglobin, as used in these experiments. Solid line, oxyleghaemoglobin; dashed line, completely deoxygenated ferrous leghaemoglobin. The Soret region was recorded with a 1:10 dilution.

species of legumes, was located in the cytoplasm of the nodule cells. Bergersen and Goodchild [10] set out to resolve the discrepancy by studying the effects of fixation procedures on 0.4 mM solutions of pure leghaemoglobin, before developing fixation and staining methods which would minimise migration and loss of leghaemoglobin from the small blocks of soybean nodule tissue which were used. Also, sections of unfixed nodules, stained with oxidised diaminobenzidine to visualize the leghaemoglobin, were examined with light microscopy. The results showed that the bulk of the leghaemoglobin, if not all of it, was located within the membrane envelopes, and thus was in contact with the bacteroids. In this location, it was calculated that the functional concentration was about 1.5 mM during the most active period of N_2-fixation. Similar conclusions about the location of leghaemoglobin have been obtained by Truchet [37] for _Pisum sativum_ and by Gourret and Fernandez-Arias [24] for _Trifolium repens_.

Recently the controversy has been re-opened by Verma and Bal [39], who used ferritin-labelled antibody, prepared against soybean leghaemoglobin, to show the location of leghaemoglobin within soybean nodule cells.. In my opinion, their results are open to a number of criticisms which mainly arise from the prolonged washing to which the tissue was subjected during preparation. Some crucial controls were also omitted. The intensity of the immune reactions seen in their micrographs indicates that most of the leghaemoglobin was lost from the tissue. Only leghaemoglobin which was attached to cytoplasmic structures during fixation, remained. The membrane envelope space contained no leghaemoglobin.

Other workers have consistently failed to prepare membrane envelopes containing leghaemoglobin within them (Robertson, priv. comm.), and conclude therefore that the haemoprotein resides in the cytoplasm.

These results emphasize the need for more specific methods of microscopic identification and better techniques of leghaemoglobin preservation so that its location may be determined unequivocally. This is essential to proper validation of experiments on the physiological function of leghaemoglobin. In the meantime, the weight of evidence indicates that leghaemoglobins are probably located within the membrane envelopes [10,20,24].

EVIDENCE FOR AN O_2-BINDING _IN VIVO_ ROLE FOR LEGHAEMOGLOBIN

Spectrophotometry of intact soybean nodule tissue showed that leghaemoglobin _in vivo_ was normally 20-25% oxygenated, the balance was in the non-oxygenated, reduced form [2]. Nash and Schulman [31] found no evidence of any other form of leghaemoglobin during the functional life of soybeans. In intact tissue, oxygenation increased when pure O_2 was substituted for air [2].

Tjepkema and Yocum [36] reported that the respiration of thin slices of soybean nodules was halved when carbon monoxide blocked O_2-binding by leghaemoglobin. These findings provide clear evidence for an O_2-binding function _in vivo_, but the evidence that this function could be related to N_2-fixation rested only on the correlation between N_2-fixation capacity and leghaemoglobin concentration [41]. Bacteroids which had been washed free of leghaemoglobin and other nodule components fixed N_2 [11], so the role of leghaemoglobin is seen as a supporting one.

The use of H_2-evolution by nitrogenase, as a CO-insensitive assay, enabled the demonstration of the O_2-binding function of leghaemoglobin in relation to nitrogenase (ref. 14; Table 1).

TABLE 1

A role for leghaemoglobin in nodule tissue.
Data adapted from Bergersen _et al_. [14], showing that intact soybean nodules have a carbon monoxide sensitive component to their hydrogen evolving nitrogenase activity. This sensitive component was absent from washed bacteroids, but was re-established when leghaemoglobin was added to dense suspension of bacteroids.

System	pO_2 (mm Hg)	pCO (mm Hg)	H_2 evolved (n mol)
Intact nodules	70	0	1557 /h/g fr.wt.
	70	14	433
	140	0	2050
	140	14	194
	210	0	2674
	210	14	137
Washed bacteroids (no leghaemoglobin)	27	0	17.1 /h/mg dry wt.
	27	14	14.5
	62	0	42.9
	62	14	39.2
	111	0	24.3
	111	14	23.7
Washed bacteroids leghaemoglobin (mM)			
0	84	0	58 /assay
0.2	84	0	219
0.2	84	1	134
0.2	84	5	81
0.2	84	25	58

In intact nodules CO inhibited H_2 evolution, whereas in washed bacteroid suspensions it did not. Addition of oxyleghaemoglobin to dense suspensions of bacteroids stimulated H_2-evolution and the stimulation was abolished by CO. Carbon monoxide had blocked the O_2-binding function of leghaemoglobin and thus indirectly diminished nitrogenase activity. Subsequent experiments have elucidated some features of the mechanisms involved in the supporting role of leghaemoglobin.

BACTEROIDS, LEGHAEMOGLOBIN AND GASEOUS O_2

The stimulation of the nitrogenase activity in bacteroid suspensions to which oxyleghaemoglobin was added and shaken with a low gas phase concentration of O_2, was confirmed by C_2H_2-reduction and $^{15}N_2$ experiments [14]. Further studies of the respiration and nitrogenase activities of bacteroids in a similar experimental system [45] showed that the stimulation was produced only by ferrous-oxyleghaemoglobin. Ferric leghaemoglobin and other ferric haemoproteins were ineffective. However, other O_2-binding proteins produced some stimulation (but less effectively than oxyleghaemoglobin). The order of their effectiveness was approximately the order of their O_2-affinities. The stimulation produced by oxyleghaemoglobin was not affected by superoxide dismutase and catalase and the mechanism therefore did not involve a superoxide radical. The effects of O_2 concentration in the gas-phase and of Lb concentration were consistent with a 'facilitated diffusion' function for leghaemoglobin. Facilitated diffusion [42] is the term given to the phenomenon wherein an O_2-carrier such as leghaemoglobin brings about an increment of O_2 flux through a layer of solution if the carrier is substantially more oxygenated on the high O_2 concentration side of the layer than at the low O_2 concentration side.

$$\text{Flux}_{\text{total}} = \text{flux}_{\text{facilitated}} + \text{flux}_{\text{free } O_2}$$

$$\text{Flux}_{\text{facilitated}} = m\,C\,D\,\frac{Y_o - Y_\ell}{\ell} \qquad [46].$$

where m = no. of O_2 binding sites/molecule (1 for leghaemoglobin)
C = concentration protein
D = diffusion coefficient for the protein
Y_o & Y_ℓ are the fraction of the protein oxygenated at the high and low boundaries of a diffusion path of length ℓ.

Wittenberg <u>et al</u>. [45] reasoned that, with bacteroids suspended in a solution of leghaemoglobin in their experiments, an unstirred layer surrounded each organism. Leghaemoglobin facilitated an increment of O_2 flux across this layer. This relatively small increment of O_2 uptake was associated with a disproportionate stimulation of nitrogenase activity (Fig. 2). Appleby <u>et al</u>. [5] used the same experimental system and showed that nitrogenase was stimulated

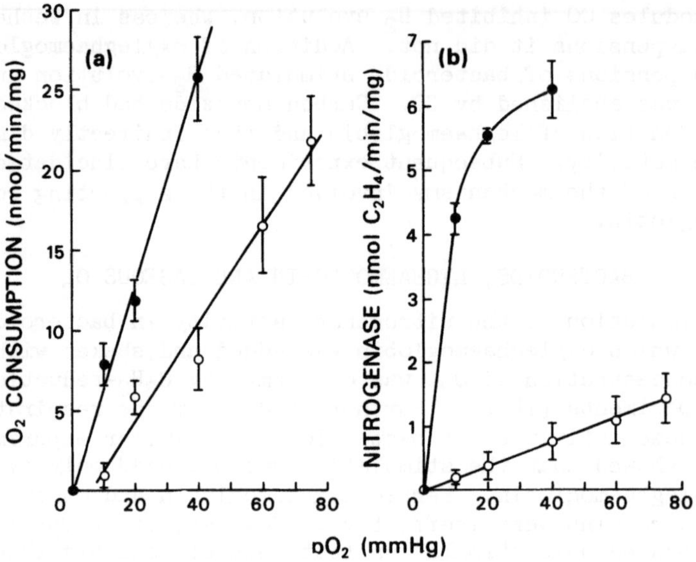

Fig. 2. Effects of leghaemoglobin on (a) respiration of bacteroids and (b) nitrogenase activity. Data from shaken assays with gas-phase O_2 concentrations as indicated. Open circles, assays with no leghaemoglobin; closed circles, assays containing 0.5 mM oxyleghaemoglobin. (Redrawn from ref. 12).

because this small increment of O_2 consumption by the bacteroids produced ATP very efficiently in a pathway involving cytochrome P-450 (Fig. 3).

These were satisfying results, but there were some inconsistencies in details of the interpretation. Furthermore, Stokes [35] calculated that effects other than facilitated diffusion of O_2 could be large in experimental systems such as these. He also calculated (unpublished results) that a major effect of leghaemoglobin, in these shaken assays with a gas phase, would be to establish a zone of stable free O_2 concentration within the volume of liquid. Such an effect may have no relevance to the effect of leghaemoglobin in vivo.

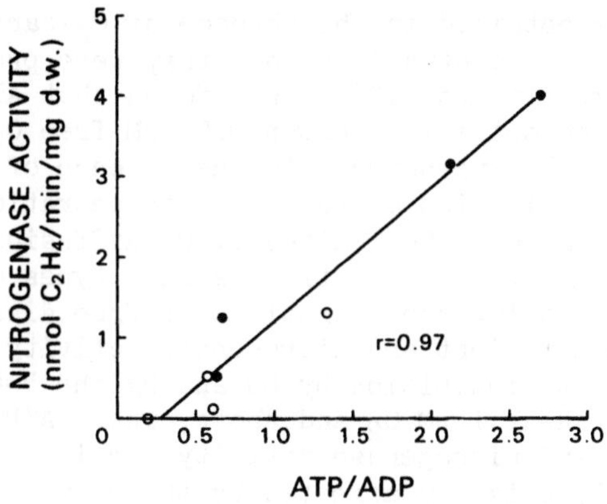

Fig. 3. The correlation between bacteroid nitrogenase activity and ATP/ADP ratios in shaken assays with (closed circles) and without (open circles) 0.5 mM oxyleghaemoglobin. Decreasing nitrogenase rates were obtained with increasing concentrations of N-phenylimidazole, a specific Cytochrome P-450 inhibitor. (Data redrawn from ref. 5).

BACTEROIDS AND OXYLEGHAEMOGLOBIN - NO GAS PHASE EXPERIMENTS

We therefore developed techniques for studying the effects of Lb on bacteroids, in experiments with no gas phase [12,13]. In these experiments, bacteroid respiration was measured electrically, or by monitoring rates of change of leghaemoglobin oxygenation spectrophotometrically. In the latter, we used the well-established O_2 equilibrium constant [44] to calculate concentrations of free O_2 from oxygenation values. Nitrogenase activity was measured by C_2H_2-reduction, using reaction solutions which were initially equilibrated with gas mixtures containing 0.15-0.2 atm. of C_2H_2 and 0.01 atm. O_2. The C_2H_4 produced was recovered by injecting reaction mixtures into evacuated, capped tubes and collecting the gas by rapid displacement. This was facilitated by the high solubility of C_2H_2 which enabled recovery of more than 1 ml of dissolved gas from 4.5 ml of reaction mixture. In some experiments oxymyoglobin replaced oxyleghaemoglobin as the source of O_2.

The first series of experiments [12] showed that nitrogenase activity was greater during maximum rates of deoxygenation of oxyleghaemoglobin than during deoxygenation of myoglobin and that only low rates were obtained in the absence of O_2-carrying proteins or at higher free O_2 concentrations when they were present. In the second series of experiments [13] it was found that ATP production in the bacteroids was optimal at about 0.05 µM free O_2 where leghaemoglobin is about ½ oxygenated. In the absence of oxyleghaemoglobin, bacteroid respiration at this low concentration of free O_2 is very low and appeared to be limited by O_2 diffusion. The experiments also showed the presence of two distinct types of respiration in the bacteroids. At low concentrations of free dissolved O_2 where ATP production was efficient and nitrogenase activity high, respiration was sensitive to inhibition by CO and by the P-450 inhibitor, N-phenylimidazole. Near 1 µM O_2 and above, where ATP production was less efficient and nitrogenase activity was low, bacteroid respiration was only slightly affected by these inhibitors. These results, which have been only briefly outlined here, led to the following conclusions [9]:

1. In soybean nodule bacteroids, formation of ATP from ADP is more efficient in a respiratory pathway(s) whose optimum activity occurs when the concentration of free dissolved O_2 in the surrounding solution is between 0.01 and 0.1 µM, than in pathways which operate at higher concentrations of O_2. Consequently, nitrogenase activity is greatest between 0.01 and 0.1 µM free O_2.
2. In the absence of leghaemoglobin, bacteroid respiration is very restricted because of limited diffusion of free O_2 below 0.1 µM.
3. Addition of leghaemoglobin to such diffusion limited bacteroids removes the restriction, because of macromolecular carrier-facilitated diffusion of O_2. Higher ATP concentrations in the bacteroids are attained and nitrogenase activity consequently increases.
4. The equilibrium constant for oxygenation ensures that bacteroids suspended in a solution of partially oxygenated leghaemoglobin are exposed to free O_2 concentrations suitable for optimum production of ATP and nitrogenase activity.
5. In nodule tissue, leghaemoglobin concentration is of the order of 10^{-3}M while free O_2 is of the order of 10^{-8}M. Partially oxygenated leghaemoglobin therefore 'buffers' the bacteroids against fluctuations in free O_2 concentration, which may occur due to external factors or to fluctuations in the supply of energy-yielding substrates.

COMPONENTS OF BACTEROID RESPIRATION - RELATIONSHIPS WITH NITROGENASE

In early experiments with soybean bacteroids, respiration and N_2-fixation were usually greater with succinate or fumarate as substrate than with malate, pyruvate or endogenous substrates. Sugars like glucose were also poor energy sources [8,11,32]. Consequently, all of the experiments described so far in this paper, employed succinate as an exogenous substrate. The results of the no gas phase experiments prompted closer examination of the properties of bacteroid respiration at low concentrations of free, dissolved O_2. Full details of this work will be published elsewhere. The main findings were as follows.

For simplicity it is assumed that respiration due to exogenous substrates is additive to respiration due to unknown endogenous substrates. Two series of experiments were conducted. The first used an amplifier and O_2 electrode to study respiration of bacteroids, (R. japonicum strain CB1809) from Lincoln soybean nodules, in dilute suspensions, with no added Lb. The interactions between respiration rates, concentration of free O_2, exogenous substrates and carbon monoxide, azide and cyanide as inhibitors were studied. Cyanide (20 μM) inhibited all respiration irrespective of O_2 concentration, but CO (1 mM) and azide (1 mM) were more effective inhibitors at low O_2 concentration. Endogenous respiration was mostly insensitive to CO and azide but a component of endogenous respiration, which was more sensitive to these inhibitors, was recognized below 0.5 μMO_2. Respiration due to added succinate and fumarate was completely insensitive to CO and azide above 5 μMO_2 but was very sensitive near 0.5 μMO_2. These experiments were complicated by the linear relationship between O_2 concentration and respiration rate which developed below about 0.3 μMO_2 (cf. ref. 13). With 1 mM CO as inhibitor, there was no effect of added substrate. With no inhibitor present the linear relationship of respiration rate covered a smaller range of O_2 concentration (Fig. 4). These results may indicate the presence of a CO-sensitive O_2-carrier in the bacteroids. With oxyleghaemoglobin present, this carrier could be more fully loaded with O_2 so that its diffusive flux no longer limited respiratory rate.

In the second series of experiments oxyleghaemoglobin was used both as a source of O_2 and to allow study of O_2 consumption rates by measuring rates of change of oxygenation spectrophotometrically. When nitrogenase activities were to be measured also, we used a system in which a reservoir of reaction mixture (bacteroids suspended in a stirred, pre-equilibrated solution containing leghaemoglobin and energy source) was connected through a spectrophotometer flow-cell and a pump, to 5 ml syringes containing 0.5 ml of 10% trichloracetic acid. Deoxygenation of oxyleghaemoglobin was recorded continuously and at suitable intervals 4.5 ml samples of reaction mixture were pumped into a syringe to terminate the reaction. Dissolved C_2H_4 in these samples was measured as before [12]. These experiments are not yet complete, but they have revealed differences between carbon sources in the efficiency with which they support nitrogenase activity at low concentrations of free O_2 (Table 2). Thus, while succinate doubled O_2 consumption at 0.01-0.1 μM free O_2, nitrogenase activity was not stimulated to the same extent. It remains to be determined whether this is due to more efficient ATP formation by respiratory systems fed by endogenous electron donors or by substrates like gluconate, or whether ATP production with succinate is in excess of that required for optimum nitrogenase activity.

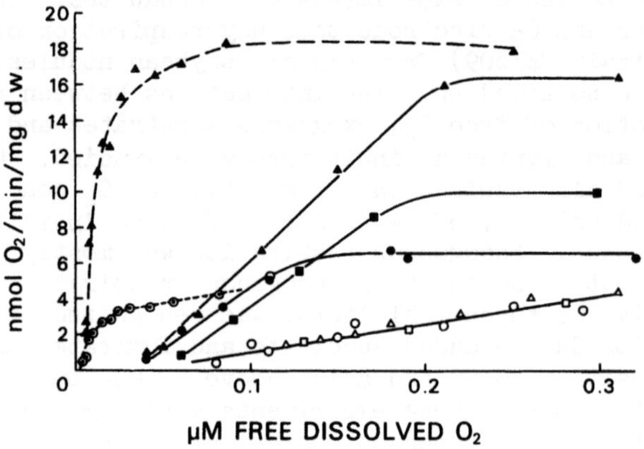

Fig. 4. Effects of substrates and carbon monoxide on respiration of soybean bacteroids at low concentrations of free dissolved O_2; results from experiments with no gas phase. Closed symbols no inhibitor, open symbols 1 mM CO. Succinate (10 mM, ▲,△,◬); gluconate (10 mM,■, □); endogenous respiration (●,○,◉). Data from assays containing 0.1 mM leghaemoglobin are shown for comparison (◬,◉). The linear relationships between respiration rates and O_2 concentration in the bulk of the solution, indicate that respiration without leghaemoglobin is limited by diffusion of O_2. Carbon monoxide accentuates this limitation and abolishes substrate effects. Leghaemoglobin abolishes the diffusion limitation, allowing respiration to proceed at much lower concentrations of O_2 in the bulk of the solution.

In early experiments with a gas phase the large stimulation of nitrogenase activity by succinate was probably connected with the generation of favourable concentrations of O_2 within the shaken suspension, rather than being due to a more direct effect on nitrogenase activity.

The results of these experiments emphasize the importance of understanding the nature of the carbon substrates supplied in vivo to bacteroids within nodule cells, especially when considering

TABLE 2

Effects of carbon source upon respiration and nitrogenase activity of soybean bacteroids during deoxygenation of oxyleghaemoglobin.

Substrate	free O_2* (μM)	Y*	respiration (n mol O_2/min/mg d.w)**	nitrogenase (n mol C_2H_4/min/mg d.w)	O_2/C_2H_4
10 mM succinate	0.70	0.95	19.2	2.2	8.8
	0.11	0.75	27.5	5.2	5.3
	0.029	0.44	23.1	5.3	4.4
	0.014	0.27	17.2	5.5	3.1
	0.007	0.16	8.9	1.4	6.2
10 mM gluconate	1.24	0.97	6.0	0.8	7.1
	0.14	0.79	13.4	3.3	4.0
	0.021	0.36	12.0	3.7	3.3
	0.010	0.21	10.7	5.2	2.1
	0.005	0.12	5.1	0	-
endogenous	0.78	0.95	5.0	0.6	7.7
	0.16	0.81	13.0	2.8	4.7
	0.024	0.39	12.3	3.6	3.5
	0.009	0.19	10.4	2.7	3.9
	0.004	0.10	5.2	2.1	2.5

*The average concentration of free O_2 and average oxygenation of leghaemoglobin (Y) prevailing during each measurement of nitrogenase activity.

**Mean of 4-7 measurements of O_2 consumption rates made during each measurement of nitrogenase activity.

the energetic efficiency of N_2-fixation. It is of interest to observe that the most efficient nitrogenase activity (in terms of O_2 consumption) occurs in the range of leghaemoglobin oxygenation prevailing in soybean nodules in air (Table 2; ref. 2).

The assignment of the various respiratory pigments which have been identified in soybean bacteroids [3] as components of electron transport systems of varying phosphorylating efficiency and O_2-affinity will be a challenge for future research. So also will be the identification of the mechanisms which cause electron flow to switch from one electron transport system to another.

Acknowledgements:

G.L. Turner has been a valued collaborator throughout and in the recent, no gas phase experiments in particular. Since 1973, C.A. Appleby and D.J. Goodchild of this Division and Professors B.A. and J.B. Wittenberg, Albert Einstein College of Medicine,

New York have collaborated with us in a way which has made the work truly interdisciplinary and international. We also thank J. Robertson and D.P. Verma for making results of experiments available to us in advance of publication.

The provision of funds for travel to this symposium, by CNPq and EMBRAPA is also gratefully acknowledged.

REFERENCES

1. Allison, F.E., Ludwig, C.A., Hoover, S.R., and Minor, F.W. (1940) Bot. Gaz. 101, 513.
2. Appleby, C.A. (1969a) Biochim. Biophys. Acta 172, 88.
3. Appleby, C.A. (1969b) Biochim. Biophys. Acta 172, 71.
4. Appleby, C.A. (1974) Leghemoglobin. In 'The Biology of Nitrogen Fixation'. p. 521 (Quispel, A. ed) North Holland, Amsterdam.
5. Appleby, C.A., Turner, G.L., and Macnicol, P.K. (1975) Biochim. Biophys. Acta 387, 461.
6. Appleby, C.A., Blumberg, W.E., Peisach, J., Wittenberg, B.A., and Wittenberg, J.B. (1976) J. Biol. Chem. 251, 6090.
7. Appleby, C.A., Bergersen, F.J., Macnicol, P.K., Turner, G.L., Wittenberg, B.A. and Wittenberg, J.B. (1976) The role of leghemoglobin in symbiotic N_2 fixation. Proc. 1st Int. Symp. Nitrogen Fixation Vol. 1. p.274. (Newton, W.E. and Nyman, C.J. Eds). Washington State Univ. Press.
8. Bergersen, F.J. (1958) J. gen. Microbiol. 19, 312.
9. Bergersen, F.J. (1977) Legume symbiosis. In 'Interactions Between Soil Microorganisms and Plants'. (Dommergues, Y. and Krupa, S.V., Eds) Elsevier, Amsterdam (in press).
10. Bergersen, F.J., and Goodchild, D.J. (1973) Aust. J. Biol. 26, 741.
11. Bergersen, F.J., and Turner, G.L. (1967) Biochim. Biophys. Acta 141, 507.
12. Bergersen, F.J., and Turner, G.L. (1975a) J. gen. Microbiol. 89, 31.
13. Bergersen, F.J., and Turner, G.L. (1975b) J. gen. Microbiol. 91, 345.
14. Bergersen, F.J., Turner, G.L., and Appleby, C.A. (1973) Biochim. Biophys. Acta 292, 271.
15. Broughton, W.J., and Dilworth, M.J. (1971) Biochem. J. 125, 1075.
16. Cutting, J.A., and Schulman, H.M. (1969) Biochim. Biophys. Acta 192, 486.
17. Dart, P.J. (1968) Proc. 4th Eur. Reg. Conf. Electron Microscopy, Rome, p. 69.
18. Dart, P.J., and Chandler, M. (1971) Ann. Rep. Rothamsted Exp. Sta. Part 1, p. 99.

19. Dilworth, M.J. (1969) Biochim. Biophys. Acta 184, 432.
20. Dilworth, M.J., and Kidby, D.J. (1968) Exptl. Cell Res. 49, 148.
21. Ellfolk, N. (1972). Endeavour 31, 139.
22. Fuchsman, W.H., Barton, C.R., Stein, M.M., Thompson, J.T., and Willett, R.M. (1976) Biochem. Biophys. Res. Comm. 68, 387.
23. Godfrey, C.A., and Dilworth, M.J. (1971) J. gen. Microbiol. 69, 385.
24. Gourret, J-P., and Fernandez-Arias, H. (1974) Canad. J. Microbiol. 20, 1169.
25. Imamura, T., Riggs, A., and Gibson, Q.H. (1972) J. Biol. Chem. 247, 521.
26. Keilen, D., and Smith, J.D. (1947) Nature 159, 692.
27. Keilin, D., and Wang, Y.L. (1945) Nature 155, 227.
28. Kubo, H. (1939) Acta Phytochim. 11, 195.
29. Lehtovaara, P., and Ellfolk, N. (1974). FEBS Letters 43, 239.
30. Nash, D.T., and Schulman, H.M. (1976a) Canad. J. Bot. 54, 2790.
31. Nash, D.T., and Schulman, H.M. (1976b) Biochem. Biophys. Res. Comm. 68, 781.
32. Rigaud, J., Bergersen, F.J., Turner, G.L., and Daniel, R.M. (1973) J. gen. Microbiol. 77, 137.
33. Richardson, M., Dilworth, M.J., and Scawen, M.D. (1975) FEBS Letters, 51, 33.
34. Smith, J.D. (1949) Biochem. J. 44, 591.
35. Stokes, A.N. (1975) J. Theor. Biol. 52, 285.
36. Tjepkema, J.D., and Yocum, C.S. (1973) Planta (Berl.) 115, 59.
37. Truchet, G. (1972) Comptes Rend. Acad. Sci. Paris, 274, 1290.
38. Vainshtein, B.K., Harutyunyan, E.H., Kuranova, I.P., Borisov, V.V., Sosfenov, N.I., Pavlovsky, A.G., Grebenko, A.I., Kondreva, N.V. (1975) Nature 254, 163.
39. Verma, D.P.S., and Bal, A.K. (1976) Proc. Natl. Acad. Sci. USA. 73, 3843.
40. Verma, D.P.S., Nash, D.T., and Schulman, H.M. (1974) Nature 251, 74.
41. Virtanen, A.I. (1947) Biol. Rev. 22, 239.
42. Wittenberg, J.B. (1970) Physiol. Rev. 50, 559.
43. Wittenberg, J.B. (1976) Facilitation of oxygen diffusion by intracellular leghaemoglobin and myoglobin. In 'Oxygen and Physiological Function'. Professional Information Library, Dallas, Texas (in press).
44. Wittenberg, J.B., Appleby, C.A., and Wittenberg, B.A. (1972) J. Biol. Chem. 247, 521.
45. Wittenberg, J.B., Bergersen, F.J., Appleby, C.A., and Turner, G.L. (1974) J. Biol. Chem. 249, 4057.
46. Wyman, J. (1966) J. Biol. Chem. 241, 115.

NITROGEN FIXATION BY *RHIZOBIUM* SPP. IN LABORATORY CULTURE MEDIA

F.J. Bergersen and A.H. Gibson

Division of Plant Industry, CSIRO, P.O. Box 1600

Canberra, 2601, Australia

SUMMARY

Nitrogenase activity has been detected in agar cultures of 37 out of 80 strains of cowpea rhizobia and *R. japonicum*. No activity has been detected in similar cultures of 20 strains of *R. meliloti*, *R. leguminosarum*, *R. trifoli* and *R. lupini*, in our laboratory. In only a few strains are nitrogenase activities sufficient for reliable physiological studies. Results indicate that a zone of restricted O_2 concentration develops within the colonial mass on agar cultures. Nitrogenase activity within this zone is influenced by gaseous O_2 concentration, carbon substrates and source of combined-N. Published results from other laboratories reporting studies with agar cultures and batch liquid cultures are outlined.

In our laboratory, continuous cultures of two strains of cowpea rhizobia have been used to define some physiological parameters governing nitrogenase activity. Oxygen-limited cultures have nitrogenase activities similar to those of bacteroids of the same strain, isolated anaerobically from cowpea nodules and assayed in the same way. However, these activities are insufficient to sustain exponential growth without additional combined-N. Fixed nitrogen is initially in equilibrium with NH_4^+ in the medium, from where it is assimilated. Relationships between pathways of NH_4^+ assimilation, O_2 availability and control of nitrogenase, are described for various culture steady states and during transitions between them. These results are considered in relation to known properties of N_2-fixation in nodule cells.

For many years researchers failed in attempts to demonstrate nitrogen fixation in cultures of the legume root nodule bacteria (*Rhizobium* spp.). This restricted physiological and genetic studies and it was believed that the host plant played an indispensible part in the induction of nitrogenase activity of the bacteroids in nodule tissue cells [10]. Indeed, recent work has shown that plant cells produce heat stable, diffusible factors which stimulate the development of nitrogenase activity in some cultured rhizobia [24, 32]. However, these factors are not indispensible. In 1975, five laboratories independently detected nitrogenase activity in cultures of the cowpea-type *Rhizobium* sp., strain 32H1; other cowpea strains, strains of *R. japonicum* and one strain of *R. leguminosarum* also produced detectable nitrogenase in cultures. These reports were reviewed by Gibson *et al.* [14].

These results with cultures of various rhizobia have opened the way for many physiological and genetic studies of nitrogen fixation by rhizobia and permit comparisons with the bacteroid forms of these bacteria. Defined cultural conditions will undoubtedly shed light on some crucial aspects of the induction and functioning of nitrogenase in nodules. In this paper, some recent progress will be outlined and implications for some of the factors involved in the *in vivo* expression of nitrogenase activity in bacteroids, will be explained.

NITROGENASE IN AGAR AND LIQUID CULTURES

Cultural Conditions

As initially reported, nitrogenase was detected in cultures of 32H1 on a variety of culture media, and with a variety of growth conditions. Three reports concerned activity on agar media in air. Two concerned liquid cultures grown under reduced pO_2.

Carbon sources. Carbon sources varied, but best activity on agar was obtained with arabinose plus succinate. Subsequent investigations showed that a range of hexoses and pentoses, mannitol and glycerol were suitable major carbon sources, whilst with succinate, 2-oxoglutarate, malate, fumarate or pyruvate as supplements, all supported high activity [13]. In liquid media, gluconate and mannitol [15] or malate [28] were used.

Nitrogen sources. All five initial reports about 32H1 described media containing a source of combined nitrogen. Although small amounts of combined nitrogen are sometimes included in media for batch culture of N_2-fixing bacteria, there is a wealth of literature showing that many forms of combined nitrogen, when supplied at reasonable concentrations, usually inhibit nitrogen fixation. Within nodule cells, ammonia, amides and amino acids are always present, and it seems reasonable that rhizobia should be able to produce nitrogenase in the presence of these sources of combined nitrogen. In agar cultures grown in air, nitrogenase was detected only after a mass of growth developed [13]; consequently, sufficient combined nitrogen must be provided to produce the colonial mass. For such cultures, glutamine, urea $(NH_4)_2SO_4$, asparagine, casamino acids and KNO_3 have been used successfully [13].

Cultures initially grown on millipore discs placed on agar medium containing combined nitrogen, and then transferred to the surface of nitrogen-free medium, retained their nitrogenase activity, and the dry weight and protein content of the cells increased slowly for 6 weeks [14].

In liquid cultures, nitrogenase activity is produced after exponential growth ceases, and there is some evidence for specific effects of nitrogen source in relation to induction of nitrogenase and assimilation of the products of N_2-fixation [21,30].

Concentration of O_2. Undoubtedly a major factor influencing development of nitrogenase activity in cultured rhizobia is the need for a low concentration of O_2 near the cells [15,28]. On agar, mechanical disturbance of the culture, altered O_2 diffusion rates following temperature changes or altered O_2 demand following changed nutritional conditions, all affect nitrogenase activity [14]. We conclude that nitrogenase activity, in agar cultures, is confined to a zone, within the colonial mass, where the concentration and flux of O_2 are appropriate for induction and maintenance of activity. As the culture develops it is likely that the zone of suitable conditions changes position within the colony. Polysaccharide composition and cell density within the colony are likely to be important factors determining whether a strain of rhizobia will develop nitrogenase activity in agar cultures.

Applications

Cultures of rhizobia in shaken or static liquid culture have been used for several physiological studies of the induction of nitrogenase [15,28] and regulation of nitrogenase synthesis [18,21, 30]. Sometimes conflicting results have been reported (e.g. Evans and Keister[12], Keister and Evans [16], obtained differing effects of NH_4^+ in stagnant and shaken cultures). Agar cultures have also been used for studies of nitrogenase regulation [25].

Strains Other Than 32H1

We have screened 100 strains of rhizobia from 6 species, for nitrogenase activity on CS7 agar [22]. Only one strain, CB756 (cowpea type) showed activity as high as 32H1. These two strains are indistinguishable serologically, and both show no growth response to vitamins (unlike most of the cowpea strains tested). However their colony appearances are different and CB756 produces nitrogenase activity on yeast extract - mannitol agar, while 32H1 does not.

Sixty four cowpea strains were tested. Only 27 produced nitrogenase activity on CS7 agar (Table 1). A greater proportion of R. japonicum strains were active but all had relatively low activities. Keister [15] also reported that some R. japonicum strains produced no nitrogenase activity in conditions where 32H1 would have been active.

TABLE 1

Nitrogenase activity of 100 strains of six species of Rhizobium cultured on CS7 agar [22]. Strains are grouped according to the highest activity recorded between 3 and 9 days after inoculation.

	No. of strains tested	No. of strains in classes of nitrogenase activity (nmol C_2H_4/h/culture): (<0.1) (not detected)	0.1-1.0	1.0-10	>10.0
Rhizobium sp. (cowpea strains)	64	37*	11	11	5
R. japonicum	16	6*	6	4	0
R. trifolii	5	5	0	0	0
R. meliloti	5	5	0	0	0
R. leguminosarum	6	6	0	0	0
R. lupini	4	4	0	0	0

*One of the cowpea strains and 2 R. japonicum strains were active on yeast extract-mannitol agar, although negative on CS7 agar.

From a consideration of data from several experiments with cowpea and R. japonicum strains, it is obvious that they differ greatly. Some produce nitrogenase activity when grown on yeast extract-mannitol agar, but not on CS7 agar, although growth on both media is similar. Others produce activity only on CS7 agar. Activity also occurs at various intervals after inoculation and varies in the time for which it persists.

So far, there is only one report of a 'fast-growing' strain producing nitrogenase activity in culture (R. leguminosarum strain TA101; ref. 17). In our laboratory, all attempts have failed to induce nitrogenase activity in this strain and in other strains of

R. leguminosarum, R. meliloti, R. trifolii and R. lupini. Some fast-growing cowpea strains have been tested, and these were negative also. Variations to the procedure successfully used for 32H1, have included growth at lower temperatures, use of low gas-phase O_2 concentrations (0.01-0.03 atm), the use of semi-solid media (as successfully used for Spirillum lipoferum, ref. 11) and the transfer of cultures, grown on millipore discs on yeast extract-mannitol media, to media lacking combined N, or containing various tricarboxylic acid cycle intermediates. All failed with fast-growing rhizobia. There is a daunting number of combinations of conditions which could be tried and it is very possible that appropriate conditions have not been tested yet. However, it is also possible that the production of nitrogenase activity in culture will be found to be the exception, rather than the rule.

NITROGENASE IN CONTINUOUS CULTURES OF RHIZOBIA

The agar culture system proved difficult for the interpretation of physiological experiments, especially in relation to control of nitrogenase [25]. The more precise experiments with other nitrogen-fixing bacteria grown in continuous culture (e.g. ref. 31) indicated that similar methods might resolve some puzzling features of results obtained with agar cultures of 32H1 and other rhizobia.

The Chemostat

The continuous culture apparatus (chemostat) used for this work was similar to that described by Baker [1]. It was fitted with an immersed teflon coil, through which air or O_2 could be passed to increase rates of solution of O_2 in the culture. Stirring rates used did not lead to entrapment of gas bubbles. An immersed O_2 electrode, connected to a d.c. amplifier measured the concentration of dissolved O_2 and in later experiments, this system was connected to a feed-back controller, which regulated the rate at which the culture was stirred, thus controlling rates of solution of O_2. Control of pH by the automatic addition of NaOH and H_2SO_4 was used in some experiments.

N-limited Cultures

Batch and continuous cultures of many nitrogen-fixing bacteria produce active nitrogenase when combined-N in the medium becomes growth-limiting (reviewed in ref. 20). In the first series of experiments, we therefore used glutamine-limited cultures of the cowpea rhizobia, strain 32H1, grown in a simple medium based on that used for the agar cultures. In most experiments, the medium contained 43 mM glycerol. 10-25 mM Na-succinate, 1-2 mM glutamine, 30 mM K-phosphate, pH 6.25, and salts. We also developed a convenient assay which used 1.75 ml samples of culture, withdrawn from the chemostat, incubated with 0.25 ml of 2 mM oxyleghaemoglobin as a source of O_2 at low free, dissolved concentrations. It was difficult to produce these concentrations reliably using gas-phase O_2, for cultures with variable cell concentrations. Some summarized results obtained with these cultures and assays, are given below.

Low O_2 and succinate. Glutamine-limited continuous cultures produced nitrogenase (60-100 nmol C_2H_4/h/mg d.w.) when the dissolved O_2 in the culture was near 1 µM or less [6,7]. When succinate was omitted from the medium, these low concentrations of dissolved O_2 could not be maintained and nitrogenase activity disappeared rapidly. Restoration of the succinate supply led to a fall in dissolved O_2 from 24 µM to 0.6 µM within 40 min, and nitrogenase was restored over a period of about 40 h. (Note: in Canberra, elevation 600 m, air-saturated H_2O contains 224 µMO_2 at 30°C).

Repression by NH_4^+. When ammonium sulphate was added to the culture and to the inflowing medium, to a concentration of 5 mM NH_4^+, nitrogenase activities of glutamine-limited cultures declined rapidly from initial values of 60-80 nmol C_2H_4/h/mg d.w., reaching equilibrium activities of 20-30 nmol C_2H_4/h/mg d.w. after 5-8 h. [7]. When the NH_4^+ supply was stopped, nitrogenase activity was restored during 40 h following the disappearance of residual NH_4^+ in the culture. When NH_4^+ was the sole source of combined-N, and was the limiting nutrient, strains 32H1 and CB756 had nitrogenase activities of 70-100 nmol C_2H_4/h/mg d.w.

O_2-limited Cultures

When rates of solution of O_2 were reduced by slower stirring and/or reduced pO_2 in the gas phase, cultures supplied with 2 mM glutamine became O_2-limited and residual glutamine in the culture rose to about 1 mM. Nitrogenase activities in such cultures were usually much greater than in N-limited cultures, commonly reaching 300 and occasionally 400 nmol C_2H_4/h/mg d.w. This level of activity is similar to the activity of bacteroids of the same strain isolated anaerobically from cowpea nodules and assayed in the same way [3].

Need for combined N. Although these rates of nitrogenase activity were relatively high, they did not seem to be great enough to sustain even the relatively slow exponential growth of 32H1. This was confirmed by analysis of N in inflowing medium, effluent culture, centrifuged bacteria and culture supernatant (Table 2). The total increment of fixed N_2 was only about 25% of the cell N. Further, when the combined N was withdrawn from the continuous culture, nitrogenase activity did not increase as would be expected for continuous cultures of other N_2-fixing bacteria. Instead, it fell to the lower levels characteristic of N-limited cultures. The bacterial growth-rate then declined and the culture began to wash out. It was concluded that these bacteria could only supplement combined-N with fixed N_2; they could not grow on N_2 alone [3]. Similar results were obtained with $(NH_4)_2SO_4$ as a source of combined N for strain CB756.

TABLE 2

Nitrogen analysis of an O_2-limited culture of 32H1 with nitrogenase activity of 201 nmol C_2H_4/h/mg d.w. (adapted from ref. 3).

Fraction analysed	Total N (μg/ml)
medium in*	50.1
culture out	56.9
N_2 fixed	6.8
culture supernatant	30.8
washed cells	27.2
(N_2 fixed/ml washed cells	0.25)

*The medium contained 1.8 mM glutamine and was supplied at 32.7 ml/h. The effluent culture contained 0.23 mg d.w./ml of washed cells.

Assimilation of fixed N_2. O'Gara and Shanmugan [21] reported that N_2 fixed by batch cultures of 32H1 and *R. japonicum* strain CB1809, was exported into the medium. The continuous cultures used in our work were much more active, but samples from them used for $^{15}N_2$ experiments, showed that newly-fixed N_2 was apparently in equilibrium with NH_4^+ in the medium. In Table 3 it can be seen that cells from cultures containing residual NH_4^+ contained only about 1/10 of the ^{15}N enrichment obtained in similar experiments with samples of culture supplied with glutamine. Both cultures had similar nitrogenase activities. The enrichment in the cells reflected the enrichment of NH_4^+ in the culture medium, showing that the fixed N_2 was assimilated from the medium, and not directly from the site of fixation within the cells. Thus, these cultures of rhizobia resemble bacteroids from soybean nodules, whose fixed N also appears as NH_4^+ in the surrounding medium [4].

The glutamine synthetase/glutamate synthase pathway seems to be the major route of assimilation of NH_4^+ in rhizobia in NH_4^+-limited continuous cultures. In glucose-limited cultures with an excess of inorganic-N in the medium, NH_4^+ was assimilated via glutamate dehydrogenase [9]. In our O_2-limited cultures, with active nitrogenase, glutamine synthetase and glutamate synthase were present with similar activities, which were adequate for the measured rates of nitrogen assimilation. Glutamate dehydrogenase appeared in the presence of excess NH_4^+ only when the O_2-limitation was relieved and nitrogenase activity was low or absent (Table 4).

TABLE 3

Assimilation of fixed $^{15}N_2$ by samples from two O_2-limited continuous cultures of CB756, each with similar nitrogenase activity. (Adapted from ref. 6).

Component analysed	^{15}N enrichment (atom % excess)*	N_2-fixed (nmol NH_4^+/h)*
Culture A**		
total N:cells	0.083	1.44
NH_4^+ : medium	1.651	13.90
total N_2 fixed		15.34
Culture B**		
total N : cells	0.008	0.095
NH_4^+ : medium	0.160	14.555
total N_2 fixed		14.65

*Means of duplicate analyses, each combined from four standard assays with oxyleghaemoglobin.

**Culture A was supplied with 2 mM glutamine and each assay contained 0.28 mg d.w. of bacteria.

Culture B was supplied with 10 mM NH_4^+ and assays contained 0.23 mg d.w. of bacteria.

Relationships between glutamine synthetase and nitrogenase. When 5 mM NH_4^+ was added to O_2-limited cultures of CB756 supplied with 2 mM glutamine, nitrogenase activity was repressed only slightly or not at all, whereas in glutamine-limited cultures, nitrogenase was repressed by NH_4^+. Continuous O_2-limited cultures, supplied with up to 34 mM NH_4^+ as sole source of combined N, had nitrogenase activities almost as high as those of cultures whose source of combined-N was 2 mM glutamine [5]. Similarly Scowcroft et al. [25] failed to obtain repression of nitrogenase by added NH_4^+ in agar cultures of 32H1 which were O_2-limited. In continuous cultures of CB756, the failure to repress nitrogenase in the presence of excess NH_4^+ has been found to be due to the O_2-limitation imposed. To explain this, it is necessary to outline some of the theories about the control of nitrogenase synthesis.

In Klebsiella pneumoniae, glutamine synthetase appears to act as a positive control element for nitrogenase synthesis [29]. Recent work with 32H1 has shown that glutamine synthetase-deficient mutants produced no nitrogenase in culture and nodules were ineffective [18]; these results are similar to results with

TABLE 4

Effects of O_2 availability on CB756 in O_2-limited continuous cultures supplied with 10 mM NH_4^+.

For culture A effluent culture was collected in ice and the cells centrifuged and stored in liq.N_2 before cell-free extracts were prepared for assay. Nitrogenase was measured in a sample taken during effluent collection. For culture B, a 70 ml sample was collected from the culture, a portion assayed for nitrogenase and cell-free extracts prepared from cells centrifuged from the remainder immediately. For details of the assays see Bergersen and Turner (5,6). The cell yield of the cultures is given as a measure of O_2 availability.

Cell yield (mg/h)	Nitrogenase (nmol C_2H_4/ h/mg d.w.)	Specific activity (nmol/min/mg protein)					
		glutamine synthetase				glutamate synthase	glutamate dehydro-genase
		γGT* (Mn^{2+})	γGT* (Mn^{2+} + Mg^{2+})	(RA)*	GS*		
A.							
3.5	156	622	449	(1.4)	13.9	-	-
6.9	3	442	204	(2.2)	6.0	-	-
B.							
3.3	154	619	374	(1.7)	-	16.6	< 0.2
7.9	0	577	112	(5.2)	-	15.0	1.8

*γGT = Reverse γ-glutamyl transferase assays with 0.3 mM $MnCl_2$ or $MnCl_2$ + 60 mM $MgCl_2$ [27]. The ratio of these assays is assumed to measure the relative adenylylation (RA) of the glutamine synthetase.

GS = assays for the forward, biosynthetic activity of glutamine synthetase (activities in GS and γGT (Mn^{2+} + Mg^{2+}) assays are strongly correlated; ref. 6).

K. pneumoniae. In *E. coli*, control of glutamine synthetase activity and synthesis involves addition (adenylylation) or removal (deadenylylation) of adenosine-5-monophosphate moieties to subunits of the enzyme [27]. There is evidence that this system of control is widespread amongst gram-negative bacteria including N_2-fixing species although there are differences in detail [2,26]. A similar system seems to operate in rhizobia [8].

In our O_2-limited continuous cultures, supplied with excess NH_4^+ (10 mM), the biosynthetic activities of cell-free extracts were positively correlated with nitrogenase activities of the cultures from which they were prepared. With controlled increase of O_2 supply and little change in steady-state concentration of dissolved O_2, cell-yield of the cultures increased, nitrogenase activity decreased and apparent adenylylation of glutamine synthetase increased (Table 4; refs. 5,6). These results were obtained from steady-state measurements. It was also possible to follow changes during transitions between steady-states using the dissolved O_2 feed-back stirrer controller to maintain O_2 availability at pre-set levels. During de-repression, following decreased O_2-supply, a decline in apparent adenylylation of glutamine synthetase preceeded increased nitrogenase activity.

From these results we conclude that repression of nitrogenase in the presence of excess NH_4^+ involves adenylylation of glutamine synthetase. Under conditions of severe O_2-limitation, 10 mM NH_4^+ being supplied, adenylylation is inhibited, probably because the ATP required for this reaction is in short supply. When the supply of O_2 increases, cell-yield increases as a result of increased ATP availability and adenylylation of glutamine synthetase increases also. Repression of nitrogenase follows. When O_2-limited cultures are grown with 1-2 mM glutamine as sole source of combined-N, changes in O_2 supply such as those described, produce only small changes in nitrogenase activity. (Details of these experiments are presented elsewhere; refs. 5,6).

Comparison with conditions in nodules. There are significant analogies between the results from O_2-limited continuous cultures and the properties of bacteroids. For the latter, some data come from soybean nodules, but cowpea nodules are essentially similar. Bacteroids and cultures have similar nitrogenase activities; both are produced in O_2-limited systems (see paper by Bergersen - this symposium). Both types of cells have best nitrogenase activities at the low free O_2-concentrations which prevail during deoxygenation of oxyleghaemoglobin. Newly-fixed N_2 appears in the surrounding medium as NH_4^+, being assimilated from there by host-cell glutamine synthetase in nodules [19] and by bacterial glutamine synthetase in cultures. In both it seems to be an advantage that the O_2-limited conditions tend to prevent adenylylation of glutamine synthetase in the bacteria, thus preventing repression of nitrogenase in the presence of products of N_2-fixation.

Acknowledgements:

J.J. Child collaborated in the examination of strains for nitrogenase activity and G.L. Turner collaborated in the continuous culture experiments. Technical assistance was provided by Marilyn Ashby and Pat Atkinson. The provision of funds by CNPq and EMBRAPA for travel of FJB to this symposium is gratefully acknowledged.

REFERENCES

1. Baker, K. (1968) Lab. Pract. 17, 817.
2. Bender, R.A., Janssen, K.A., Resnick, A.D., Blumenberg, M., Foor, F., and Magasanik, B. (1977) J. Bacteriol. 129, 1001.
3. Bergersen, F.J. (1977) Nitrogenase in chemostat cultures of rhizobia. In 'Proc II. Int. Symp. Nitrogen Fix.' Salamanca. (Postgate, J., Newton, W.E., and Rodriguez Barrueco, C. Eds.) Academic Press, New York (in press).
4. Bergersen, F.J., and Turner, G.L. (1967) Biochem. Biophys. Acta 141, 507.
5. Bergersen, F.J., and Turner, G.L. (1976) Biochem. Biophys. Res. Comm. 73, 524.
6. Bergersen, F.J., and Turner, G.L. (1977) Biochem. Biophys. Acta (in press).
7. Bergersen, F.J., Turner, G.L., Gibson, A.H., and Dudman, W.F. (1976) Biochem. Biophys. Acta 444, 164.
8. Bishop, P.E., Guevara, J.G., Engelke, J.A., and Evans, H.J. (1976) Pl. Physiol. 57, 542.
9. Brown, C.M., and Dilworth, M.J. (1975) J. gen. Microbiol. 86, 39.
10. Dilworth, M.J., and Parker, C.A. (1969) J. Theor. Biol. 25, 208.
11. Döbereiner, J., and Day, J.M. (1976) Associative symbioses in tropical grasses: Characterisation of microorganisms and dinitrogen fixing sites. In 'Proc. I Int. Symp. Nitrogen Fix.' Pullman. p. 518. (Newton, W.E., and Nyman, C.J., Eds.) Washington State Univ. Press.
12. Evans, W.R., and Keister, D.L. (1976) Can. J. Microbiol. 22, 949.
13. Gibson, A.H., Scowcroft, W.R., Child, J.J., and Pagan, J.D. (1976) Arch. Microbiol. 108, 45.
14. Gibson, A.H., Scowcroft, W.R., and Pagan, J.D. (1977) Nitrogen fixation in plants: an expanding horizon? In 'Proc II Int. Symp. Nitrogen Fix.' Salamanca. (Postgate, J.R., Newton, W.E., and Rodriguez-Barrueco, C., Eds.). Academic Press. New York. (in press).
15. Keister, D.L. (1975) J. Bacteriol. 123, 1205.
16. Keister, D.L., and Evans, W.R. (1976) J. Bacteriol. 129, 149.
17. Kurz, W.G.W., and LaRue, T.A. (1975) Nature 256, 407.
18. Ludwig, R.A., and Signer, E.R. (1977) Nature 267, 245.
19. McParland, R.H., Guevara, J.G., Becker, R.R., and Evans, H.J. (1976) Biochem. J. 153, 597.
20. Mulder, E.G., and Brotonegoro, S. (1974) Free-living heterotrophic bacteria. In 'Biology of Nitrogen Fixation'. p. 38. (Quispel, A. Ed) North Holland, Amsterdam.
21. O'Gara, F., and Shanmugam, K.T. (1976) Biochim. Biophys. Acta 437, 313.
22. Pagan, J.D., Child, J.J., Scowcroft, W.R., and Gibson, A.H. (1975) Nature 256, 406.

23. Pagan, J.D., Scowcroft, W.R., Dudman, W.F., and Gibson, A.H. (1977) J. Bacteriol. 129, 718.
24. Reporter, M. (1976) Pl. Physiol. 57, 651.
25. Scowcroft, W.R., Gibson, A.H., and Pagan, J.D. (1976) Biochem. Biophys. Res. Comm. 73, 516.
26. Senior, P.J. (1975) J. Bacteriol. 123, 407.
27. Stadtman, E.R., Ginsberg, A., Ciardi, J.E., Yeh, J., Hernig, S., and Shapiro, B.M. (1970) Adv. Enz. Reg. 8, 99.
28. Tjepkema, J.D., and Evans, H.J. (1975) Biochem. Biophys. Res. Comm. 65, 625.
29. Tubb, R.S. (1974) Nature 251, 481.
30. Tubb, R.S. (1976) Appl. Env. Microbiol. 32, 483.
31. Tubb, R.S., and Postgate, J.R. (1973) J. gen. Microbiol. 79, 103.
32. Werner, D. (1976) Ber. Deutsch. Bot. Ges. 89, 563.

LIMITING FACTORS IN GRASS NITROGEN FIXATION

J. Balandreau (1), P. Ducerf (1), Ibtissam Hamad-Fares (2), Pierrette Weinhard (1), G. Rinaudo (3), C. Millier (4), Y. Dommergues (1,3).

(1) Centre de Pédologie C.N.R.S. B.P.5, 54500 Vandoeuvre, France
(2) Département de Botanique, Faculté des Sciences, Université de Damas, Syrie
(3) Laboratoire de Microbiologie des Sols, ORSTROM, B.P. 1386, Dakar, Sénégal
(4) Centre National de Recherches Forestières, Champenoux, 54280 Seichamps, France

INTRODUCTION

Demonstration of nitrogen fixation by bacteria associated with the roots of grasses was achieved only recently by PARKER (1957) about a Lolium rigidum cover, YANAGISAWA and TAGAHASHI (1964) then YOSHIDA (1968) in rice fields, DAY *et al.* (1973) in a temperate fallow at Rothamsted Experimental Station. These studies were conducted using the nitrogen budget method over one to fifty year periods owing to the low sensitivity of the method and the relatively low activity of grass rhizospheres.

The availability of the acetylene method gave the opportunity of a complete change of time scale making possible a measure of nitrogen fixation in a matter of hours or even minutes. Some problems did appear about the conditions of use of acetylene but most authors came to the conclusion that the less disturbing these conditions are, the better the results. It was then possible to study nitrogenase activity in the rhizosphere in different natural conditions and to investigate the effects of some factors on it. This activity was shown to vary largely and rapidly *in situ* owing to :

- Light level (BALANDREAU *et al.*, 1971; DÖBEREINER and DAY, 1974).

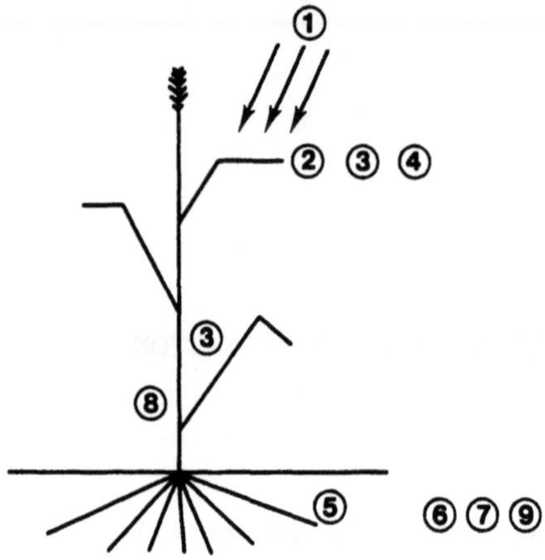

Fig. 1 - Main factors acting on Nitrogenase Activity in the rhizosphere of grasses; numbers refer to paragraphs of Part II of this paper:
1. Light intensity, 2. Potential of the phosynthetic apparatus, 3. Air temperature, 4. Air water deficit, 5. Soil temperature, 6. Soil redox and water contents, 7. Soil factors, 8. Plant genetic factors, 9. Bacteria.

- Plant age (RINAUDO et al., 1971), variety (HAMAD-FARES, 1976, Von BÜLOW and DÖBEREINER, 1975) or potential for photosynthesis (BALANDREAU et al., 1976).

- Air temperature and water deficit (BALANDREAU et al., 1976, HAMAD-FARES, 1976).

- Soil temperature (BALANDREAU et al., 1976), redox and closely related water content (HAUKE-PACEWICZOWA et al., 1970), soil inorganic N content (BALANDREAU et al., 1975) or soil type (RINAUDO et al. 1971).

- The nature and number of diazotrophs associated with grass roots (Table I) has not been very closely investigated but the number of inoculation trials (BURRIS, 1976) shows that authors regard it as an important factor which could limit N fixation.

These factors (Fig 1) are likely to act at different levels through photosynthesis (light, potential for photosynthesis, air temperature and water deficit), downwards translocation of assimilates (air temperature), exudation (soil water content) or directly on the N_2 fixers (soil temperature, soil inorganic N content, redox potential). We know that the plant can benefit from the fixed nitrogen (RUSHEL et al., 1975, De POLLI et al., 1977), but nothing is known about the way it gets it or transports it upwards and the factors acting on these processes.

Table I - Nitrogen fixing bacteria in the rhizosphere of higher plants. R/S is the rhizosphere effect on their population (ratio of MPN in and outside the root zone). Dashes stand for not determined values.

PART I : AEROBES

Bacterium	Plant	Reference	R/S
Azotobacter	Cereals	Katznelson, 1965 Clark, 1969 Krasil'nikov, 1958	about 1
	Ammophila arenaria	Hassouna and Wareing, 1964	10
	Legumes	Clark, 1969 Vancura *et al.*, 1965 Krasil'Nikov, 1958	about 10
Azotobacter paspali	*Paspalum notatum cv. Batatais*	Döbereiner, 1966	> 1
	Cynodon dactylon Hyparrhenia rufa	Machado & Döbereiner 1969	< 1
Beijerinckia	*Medicago sativa*	Vancura *et al.*, 1965	10
	Sugar cane	Döbereiner, 1961	1 to 56
	Cynodon dactylon	Ruschel & Döbereiner, 1965	20
	Digitaria decumbens	" " " "	19
	Rice	Balandreau *et al.*, 1975	19
		Yoshida, 1970	-
		Döbereiner & Day, 1974	1,4 to 2,2
	Cyperus, forage grasses (tropical)	Ruschel & Britto, 1966	-
Derxia	Mainly grasses (tropical)	Campelo & Döbereiner, 1970	-
Spirillum	Most tropical forage grasses including maize	Döbereiner *et al.*, 1976 Lakshmi Kumari *et al.*, 1976	- -

Table I (cont.) PART II : FACULTATIVE ANAEROBES

Bacterium	Plant	Reference	R/S
Bacillus (polymixa and macerans)	Various field crops	Jurgensen & Davey, 1970	-
	Wheat	Nelson *et al.*, 1976	-
	Agrostis tenuis, Festuca sp.	Nelson *et al.*, 1976	-
	Ammophila arenaria	Wahab, 1975	-
Enterobacter (cloacae and aerogenes)	Zea mays	Raju *et al.*,	-
	Wheat	Nelson *et al.*, 1976	-
	Andropogon gerardii, Panicum virgatum Schizachirium scoparium	Kaputska & Rice, 1976	-
	Rice	Balandreau, *et al.*, 1975	-
		Watanabe *et al.*, 1977	-
	Juncus balticus Agrostis tenuis	Nelson *et al.*, 1976	-
	Panicum maximum Chloris divaricata	Koch & Oya, 1974	-
Klebsiella cf. pneumoniae	Legumes	Evans *et al.*, 1972	-
	Panicum maximum Chloris divaricata	Koch & Oya, 1974	-
Rhodopseudomonas capsulatus	Rice	Kobayashi & Hague, 1971,	-

Table I (concl.) PART III: ANAEROBES

Bacterium	Plant	Reference	R/S
Clostridium	Grape Vine	Vidal & Leborgne, 1963	-
	Wheat	Katznelson, 1965	-
	Maize	Balandreau, *et al.*, 1975	7,9
	Rice	Villemin *et al.*, 1974	1,6
	Grassland	Vlassak *et al.*, 1973	-
	Digitaria smutzii	Tow and White, 1976	5,4
Sulfate reducers (*Desulfovibrio* and others)	*Diplanthera wrightii*	Patriquin & Knowles, 1972	10
	Syringodium filiforme		1
	Thalassia testudinum		96
	Zostera marina		22
	Mangrove trees	Silver *et al.*, 1977	-

Indeed, some cautions must be taken regarding the use of the acetylene method (conversion factor from acetylene reduction to nitrogen fixation, problems of diffusion of gases in waterlogged stands) but it is not our purpose to discuss them here; we shall only present the way we have tried to understand the observed variations in nitrogenase activity.

In some instances the experiments were conducted in expensive growth cabinets but most of the results have been obtained in the field; in the latter case a close analysis of the results gave us as much relevant information as in the former case about the action of an individual factor and extra information: how much and when this factor is limiting nitrogenase activity.

I - MATERIAL AND METHODS

1. Measure of nitrogenase activity

Only results obtained with non-disturbing devices are taken into account in this paper. In the laboratory entire plant + soil systems in their containers were enclosed in a gas-tight device and 10% C_2H_2 injected (V/V); while starting incubations and during them, plants were left under previous conditions of light, temperature and air humidity. Whenever the volume of the incubation chamber was unknown, a known amount of propane had to be injected: its dilution provided a simple means to work out the volume of the internal atmosphere. Incubations were as short as possible compatible with producing a measurable amount of ethylene. For each replicate two successive samplings were made, usually at one hour intervals, and ethylene and propane measured by gas chromatography using a 2 m 1/8" stainless steel column packed with 100-200 mesh Spherosil XOB 075 (Pechiney Saint-Gobain, Departement Commercial Produits Chimiques, 92200 Neuilly-sur-Seine, France); from this column propane is eluted before acetylene and with a better definition than in the case of Porapak.

Whenever possible field measurements (BALANDREAU and COMMERGUES, 1973) were preferred using the device shown on Fig. 2.

2. Measure of environmental parameters

Air temperature and humidity were measured in the usual way under standard meteorological shelter. Available light energy was measured with a thermolinear pyranometer and expressed as J cm^{-2}. Soil temperature was determined using thermometric probes buried at several depths.

3. Calculations

Observations of five consecutive days are grouped in two samples, one for daytime and another for night.

On each of these samples, the correlations between environmental factors and nitrogenase activity are computed. Some delayed influence of these factors on activity can be expected because of system inertia; owing to that, environmental factors collected at the very moment of measuring N.A. and one (two, three) hour before are introduced and correlated with N.A. (Fig. 3).

But for light intensity which shows visually a delay of one or two hours, other factors produce better correlations instantaneously with N.A. (Nitrogenase Activity).

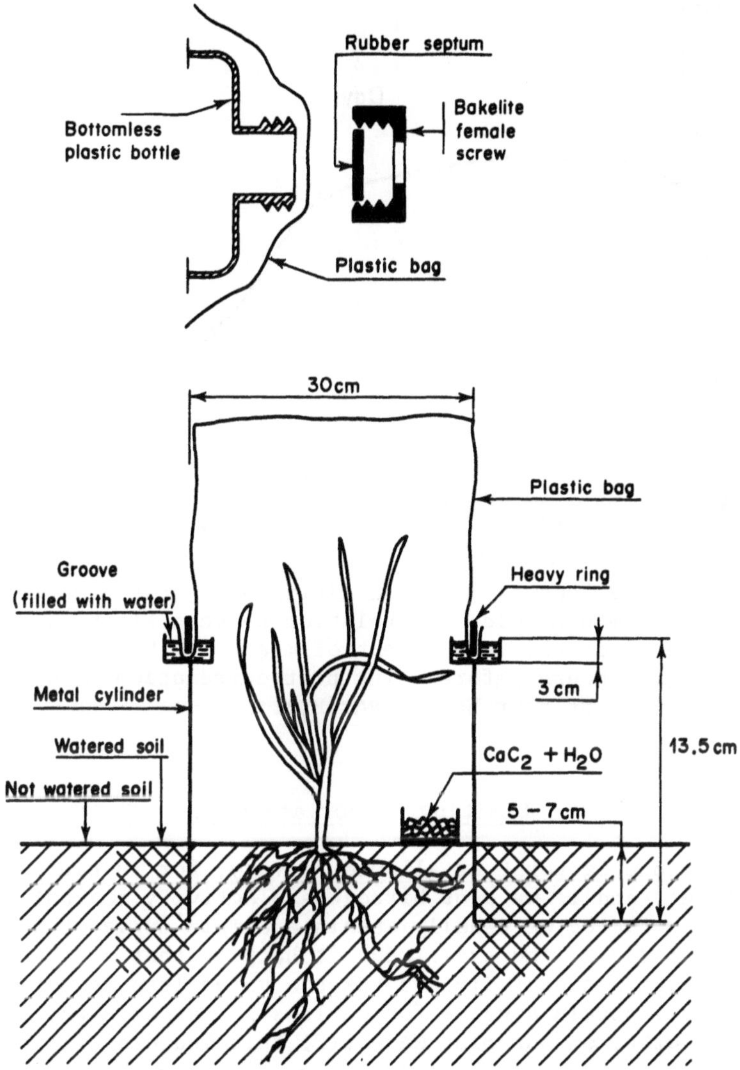

Fig. 2 - Cross section of the device used for measuring Nitrogenase Activity in the field ; top : details of the sampling port.

Use of concomitant information upon correlation between enviromental factors and interpolation of the correlation provide a crude hierarchy on the explanation power of the different factors.

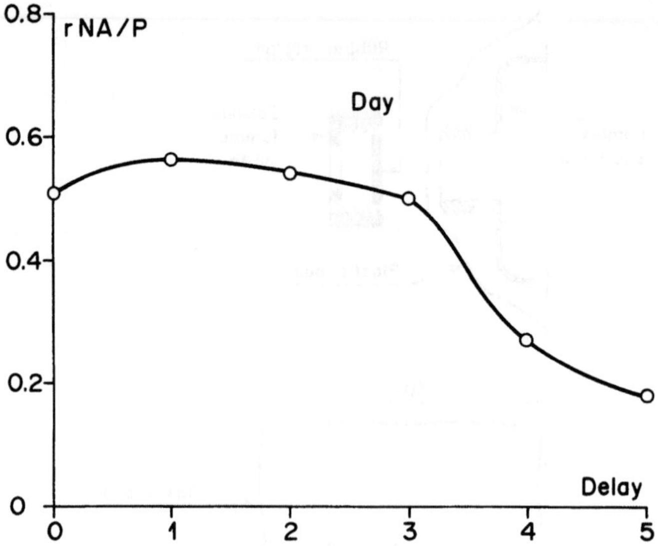

Fig. 3 - Delayed effect of light on Nitrogenase Acitivty (NA). Maize 2 months old. Correlation between NA and light measured at the very moment of measuring NA and one (two, three..) hours before. The best correlation is with light measured one hour before NA (P_1).

4. Limiting factor analysis

When only one factor F is allowed to vary, for instance in a growth cabinet, and if we plot Nitrogenase Activity against the values of this factor we usually obtain a simple curve of action NA = f(F). If during the experiment it happens that another factor F' varies unexpectedly and takes a value less favourable to NA, we shall obtain a value for NA less than expected owing to the value of F; on our plot NA = f(F) this measure will be represented by a point below the curve. So, the curve NA = f(F) represents all the measures made when no other factor than F was detrimental to NA: F was the only limiting factor. The points below the curve represents measures made when factors other than F were limiting.

The same applies in field situations where several factors act simultaneously on NA. Each plot of NA against any individual active factor is a cloud of points whose upper boundary is the curve of action of this particular factor, and its points represent the measures when this factor was the only one limiting NA. Points below this curve represent measures when another factor was limiting.

It must be stressed that many environmental parameters are closely related (for instance light level and air temperature or air temperature and soil temperature) and it is sometimes difficult to discriminate between two possible limiting factors: This is the case when the representative point belongs to two different boundary curves NA = f(F) and NA = f(F'). Another point is that it is necessary to have well scattered points to draw the boundary curve with enough certainty. It is worth noting also that this type of analysis must be conducted on comparable plant + soil samples.

5. Soils and plants

Measures reported here were made :

- In Ivory Coast in the Lamto Savannah: under a light cover of palm trees (*Borassus aethiopium*); it consists mainly of dense nearly pure populations of *Hyparrhenia* (essentially *H. diplandra* and *dissoluta*) or *Loudetia simplex*. The soil is a very poor sandy soil.

- In France near Nancy: in a maize plot on a clay soil. Maize cultivars were INRA 260 and INRA F_7xF_2. At the same location a *Dactylis glomerata* (cultivar Floreal) plot has been occasionally studied.

- Rice studies were conducted in growth cabinet using a soil from a rice field in the south of France. Rice cultivar was IR8 except if otherwise stated.

Some characteristics of the soils are shown on Table II.

1973 field data have been obtained during a five day period of continuous measures (day and night). 1975 field data have been obtained throughout June, July and August, with one day of continuous measures every week.

TABLE II : Some characteristics of soils used in this study

	texture	C%	N%	pH
Rice field (alluvial, southern France)	clay and silt	1.2	0.150	7.8
Maize plot (North east of France)	clay and silt	1.6	0.155	6.8
Savannah (Ivory Coast, Lamto)	very sandy	0.83	0.07	6.0

II - RESULTS

1. Light intensity

In growth cabinets, the effect of illumination is examplified by Fig. 4.

In field conditions available light energy exerts a comparable effect. Fig. 5 shows some results obtained with *Dactylis*. The same applies for maize: in both cases a straight line fits well the boundary, in the studied range of values.

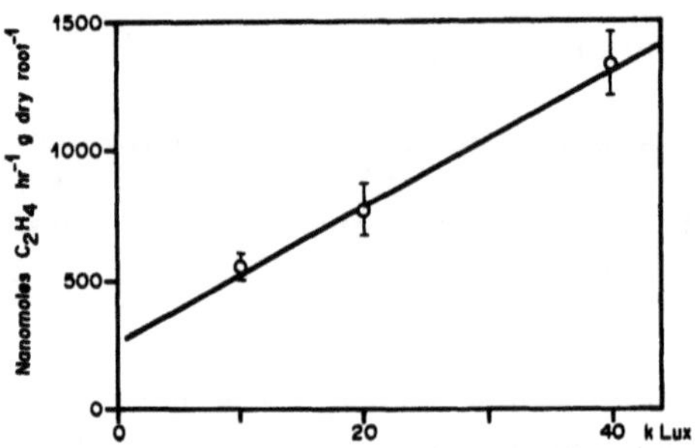

Fig. 4: Action of illumination on Nitrogenase Activity in the rhizosphere of Rice IR8, growth cabinet, 22 day old seedlings, 5 replicates.

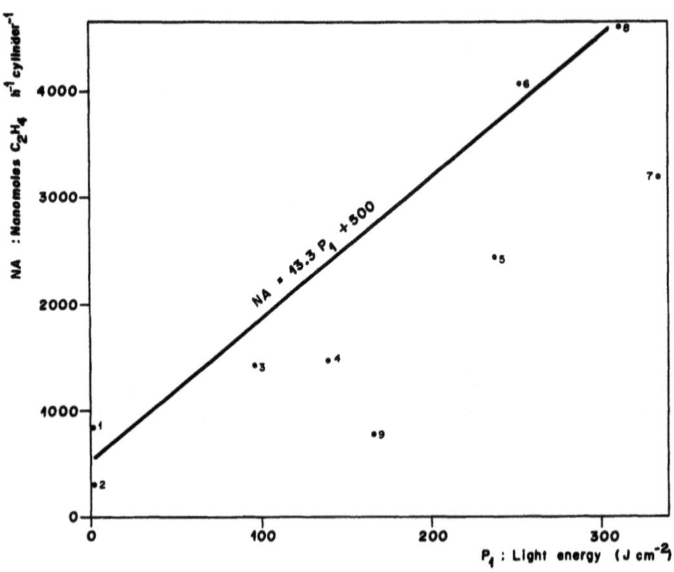

Fig. 5: *Dactylis glomerata* in the field: effect of light energy on Rhizospherical Nitrogenase Activity. Measures throughout one single day.

2. Potential of the photosynthetic apparatus

The quantity of carbonaceous compounds synthesized for a given light energy depends primarily upon the potential of the photosynthetic apparatus. This potential can be represented in several ways, either by plant age (RINAUDO *et al.*, 1971), dry weight (Fig. 6) or leaf area (L.A.I.), or any related parameter. For the maize we have studied, nitrogenase activity expressed as 10^{-9} moles C_2H_4 hr^{-1} $plant^{-1}$ can reach one hundred times the dry weight of green blade (in g.) when no other factor is limiting.

The physiological type (C_3 or C_4) of the photosynthetic apparatus is perhaps a minor factor as Rice which possesses the C_3 physiology seems able to fix more than Maize which belongs to the C_4 grasses.

3. Air temperature

This factor (BALANDREAU *et al.*, 1976) acts at two levels : on the photosynthetic process itself, and on the translocation rate of assimilates through the plant (WEST, 1971). Its first type of action is usually difficult to distinguish from the role of light, for light and air temperatures are closely linked. The effect of air temperature on translocation seems prominent by night; we have never observed high nitrogenase activity when air temperature was below 19°, which confirms a DÖBEREINER hypothesis.

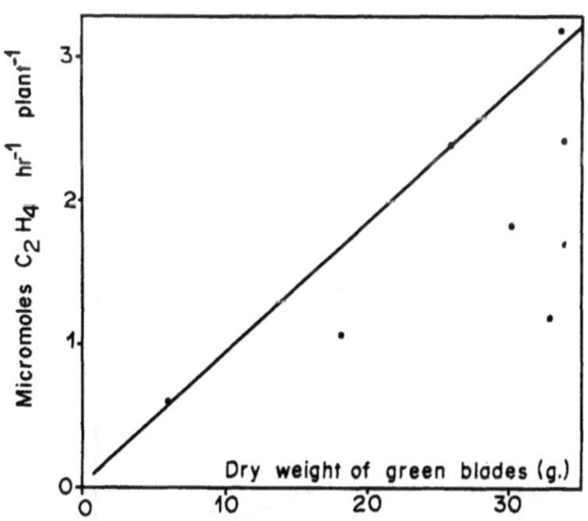

Fig. 6: Maize. Relation between daily maxima of Nitrogenase Activity and potential of the photosynthetic apparatus (exemplified by the dry weight of green blades). The straight line NAmax = 95B seems to limit the repartition of representative points and was adopted as the model for representing the action on Nitrogenase Activity of the potential of the photosynthetic apparatus.

4. Air water deficit

This factor also is likely to act at two different levels; first on stomatal aperture then on photosynthesis itself, second on the upward transpiration stream; MINCHIN and PATE (1974) have shown that in legumes this stream had a great role by removing newly synthesized nitrogenous compounds from the sites of active nitrogen fixation where they could be inhibitory; such an action is not completely unlikely in grass associative symbiosis.

Fig. 7 shows the type of action of air water deficit on rice rhizosphere nitrogenase activity; there is a marked optimum for a light deficit of about 5 mm Hg. In the field, maize seems to show the same type of optimum at a low deficit then an increase at higher deficits (fig. 8).

5. Soil temperature

Soil temperature is also a frequent limiting factor as shown on figures 9 and 10 by the large number of points on the boundary curve. Soil temperature in our maize plot was often limiting in the middle of the day especially when the soil was wet (and had a higher thermal capacity).

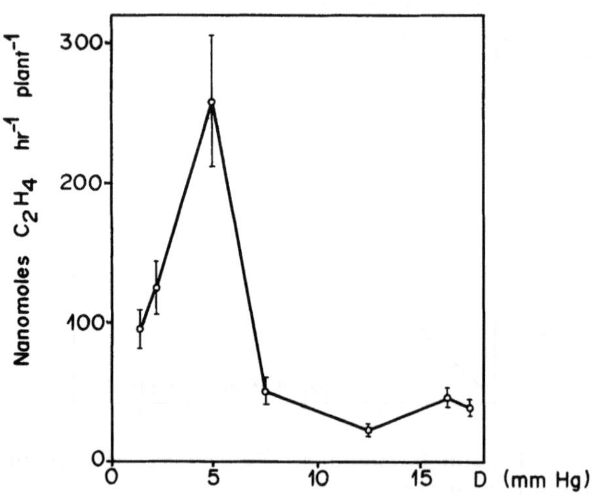

Fig. 7:Action of Air Water-Content deficit (D)on Nitrogenase Activity. Growth cabinet, Rice IR8, 11 day old seedlings, 8 replicates.

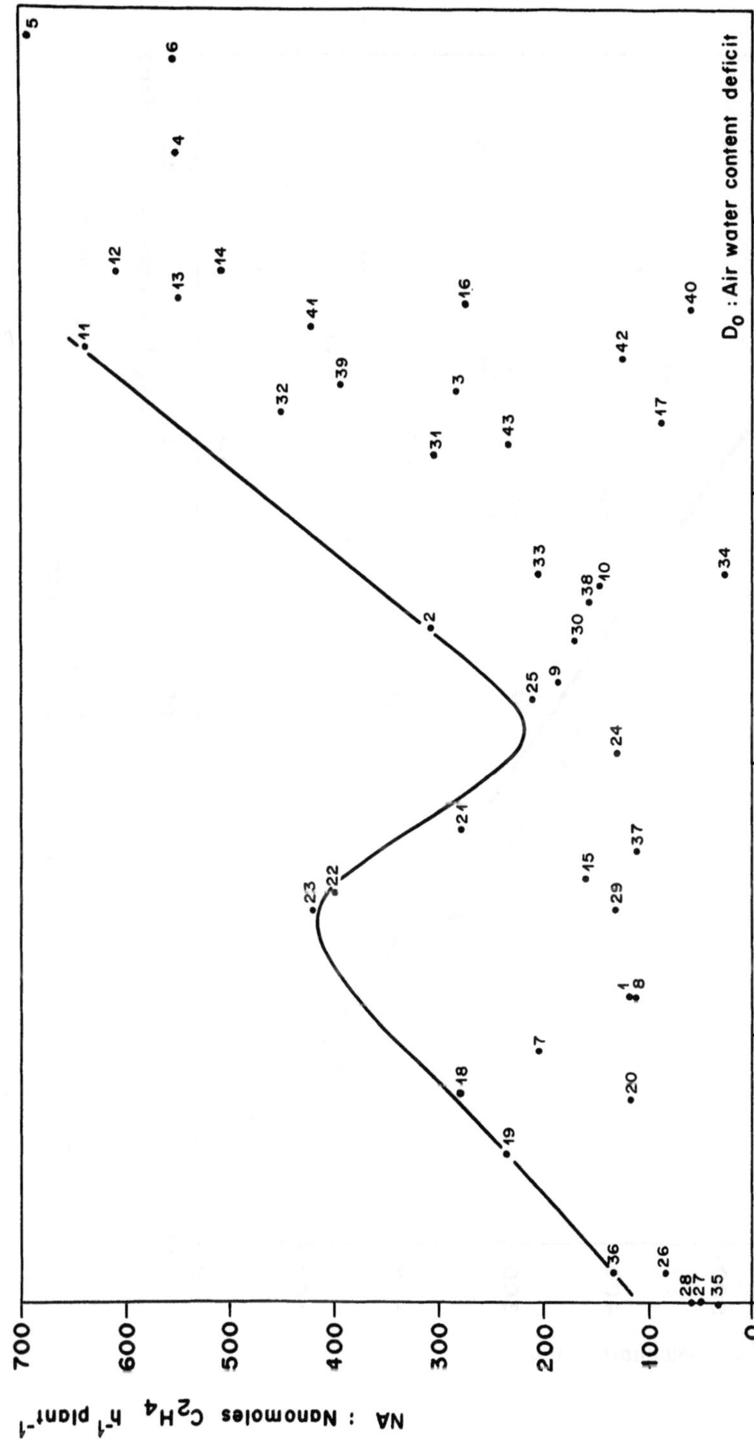

<u>Fig.8</u>: Maize in the field: Action of Air Water-Content deficit (D_O) on Nitrogenase Activity. The shape of this curve, still unexplained, has been confirmed by some recent unpublished data.

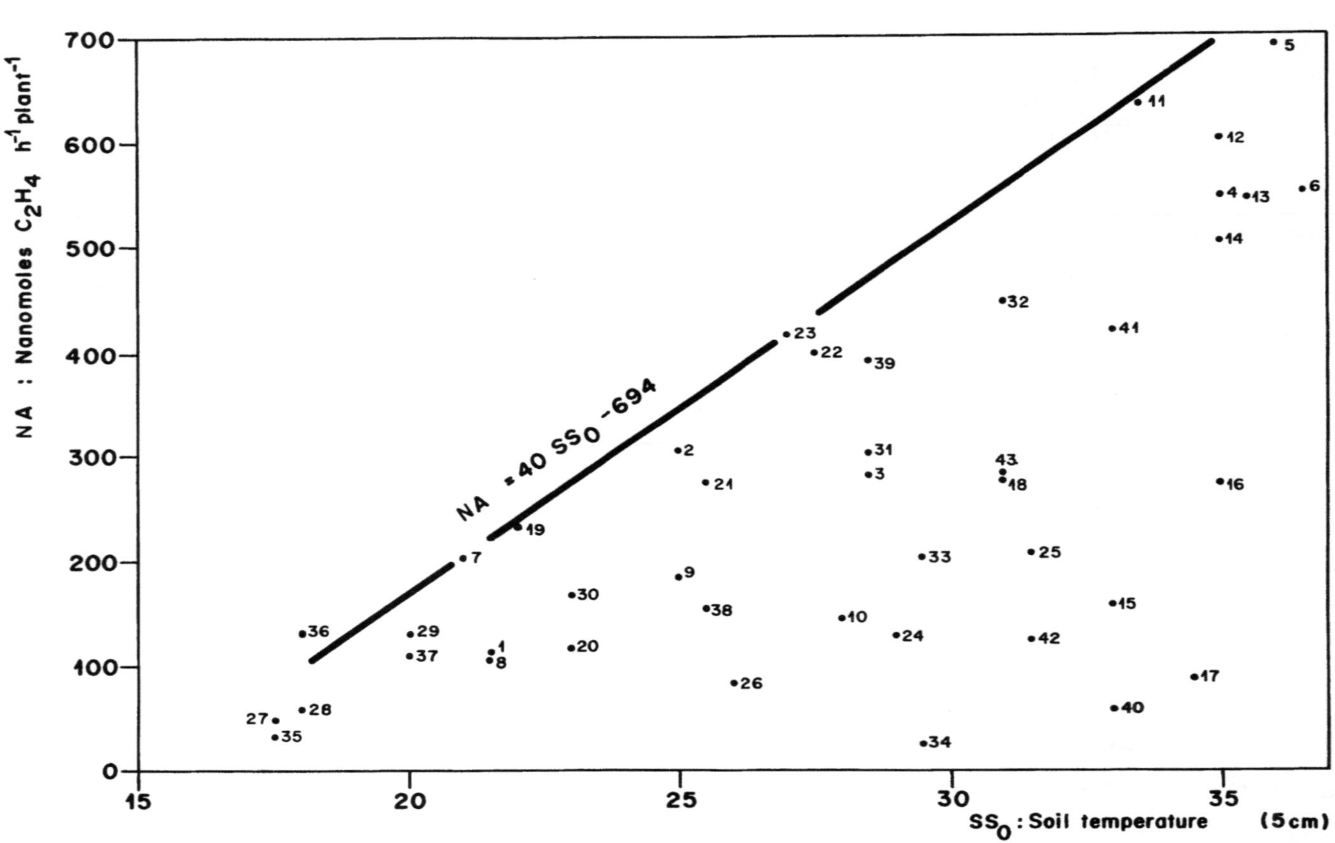

Fig. 9 : Maize in the field : Action of Surface (5 cm) Soil Temperature (SS_o) on Nitrogenase Activity.

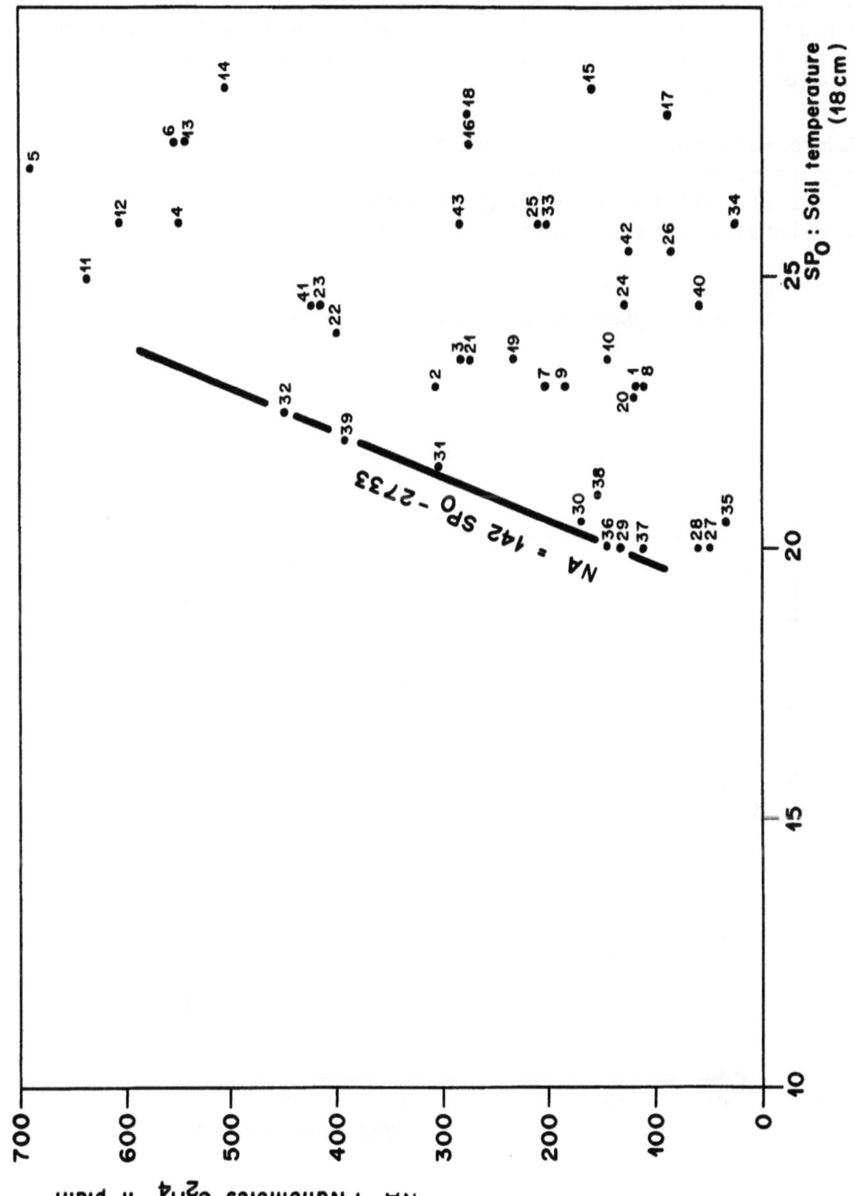

Fig. 10 : Maize in the field : Action of Deep (15 cm) Soil Temperature (SP$_o$) on Nitrogenase Activity.

6. Soil redox and water content

On the days when the observed maximum of nitrogenase activity was lower than allowed by the potential of the photosynthetic apparatus (measured by the weight of green blades) soil water content seemed to be the actual limiting factor, especially when below 18%, as shown in Fig. 11.

In Lamto Savannah the same type of analysis (Fig. 12) shows that a high level of nitrogenase activity is possible only above a certain level of soil water content, measured in that case by the cumulated rainfalls during the ten previous days.

Fig. 13 shows the two different patterns of nitrogenase activity in the rhizosphere of *Hyparrhenia* during a dry or a wet period.

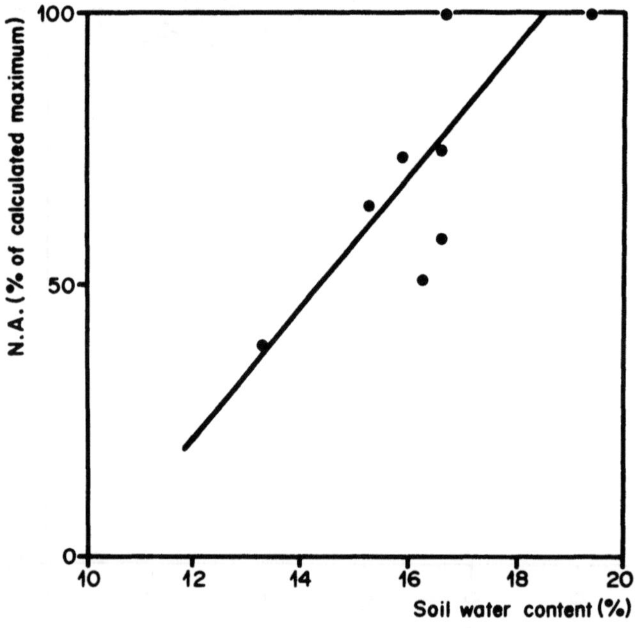

Fig. 11 : Maize in the field : Daily Maxima of Nitrogenase Activity (NAmax) in % of the Maxima calculated following NAmax = 95B (see Fig. 6). Below 18% (of soil fresh weight) there is a depressive effect on NAmax.

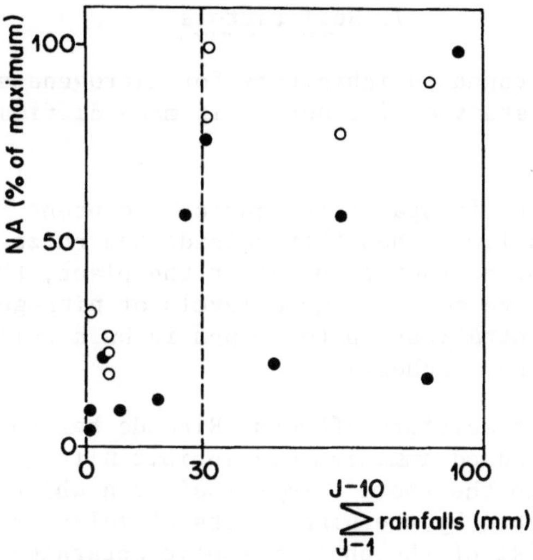

Fig.12 : Ivory Coast Savannah : effect of soil water content (represented by cumulated rainfalls during 10 days) on Nitrogenase Activity. Measures were made at the beginning of the afternoons throughout the year and results obtained on *Loudetia* and *Hyparrhenia* (expressed as % of observed maxima) were pooled together.

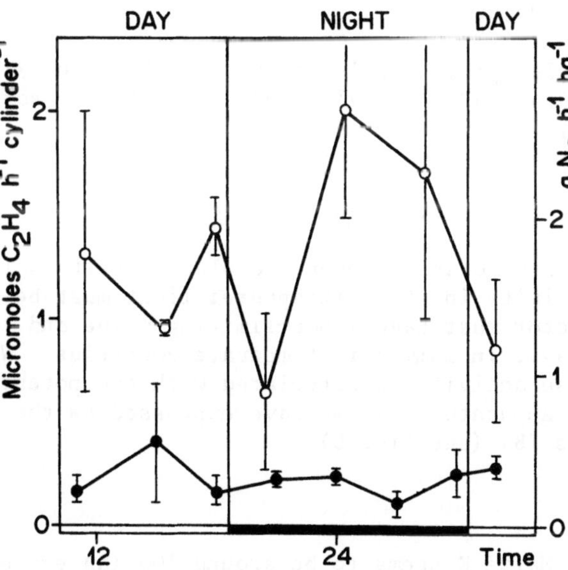

Fig. 13: *Hyparrhenia* : diurnal variations of Nitrogenase Activity on a dry day (lower curve) and on a wet day (upper curve). Note the depressive effect of drought.

7. Soil factors

Inorganic nitrogen is inhibitory for nitrogenase but in soil plant systems the study of its action is made difficult for two reasons :

Heterogeneity in space: inorganic N content in the rhizosphere can be much lower than that outside the rhizosphere because of absorption by plant roots; the older the plant, the greater this discrepancy. We have got very good levels of nitrogenase activity with nitrate concentrations up to 50 ppm in bulk soil (but less than 5 ppm in the rhizosphere).

Long term stimulatory effects: Rinaudo has shown (BALANDREAU *et al.*, 1975) that added ammonium did inhibit nitrogenase activity in rice rhizosphere in the short term, but after a while this ammonium disappeared, absorbed by the plant. Its stimulatory effect on growth and potential of the photosynthetic apparatus could result in a stimulation of nitrogenase activity.

So, the effect of inorganic nitrogen is not always simple and we need more experiments before assessing if it can be an important limiting factor in the field situation.

P content must be an important limiting factor especially on poor soils, as is the case for legumes, but we know no experimental data on the subject.

Soil type also plays a role as shown by Rinaudo (RINAUDO *et al.*, 1971, RINAUDO 1974, GARCIA *et al.*, 1974), but the reasons for it are still not clear.

8. Plant genetic factors

It is not easy to assess when the plant itself is limiting nitrogenase activity in its rhizosphere; light must be maximum and every other factor must have a non-limiting value and this happens only on fine days, in summer in temperate countries. When it is the case nitrogenase activity is correlated with the potential of the photosynthetic apparatus that we have expressed as the dry weight of green blades (B) (see fig. 6)

$$NA = KB$$

in the case of Maize K seems to be around 100 (NA expressed as 10^{-9} moles C_2H_4 hr^{-1} $plant^{-1}$ and B as grams) for F7 x F_2 variety. This K coefficient has probably a very interesting significance It is the ultimate limiting factor when all environmental factors have values favourable to nitrogen fixation; it reflects the ability of the plant to divert a part of its assimilates towards nitrogen fixing organisms, and the efficiency of utilization of these assimilates after exudation in the soil under study. Comparing the K values for different plants could be the best way to compare them for the ability to support nitrogen fixation on a given soil.

I. Fares-Hamad has tried such a comparison about Rice in a growth cabinet at maximum light level; rice seedlings were 13 days old and cultivated in the same optimal conditions, on the same soil.

Table III : Comparison of rice varieties for their ability to support Nitrogen fixation.

Cultivars	Nanomoles C_2H_4 hr^{-1} g dry leaves^{-1}
Cristal	7000 ± 277
Cesariot	6375 ± 197
IR5	6290 ± 175
Delta	6089 ± 225
Cigalon	4057 ± 223
Arlésienne	3662 ± 206
IR8	2909 ± 195

This type of result, and corresponding results obtained by Von Bülow and Döbereiner (1975), must provide a good challenge for plant geneticists: if such differences do exist between varieties, it is possible to improve nitrogen fixation by improving the plant partner of the association.

9. Bacteria

Very little is known about N_2-fixing bacteria actually responsible for nitrogen fixation in the rhizosphere of grasses. Koch's postulates have been completely fulfilled only for rice associated with *Beijerinckia* (HAMAD FARES, 1976) and sterile isolated maize roots associated with *Spirillum lipoferum* (BALANDREAU, unpublished data) and if the presence of a large number of species of N_2 fixers has been evidenced (Table I) only in some instances has the size of the population been actually measured and compared to the N_2 fixing population in bulk soil. Such a measure of the R/S ratio has been performed on aerobic N_2 fixers using the acetylene method (VILLEMIN *et al.*, 1974) by P. DUCERF (unpublished data) on our Maize plot and has evidenced a steady increase of its value throughout the growth cycle (Table IV).

Table IV : R/S values of aerobic N_2 fixers in the rhizosphere of maize (in the field). Maize had been sown in May. Ratios of MPN in and outside the rhizosphere.

Date	R/S
July 17th	3.8
August 13th	6.5
November 10th	8.8

This shows an increase of bacterial populations in the rhizosphere soil (soil adhering to the roots was ground together with the roots before diluting) but does not tell us whether the size of the population is a significant limiting factor or if it continuously adapts itself to the increasing level of exudation.

Inoculation trials are an approach to study this point but their results must be studied with much caution. (BURRIS 1976). Most of their results are not conclusive either with *Azotobacter* or with *Spirillum*, in spite of a tendency towards better growth or nitrogen content for inoculated plants. It must be stressed that indeed no effect of inoculation can be seen if a factor other than microbial population is limiting.

Nevertheless the possibility of a transient effect of such inoculations has not been sufficiently explored; in some poor tropical soils at the end of the hot season, soils have a very low microbial density and it is possible that the number of N_2 fixers is so low that it becomes limiting; inoculation could help establish Nitrogenase Activity in the rhizosphere in such conditions.

III - DISCUSSION AND CONCLUSIONS

We can use a boundary curve analysis in several ways:

First of all we can ascribe an explanation to each observed Nitrogenase Activity provided the representative point is on or near the boundary curve corresponding to a possible limiting factor. Table V shows for one particular summer day that deep soil temperature was never limiting; the main limiting factors were light intensity and air temperature with some difficulty to distinguish between the two. On some other summer days, after rain, deep soil temperature could actually become limiting.

Another possibility offered by this type of analysis was the building up of a mathematical model describing the observed variations of Nitrogenase Activity in the rhizosphere of maize, (BALANDREAU *et al.*, 1976) and using seven parameters:

- Dry weight of green blades : B(G;)
- Soil water content : W(% fresh weight)
- Air temperature: T° (°C) and water content deficit : D(mm Hg)
- Light energy received for one hour, on to one hour before the measure : P_1 (J $cm^{-2}hr^{-1}$)
- Surface (5 cm) soil temperature: SS° (°C)
- Deep (15 cm) soil temperature : SP°(°C)

In its present state this model allows the calculation of the possible maximum Nitrogenase Activity (10^{-9}moles C_2H_4 $hr^{-1}plant^{-1}$) for a given day

$$NA\ max = 95B \quad \text{if} \quad W > 18\%$$

$$NA\ max = 95B\ \frac{12W - 124}{100} \quad \text{if} \quad 10 < W < 18\%$$

the next step gives the actual value of Nitrogenase Activity (in % of the previously calculated maximum), at a given time of the day; this value is the minimum of the four values calculated with the equations

$$4.8\ T^\circ - 47$$

$$0.28\ P_1 + 20$$

$$5.7\ SS^\circ - 99$$

$$20\ SP^\circ - 390$$

These equations are those of boundary curves like those of fig. 8,9 and 10, expressed in percent of the maximum value observed during the studied period (700.10^{-9}moles C_2H_4 $h^{-1}plant^{-1}$). To those four values one must add the value graphically determined on fig. 8 to take into account the possible action of air water-content deficit. The minimum of these five values is the model estimate for Nitrogenase Activity. Fig. 14 shows agreement between calculated and measured values, for one particular day, and fig. 15 does the same for an entire growth cycle.

To end with, a few points are worth being emphasized:

1. When ignored, a limiting factor can invalidate completely two kinds of studies : comparisons between species or varieties and inoculation trials.

2. We still do not know which N_2 fixers are the most important in grass rhizospheres and it is not sure whether microbial populations are actually an important limiting factor or not.

3. In many circumstances expensive growth cabinet and computer studies can be replaced by a close study of what happens *in situ*.

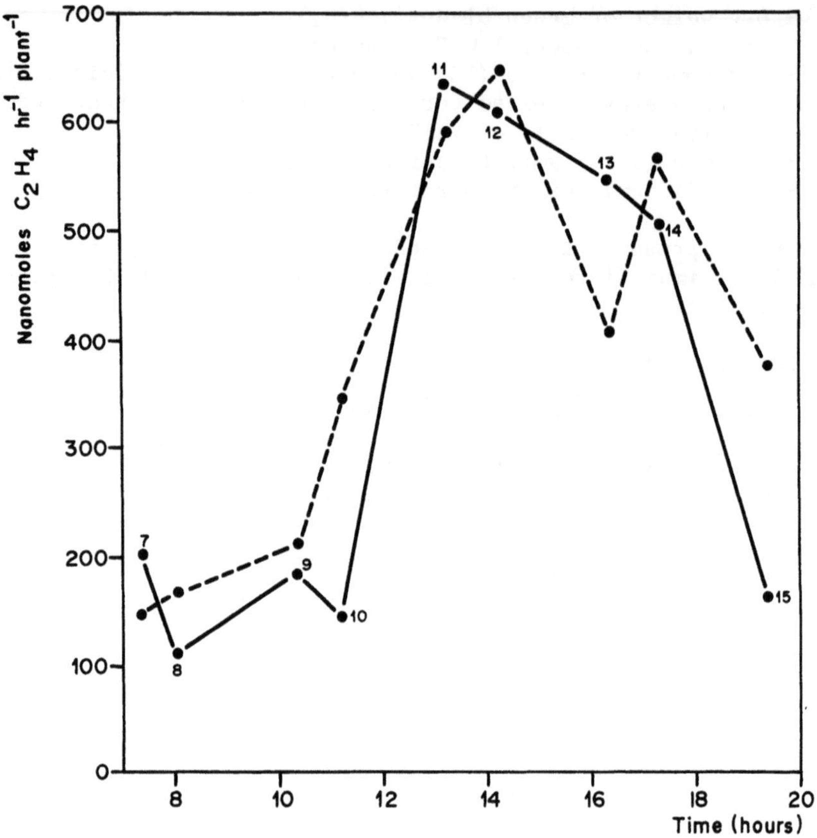

Fig.14 : Maize, in the field; agreement between observed (solid line) and calculated (dotted line) Nitrogenase Activity for one particular day (July 4th 1973); agreement is better at the beginning of the day; figures identify measures (see table V and figs.8,9,10).

4. Large discrepancies in data from literature about Nitrogenase Activity can be attributed (1) to over- and underestimates due to the methods of measurement and (2) to the interaction of overlooked limiting factors.

We hope that, in the near future, much new work on the ecology of rhizosphere dinitrogen fixation will appear so that improving the process will become possible and will bring about an

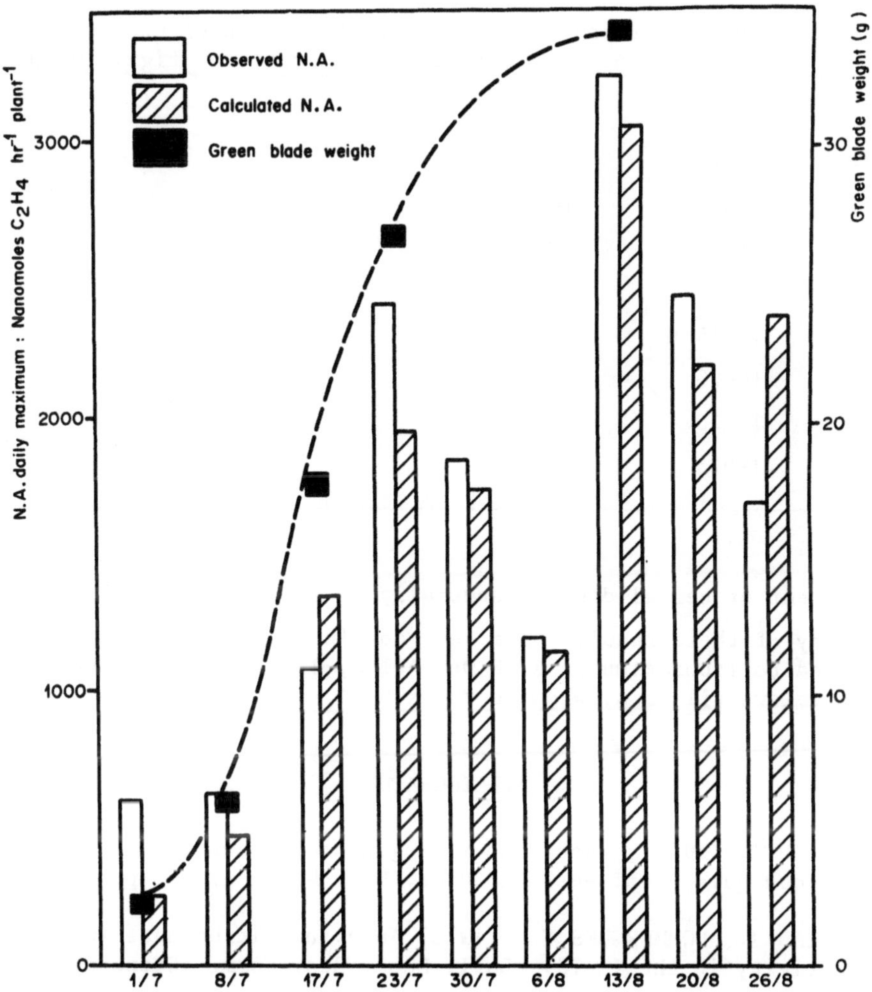

Fig. 15 : Maize, in the field: agreement between observed (white) and calculated (grey) values for daily maxima of Nitrogenase Activity throughout most of a growth cycle (after BALANDREAU *et al.*, 1976).

Table V : Possible limiting factors at different times of a day. Maize plot (sown on May 10th) July 4th 1973.

Sample N°	Time	Possible limiting factors (representative point on or near a boundary curve; see figs.8,9,10)
7	7h21	SS∘ P_1
8	8h10	P_1
9	10h20	D∘
10	11h10	D∘
11	13h20	SS∘ P_1 T∘ D∘
12	14h15	SS∘ P_1 T∘
13	16h20	P_1 T∘
14	17h20	P_1
15	19h25	?

improvement of cereal diets, especially in tropical countries.

Many of these results have been obtained with the technical skill and collaboration of R. N'Dri Allou, J.L. Renou, P. Villecourt, and Genevieve Villemin, whom we thank here.

BIBLIOGRAPHY

BALANDREAU J. (1970), Activité Nitrogénasique dans la rhizosphère de quelques graminées. Th. Doct. Etat Nancy I, 162 p.

BALANDREAU J., DOMMERGUES Y., (1973) Assaying Nitrogenase (C_2H_2) activity in the field. Bull. Ecol. Res. Comm. (Stockholm) 17, 247 - 254.

BALANDREAU, J.P., MILLIER C.R., WEINHARD Pierrette, DUCERF P., DOMMERGUES Y., (1976) A modelling approach of acetylene reducing activity of plant-rhizosphere diazotroph systems in "Proc. 2nd Intern. Symp. on Nitrogen Fixation Salamanca" (in press).

BALANDREAU J., RINAUDO G., FARES-HAMAD Ibtissam, DOMMERGUES Y., (1975) Nitrogen fixation in the rhizosphere of rice plants in "Nitrogen fixation and the biosphere" W.D.P. Stewart ed. Cambridge Univers. Press, 6, 57-70

BALANDREAU J., WEINHARD Pierrette, RINAUDO G., DOMMERGUES Y., (1971) Influence de l'intensité de l'éclairement de la plante sur la fixation non symbiotique de l'azote dans sa rhizosphere. Oecol. Plant. 6, 341-351.

Von BÜLOW, J.F.W., DÖBEREINER Johanna (1975) Potential for Nitrogen Fixation in Maize Genotypes in Brazil Proc. Natl. Acad. Sci. U.S.A., 72, 2389-2393.

BURRIS, R.H., (1976) A synthesis paper on nonleguminous N_2-fixing systems in "Proc. 2nd Intern. Symp. on Nitrogen Fixation, Salamanca" (in press).

CAMPELO, A.B., DÖBEREINER J., (1970) Ocorrencia de Derxia sp. em solos de alguns estados brasileiros. Pesq.agropec.bras. 5, 327 -332

CLARK F.E., (1969) Ecological associations among soil micro-organisms in "Soil biology, reviews of research", Paris, UNESCO, 125-161.

DAY J., HARRIS D., DART P., VAN BERKUM P., (1973) The Broadbalk experiment. An investigation of nitrogen gains from non-symbiotic fixation, in "Nitrogen fixation and the biosphere", WDP Stewart ed., Cambridge Univers. Press, 6, 71-84.

DE-POLLI H., MATSUI E., DÖBEREINER Johanna, SALATI E., (1977) Confirmation of Nitrogen fixation in two tropical grasses by $^{15}N_2$ incorporation. Soil Biol. Biochem., 9, 110-123.

DÖBEREINER Johanna (1961) Nitrogen-fixing bacteria of the genus *Beijerinckia Derx.* in the rhizosphere of sugar cane. Plant Soil 15, 211-216

(1966) *Azotobacter paspali* sp.n. a nitrogen-fixing bacterium in the rhizosphere of Paspalum. Pesq.agropec.bras. 1, 357-365.

DÖBEREINER Johanna, DAY J., (1974) Associative symbioses in tropical grasses. Characterization of microorganisms and dinitrogen fixing sites. Proc. 1 Int. Symp. Nitrogen fixation, Washington State Univ. 518-538.

(1975) Nitrogen fixation in the rhizosphere of tropical grasses in: "Nitrogen fixation and the biosphere", W.D.P. Steward ed. Cambridge Univers. Press., 6, 39-56.

DÖBEREINER Johanna, MARRIEL I.E., NERY M., (1976) Ecological distribution of *Spirillum lipoferum* Beijerinck. Can. J. Microbiol. 22, 1464-1473.

EVANS H.J., CAMPBELL N.E.R., HILL Susan, (1972)
A symbiotic nitrogen-fixing bacteria from the surfaces of nodules and roots of legumes. Can. J. of Microb. 18, 13-21.

GARCIA J.L., RAIMBAULT M., JACQ V., RINAUDO G., ROGER P., (1974)
Activités microbiennes dans les sols de rizières du Sénégal: relations avec les caractéristiques physico-chimiques et influence de la rhizosphère. Rev. d'Ecol. Biol. Sol, II 169-185.

HAMAD-FARES Ibtissam (1976)
La Fixation de l'azote dans la rhizosphère du Riz. Th. Doct. Etat. Nancy I, 137 p.

HASSOUNA M.G., WAREING P.F., (1964)
Possible role of rhizosphere bacteria in the nitrogen nutrition of *Ammophila arenaria*, Nature, 202, 467-469.

HAUKE-PACEWICZOWA Theresa, BALANDREAU J., DOMMERGUES Y., (1970)
Fixation microbienne de l'azote dans un sol salin tunisien Soil Biol. Biochem. 2, 47-53.

JURGENSEN M.F., DAVEY C.B., (1970)
Nonsymbiotic nitrogen-fixing microorganisms in acid soils and the rhizosphere. Soils and Fertilizers 33, 435-446.

KAPUSTKA L.A., RICE E.L., (1976)
Acetylene reduction (N_2 fixation) in soil and old field succession in central Oklahoma. Soil Biol. Biochem. 8, 497-503.

KATZNELSON H., (1965)
Nature and importance of the Rhizosphere in "Ecology of Soil-borne Plant Pathogens". 187-209.

KOBAYASHI M., HAGUE M.Z., (1971)
Contribution to nitrogen fixation and soil fertility by photosynthetic bacteria. Plant Soil, Spec. Vol, 443-456.

KOCH B.L., OYA J., (1974)
Non symbiotic nitrogen fixation in some Hawaïian pasture soils. Soil Biol. Biochem. 6, 363-367.

KRASIL'NIKOV N.A., (1958)
Soil microorganisms and higher plants. Acad. of Sc. of USSR, Moscow, 1958.

LAKSHMI KUMARI M., KAVIMANDAN S.K., SUBBA RAO N.S., (1976)
Occurrence of Nitrogen Fixing *Spirillum* in Roots of Rice, Sorghum, Maize and other Plants. Indian J. Exp. Biol. 14, 638-639.

MACHADO W.C., DÖBEREINER J., (1969)
Estudos complementares sobre a fisiologia de *Azotobacter paspali* e sua dependencia da planta(*Paspalum notatum*). Pesq. agropec. bras. 4, 53-58.

MINCHIN F.R., PATE J.S., (1974)
Diurnal functioning of the legume root nodule. J. of Exper. Bot. 25, 295-308.

NEAL J.L. Jr., LARSON R.I. (1976)
Acetylene reduction by bacteria isolated from the rhizosphere of wheat. Soil Biol. Biochem. 8, 151-155.

NELSON A.D., BARBER Lynn E., TJEPKEMA J., RUSSELL S.A., POWELSON R., EVANS H., SEIDLER R.J.,(1976)
Nitrogen fixation associated with grasses in Oregon. Can. J. Microbiol. 22, 523-530.

PARKER C.A., (1957)
Non-symbiotic nitrogen-fixing bacteria in soil. III - Total nitrogen changes in a field soil. J. Soil Sci. 8, 48-59.

PATRIQUIN D., KNOWLES R., (1972)
Nitrogen fixation in the rhizosphere of marine angiosperms Marine Biology (Berlin) 16, 49-58.

RAJU P.N., EVANS H.J., SEIDLER R.J., (1972)
An asymbiotic nitrogen fixing bacterium from the root environment of corn. Proc. Natl. Acd. Sci. USA, 69, 3474-3478

RINAUDO G., (1974)
Fixation biologique de l'azote dans trois types de sols de rizières de Côte d'Ivoire. Rev. d'Ecol. Biol. Sol. 11, 149-168.

RINAUDO, G., BALANDREAU J., DOMMERGUES Y., (1971)
Algal and bacterial non-symbiotic nitrogen fixation in paddy soils. Plant and Soil, Special volume 1971, 471-479.

RUSCHEL Alaides P., BRITTO D.P.,(1966)
Fixação assimbiotica de Nitrogênio atmosférico em algumas gramineas e na tiririca pelas bactérias do gênero *Beijerinckia* Derx. Pesq. agropec. bras. 1, 65-69.

RUSCHEL Alaides P., DÖBEREINER J., (1965)
Bactérias assimbioticas fixadoras de N na rizosfera de gramineas forrageiras. An. IX Congr. Int. Pastagens. Sao Paulo, 2, 1103-1107.

RUSCHEL Alaides P., HENIS Y., SALATI E., (1975)
Nitrogen-15 tracing of N-fixation with soil-grown sugar cane seedlings. Soil Biol. Biochem. 7, 181-182.

SILVER W.S., ZUBERER D.A., BABIARZ E.J., (1977)
Nitrogen fixation associated with mangroves and selected sea grasses. Translat. Intern. Sympos. on Environmental Role of Nitrogen fixing Blue-green algae and asymbiotic bacteria. (In press).

TOW P.G., WHITE G.J., (1976)
Non-symbiotic nitrogen fixation in the rhizosphere of Digitaria Smutsii Stent. Plant and Soil 45, 637-646.

VANCURA V., ABD EL MALEK Y., ZAYED M.N., (1965)
Azobacter and *Beijerinckia* in the soils and rhizosphere of plants in Egypt. Folia Microbiologica 10, 224-229.

VIDAL G., LEBORGNE L., (1963)
Recherches sur la rhizosphère de la vigne (*Vitis vinifera*). Ann. Inst. Pasteur, 105, 361-367.

VILLEMIN Geneviève, BALANDREAU J., DOMMERGUES Y., (1974)
Utilisation du test de réduction de l'acétylène pour la numération des bactéries libres fixatrices d'azote. Ann. Micr., 24, 87-94.

VLASSAK K., PAUL E.A., HARRIS R.E., (1973)
Assessment of biological nitrogen fixation in grassland and associated sites. Plant and Soil 38, 637-649.

WATANABE I., LEE K.K., ALIMAGNO B.V., SATO M., del ROSARIO D.C., de GUZMAN M.R., (1977)
Biological Nitrogen Fixation in paddy fields studied by in situ acetylene-reduction assays. IRRI Research paper series N° 3, 16 p.

WAHAB A.M.A., (1975)
Nitrogen fixation by *Bacillus* strains isolated from the rhizosphere of *Ammophila arenaria*. Plant & soil, 42, 703-708.

WEST S.H., (1971)
Biochemical mechanism of photsyntehsis and growth depression in *Digitaria decumbens* when exposed to low temperatures. Proc. XI Intl. Grassl. Cong. 514-517.

YANAGISAWA M., TAKAHASHI J., (1964)
Studies on the factors related to the productivity of paddy soil in Japan with special reference to the nutrition of rice plants. Bull. Natl. Inst. Agric. Sci. Japan, Ser. B., 14:41.

YOSHIDA T., (1968)
Soil microbiology. The IRRI An. Report, 131-146.

YOSHIDA T., (1970)
Soil microbiology. The IRRI An. Report, 47-59.

PHYSIOLOGY AND BIOCHEMISTRY OF *SPIRILLUM LIPOFERUM*

R. H. Burris, Stephen L. Albrecht and Yaacov Okon

Department of Biochemistry and Center for Studies of N_2 Fixation

College of Agricultural and Life Sciences
University of Wisconsin-Madison
Madison, Wisconsin 53706

There has been great interest in the organism *Spirillum lipoferum* because of its reported potential for fixing N_2 in association with tropical grasses (3-7, 15, 19). Our discussion will center primarily on the physiology and biochemistry of the organism, but there will be reference to growth chamber and field tests of its effectiveness in N_2 fixation. The organism has been discussed previously at these meetings in relation to a number of its properties.

Spirillum lipoferum is clearly capable of fixing N_2 (5, 10). Not only is it easy to demonstrate that the organism reduces acetylene, but it also is possible to show, as indicated in Table I, that the organism can reduce ^{15}N-enriched N_2 (10). Note that the cells acquired up to 0.453 atom % ^{15}N excess in a period of 30 min. This is 100 times the level needed for easy detection of ^{15}N enrichment. So *S. lipoferum* reduces N_2 at a substantial rate and thus can grow quite adequately on N_2. The organism grows best at a pH near neutrality (3, 10). It is commonly started at a pH near 7, and it grows well until it reaches a pH of about 7.8. Above this pH, growth and N_2 fixation are very limited.

In growing *S. lipoferum* one can choose among several effective substrates. As most strains of the organism do not use sugars well, they commonly are grown on malate (3). It also is possible to grow them effectively on succinate, lactate, or pyruvate (11, 12). Pyruvate is not a particularly convenient compound to use, but the other three organic acids are very convenient (10); they are quite

TABLE I

Fixation of $^{15}N_2$ by Spirillum lipoferum. Cultures were in 60-ml bottles at a pO_2 of 0.01 atm, pN_2 of 0.30 atm (80 atom % ^{15}N excess) and pAr of 0.69 atm. The incubation was at 30° C for a period of 30 min.

Sample	Total mg N in 12 ml sample	Atom % ^{15}N excess	$\frac{\text{nmoles } N_2 \text{ fixed*}}{\text{h x ml culture}}$
1	0.425	0.449	14.2
2	0.400	0.453	13.5
3	0.425	0.353	11.8

*Calculation:

$$\frac{\text{atom \% }^{15}\text{N excess in sample}}{0.80 \text{ fraction of } N_2 \text{ as } ^{15}\text{N}} \times \frac{\mu\text{g N in 12 ml culture}}{28 \times 12 \text{ ml}} \times \frac{60 \text{ min}}{30 \text{ min}} = \text{nmoles } N_2 \text{ fixed/(h x ml culture)}$$

inexpensive, readily available and stable to autoclaving. As the organisms use the organic acid during growth, the pH rises, and to counter too high a pH rise requires the use of a well-buffered medium or controlled addition of acid. Fortunately the organism can tolerate a rather high concentration of phosphate as buffer. We use 10 g of mixed phosphate salts per liter in the growth medium (10), and this does not appear to inhibit the organisms; it keeps the pH below 7.8 for an extended period.

Although S. lipoferum is quite capable of fixing N_2, it can grow more rapidly when supplied a source of combined nitrogen. It will grow with a generation time of about one hour when supplied ammonia as the nitrogen source (13). Under these conditions S. lipoferum grows as a typical aerobic organism. In contrast, when fixing N_2 the organism is microaerophilic. As indicated in Fig. 1, the organism grows very rapidly until it exhausts the small amount of ammonia that was transferred with the ammonia-grown inoculum. There is a very short lag period as the organism adjusts to the shift from the use of combined nitrogen to the use of N_2. Then after a short transition period, it assumes a new and slower rate of growth characteristic of its growth on N_2. The generation time increases from an hour on ammonia to 5.5 to 7 hours on N_2. When

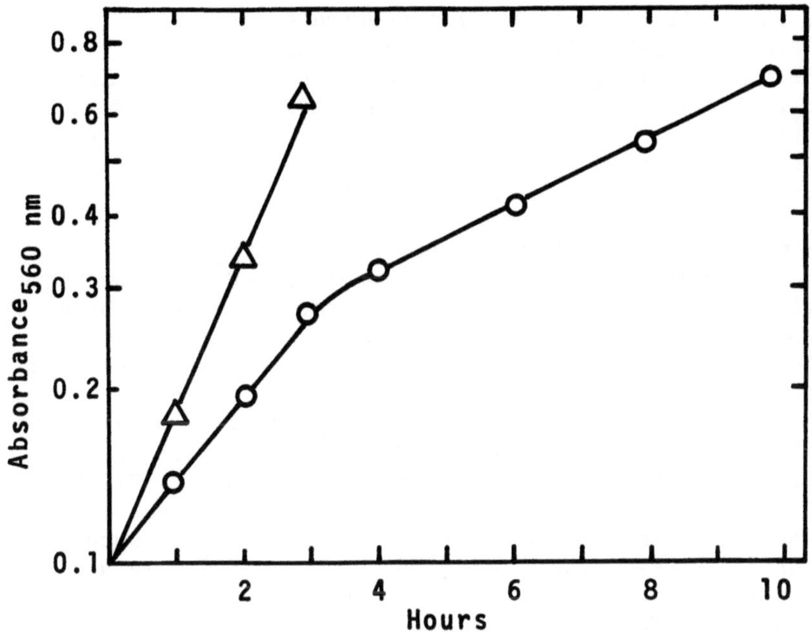

Fig. 1. Growth curves of S. lipoferum grown at 0.2 atm O_2 in the presence of 0.1 % NH_4Cl (Δ), and on N_2 in a nitrogen-free medium at a pO_2 of 0.005 to 0.007 atm and a pN_2 of about 0.99 atm (o). Residual NH_4^+ from the inoculum supported growth at the low pO_2 for about 3 h.

S. lipoferum is supplied ammonia, it is not inhibited by air sparged vigorously through the culture (Fig. 1, top line). However, when the organism is utilizing N_2, such aeration of the culture is distinctly inhibitory. Later we will discuss further the effect of the partial pressure of O_2.

Another characteristic of growth of S. lipoferum on N_2 is that it produces rather high levels of poly-β-hydroxybutyrate (11). When supplied ammonia, the levels of the polymer are minimal. Apparently there is such a marked difference in metabolism on the two sources of nitrogen that N_2-grown cells may contain 25% of their dry weight as poly-β-hydroxybutyrate, whereas ammonia-grown cells may have less than 1%.

Measurements of respiration gave responses to substrates very similar to the responses to substrates in support of growth of S. lipoferum (11). Again, most of the strains did not utilize sugars well to support uptake of oxygen, but they used organic acids effectively. If one grows the organism on a substrate such as malate, it adapts not only to this substrate but also to the comparable four carbon dicarboxylic acid, succinic acid. Likewise, when the organism is grown on lactate, it not only uses lactate as an excellent respiratory substrate, but it also uses pyruvate very effectively (11). The organic acids support a rate of oxygen uptake by S. lipoferum comparable to that observed with other bacteria on their most active substrates.

Examination of a heavy suspension of S. lipoferum with a microspectroscope reveals the presence of characteristic bands of a cytochrome system. There appear to be two type-c cytochromes as well as a type-b cytochrome and an oxidase. Characteristically, cytochrome systems are capable of operating at a low partial pressure of oxygen, and hence a cytochrome system is well adapted to supporting the respiratory activity of a microaerophilic organism such as S. lipoferum. An extract from S. lipoferum has an absorption spectrum as shown in Fig. 2. The peak at 559 nm may correspond to a b-type and the peak at 553 nm to a c-type cytochrome. The peak at 524 nm probably represents the β-band of the type-c cytochrome. The c-type cytochrome has not been purified to homogeneity, but its characteristics are those one would anticipate for a cytochrome of this type.

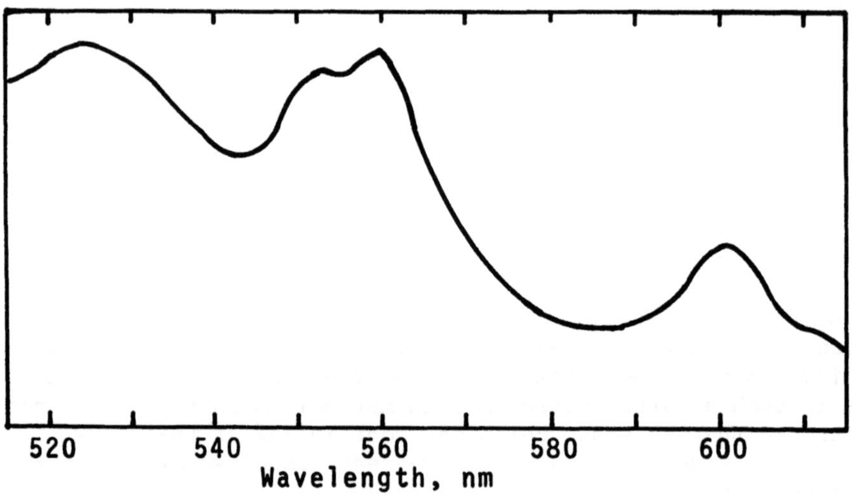

Fig. 2. Spectrum of an extract from Spirillum lipoferum that shows absorbance by components of the cytochrome system.

It has been possible to isolate a cell-free nitrogenase from *S. lipoferum* by methods rather similar to those employed for the isolation of the enzyme system from *Rhodospirillum rubrum* (11, 13). The organism can be disrupted with a French press, and after preliminary treatment the two nitrogenase components can be separated on DEAE cellulose and Sephadex. Although these preparations have not been purified to homogeneity, they have been purified to the extent that the individual fractions have no nitrogenase activity until supplemented with the complementary fraction. The properties of this nitrogenase are very similar to those of other nitrogenases. It has an absolute requirement for MgATP, and its activity must be supported by a strong reducing agent such as reduced ferredoxin or $Na_2S_2O_4$. The physical-chemical properteis of the individual proteins have not been established.

An interesting property of the Fe protein is its requirement for an activating factor (9). Among all the nitrogenases we have tested, only those from *R. rubrum* and *S. lipoferum* release an Fe protein requiring activation. The activating factors from the two organisms are interchangeable. When incubated in the presence of the activating factor plus Mn^{++} and MgATP, the inactive Fe protein is converted to a catalytically active form. The activating factor then can be removed, and the Fe protein remains active in its absence. The activating factor is a protein and operates in catalytic fashion, but it still is not clear exactly how it converts the Fe protein from an inactive to an active form. The phenomenon is of particular interest, because activation has the potential to be an important control mechanism for the enzyme system.

Examination of the nitrogen metabolism of *S. lipoferum* suggests that the organism assimilates fixed nitrogen from N_2 by way of glutamine synthetase and glutamate synthase (16, 20). In this reaction, glutamate + ammonia form glutamine catalyzed by glutamine synthetase. Then glutamate synthase catalyzes the conversion of glutamine + α-ketoglutarate + NADPH + H^+ to 2 glutamate. As a general rule, the organisms fixing N_2 have a limited concentration of ammonia available, and they utilize the glutamine synthetase reaction rather than glutamate dehydrogenase (20) to assimilate the ammonia initially. The Michaelis constant for ammonia is considerably lower for glutamine synthetase then for glutamate dehydrogenase. On the other hand, when the organisms have an abundant supply of externally furnished ammonia, they commonly utilize the glutamate dehydrogenase pathway for assimilation. In this reaction, α-ketoglutarate + ammonia + NADH + H^+ is converted to glutamate + NAD^+. An examination of the influence of methionine sulfoximine can be helpful in distinguishing the pathways of nitrogen assimilation. Ammonia represses synthesis of nitrogenase, but in the presence of methionine sulfoximine the system is derepressed and nitrogenase is formed. As indicated in Table II (11), the addition of methionine sulfoximine (or methionine sulfone) to growing cultures of *S. lipoferum*

TABLE II

Effect of various derepressing substances on acetylene reduction by NH_4^+-grown S. lipoferum*

Addition/ml	Rate of acetylene reduction as a function of time of incubation (h)			
	2	6	8	16
0.8 mg of NH_4Cl	0	0	0	0
NH_4Cl + 20 mg of methionine sulfoximine	0	1.5	3	15
NH_4Cl + 20 mg of methionine sulfone	0	0	0	20

*Data are given as nanomoles of C_2H_4 produced/(hour x ml culture). For further details see (11).

effects a derepression of nitrogenase synthesis. Switching from N_2 to NH_4^+ is accompanied by a change in balance among the ammonia assimilating enzymes, such as would be predicted for a system now operating with the glutamate dehydrogenase pathway of assimilation. The total glutamine synthetase activity from N_2-grown cells was about eight times that from ammonia-grown cells, and the response of glutamate dehydrogenase was the opposite. The glutamate synthase activity in N_2-grown cells was about twice that in ammonia-grown S. lipoferum. As indicated earlier, the culture growing on ammonia quickly shifts with a minimal lag period to the utilization of N_2 as a nitrogen source upon exhaustion of the externally supplied ammonia.

The nitrogenase from S. lipoferum is very similar to that from other organisms. As indicated in the table on cross-reactions in our paper on "Nitrogenase Systems" in this volume, it is possible to cross the individual components of nitrogenase from S. lipoferum with the components from a variety of other organisms. The MoFe protein from S. lipoferum formed active crosses with the Fe protein from all organisms tested with the exception of Clostridium pasteurianum. Likewise, the Fe protein from S. lipoferum formed active crosses with the MoFe proteins from all organisms tested except C. pasteurianum. There is variation in activity among these crosses, but there is a demonstrable homology with all the organisms except C. pasteurianum. Perhaps there is a tight but ineffective binding between the proteins of S. lipoferum and

C. pasteurianum comparable to the binding that has been observed between Azotobacter vinelandii and C. pasteurianum (8).

The interesting response of S. lipoferum to the partial pressure of O_2 has been mentioned before. The organism supplied N_2 as a source of nitrogen is microaerophilic in nature. We were interested in determining what the optimal concentration of O_2 was for the organism when it was fixing N_2. As is well known, the nitrogenase itself is sensitive to O_2, and N_2-fixing organisms have adopted various techniques for avoiding O_2 inactivation of their nitrogenase. S. lipoferum apparently has no specific mechanism; it merely ceases to grow on N_2 unless the partial pressure of O_2 is low. To determine the optimal level, we have used an oxygenstat that senses the concentration of dissolved O_2 in the medium with a sterilizable O_2 electrode (14) and calls for an additional supply of O_2 whenever there is a depletion of dissolved O_2 below the set level. Our observations, obtained during the exponential growth phase, indicated the maximum growth rate occurred at a partial pressure of O_2 between 0.005-0.007 atm. This is a rather low concentration considering that the organism grows vigorously in air when supplied ammonia as a source of fixed nitrogen.

To determine the optimal concentration of O_2 in terms of conversion of substrate to cellular nitrogen, the oxygenstat again was used to maintain specific levels of dissolved O_2 (13). The optimal pO_2 proved to be lower for maximal cell yield than for most rapid growth. The highest yield (about 12 mg cellular N/g malate used), measured during exponential growth on malate, was at a partial pressure of O_2 of about 0.002 atm. Others have reported somewhat higher efficiencies under different conditions of measurement (3, 18). Our measurements were made on organisms growing exponentially, and the content of malate in the medium was measured at the time the first sample for cellular nitrogen was taken and was compared to the level at the time the second measurement of cellular nitrogen was made. Thus, cell yield and substrate utilization were measured on the same samples during conditions of exponential growth.

There have been reports that the association between S. lipoferum and certain plants, particularly tropical grasses, may add substantial amounts of fixed nitrogen to the soil (4-6, 15, 19). Our tests on the association have been made in the field and in a facility providing controlled temperature and light intensity. The Biotron at the University of Wisconsin has a room in which the light intensity and temperature can be varied at right angles to each other. We adjusted the controls to give temperatures of 28, 32, 36, and 40°C. during the day and 10° less than this at night. At right angles to the temperature gradient, the light intensity gradient was set at 500, 1250, 2400, and 3000 ft-c. Thus in a single room there were 16 conditions of light and temperature. Maize plants were placed under the 16 conditions, and three pots of maize were inoculated with S. lipoferum and two were kept as controls in each treatment (controls received heat-killed S. lipoferum). Unfortunately, there was inadequate space for more thorough replication of the treatments.

Under these conditions, which embraced a rather broad range of temperature and light intensity, we found relatively small changes in apparent N_2 fixation. The plants differed measureably in total dry weight and percentage nitrogen from one end of each gradient to the other, but there was very little difference between the inoculated and the control plants. Periodically, roots were sampled from one of the inoculated plants and tested for their ability to reduce acetylene. Acetylene reduction often was detected, although it was not particularly vigorous. Responses were highly variable between replicates (1), so one could not rely on the data to predict the N_2-fixation under the various treatments (Table III). These observations are similar to those reported by Barber et al. (2) and by Tjepkema and Van Berkum (17). These investigators also found variable responses when roots were preincubated at a low pO_2 and then were tested for their rates of C_2H_2 reduction. The preincubated roots exhibited rates of C_2H_2 reduction were often 30 times the rates of comparable sets of roots examined in soil cores. The difficulties with the preincubated root method usually have been attributed to proliferation of N_2-fixing bacteria during preincubation (2, 12). Rates of C_2H_2 reduction between replicate soil cores are more consistent than rates between preincubated root samples, and they are more nearly comparable than rates by preincubated roots to rates of N_2 fixation measured by increase in total nitrogen of the plants.

In our opinion, greater reliance can be placed upon the total dry weight of the plant and the total nitrogen of the plant at harvest than on C_2H_2 reduction by preincubated roots. Tables IV and V show very small differences in dry weights and total nitrogen yields between the inoculated and the uninoculated maize. Calculation from the increase in total nitrogen of our maize plants suggested an increase of about 375 grams of nitrogen per hectare during a 100 day growing season. This is scarcely beyond the experimental error of the method, and certainly under these specific conditions the increase of nitrogen from "associative symbiosis" would have little practical significance to the overall growth of maize.

We also have made limited field trials with S. lipoferum as the inoculant for a variety of plants. The trials in 1975 gave a somewhat larger number of positive than negative results, but none of the positive increases was statistically significant. A

TABLE III

Rates of acetylene reduction by excised roots that had been pre-incubated at a low pO_2 before acetylene was supplied. Roots were removed from inoculated plants and were placed in an atmosphere with 0.025% O_2 for 18 hr before C_2H_2 was supplied. Rates shown in the table represent nmoles C_2H_2 formed in an h/g dry roots.

Temperature, C	Light, ft-c	Days after planting						
		48	58	67	74	81	89	94
28	500	43	7	36	7	169	0	125
28	1250	429	0	37	15	471	28	202
28	2400	112	64	7	45	867	23	866
28	3000	18	27	16	0	266	141	308
32	500	3	43	250	35	364	90	54
32	1250	109	13	9	145	1042	0	434
32	2400	15	71	64	30	1615	42	26
32	3000	8	358	48	61	711	209	65
36	500	163	22	32	6	283	2871	300
36	1250	3	34	149	92	903	1273	69
36	2400	2	517	58	45	203	0	0
36	3000	2	162	14	0	119	0	0
40	500	222	52	82	7	928	914	25
40	1250	66	7	128	4	362	1675	242
40	2400	384	492	164	16	743	1375	713
40	3000	97	566	12	80	266	0	6

TABLE IV

The dry weights of 94 day maize plants grown under different temperatures and light intensities. The data are expressed as grams dry weight of the entire above-ground portion of the maize plant (average values for 3 inoculated and 2 uninoculated plants are given).

Treatment	Temp, C	Light intensity, ft-c				
		500	1250	2400	3000	Average
Inoc	40	27.8	34.5	32.4	32.3	31.8±2.8
Control		25.0	30.8	28.3	33.8	29.5±3.7
Inoc	36	31.0	34.6	33.8	29.5	32.2±2.4
Control		26.7	35.9	31.1	36.0	32.4±4.5
Inoc	32	31.2	33.5	36.8	41.0	35.6±4.3
Control		32.6	38.6	38.8	36.2	36.6±2.9
Inoc	28	34.2	42.2	44.2	43.0	40.9±4.5
Control		30.7	43.1	44.1	43.6	40.4±6.5
Average	Inoc.	31.1±2.6	36.2±4.0	36.8±5.3	36.5±6.6	
	Cont.	28.8±3.5	37.1±5.1	35.6±7.2	37.4±4.3	

TABLE V

The total nitrogen of 94 day maize plants grown under different temperatures and light intensities. The data are expressed as grams of nitrogen in the entire above-ground portion of the maize plant (average values for 3 inoculated and 2 uninoculated control plants are given).

Treatment	Temp, C	Light intensity, ft-c 500	1250	2400	3000	Average
Inoc	40	0.41	0.36	0.32	0.33	0.36±0.04
Control		0.42	0.37	0.30	0.35	0.36±0.05
Inoc	36	0.33	0.31	0.29	0.26	0.30±0.03
Control		0.37	0.32	0.29	0.30	0.32±0.04
Inoc	32	0.34	0.27	0.25	0.31	0.29±0.04
Control		0.31	0.27	0.25	0.24	0.27±0.03
Inoc	28	0.38	0.29	0.27	0.25	0.30±0.06
Control		0.34	0.29	0.27	0.25	0.29±0.04
Average	Inoc	0.37±0.04	0.31±0.04	0.28±0.03	0.29±0.04	
	Cont	0.36±0.05	0.31±0.04	0.28±0.02	0.29±0.05	

summary of some of the data from our 1976 trials in Table VI shows that 3 different treatments gave a statistically significant increase in the total nitrogen of the plant. The plants were grown at two locations, on a sandy soil with irrigation (Hancock Experiment Station) and on a clay-loam soil (Charmany Farms). The total nitrogen of the above-ground portion of the plants was determined; data are expressed as g of total nitrogen (± the standard deviation) in the above-ground portion of single plants (popcorn, maize) or of a composite sample from 0.1 sq meter of other grasses (sorghum-sudan, wheat, German millet, white proso millet, fescue and Digitaria sanguinalis). No statistically significant increases were found in the plants grown on the clay-loam soil. On the sandy soil there was a statistically significant increase in total nitrogen of inoculated vs control inbred maize A629, wheat and German millet. The increases were not large, but they were encouraging. Among the other 11 plant varieties used, 8 showed higher total nitrogen when inoculated, one showed the same and two showed lower total nitrogen when inoculated; the differences between inoculated and control plants were not statistically significant.

Although it has been possible in very few cases for us to demonstrate a statistically significant improvement in the yield of maize and other plants in the field by inoculation with

TABLE VI

Total nitrogen (grams) in field-grown plants inoculated with *S. lipoferum* or treated with heat-killed *S. lipoferum* (control). See text for details.

Cultivar*	Sand		Clay	
	Inoc.	Control	Inoc.	Control
White cloud	1.50±0.44	1.03±0.21	1.01±0.52	1.26±0.68
Gold rush	1.45±0.45	1.12±0.45	1.73±0.78	1.66±0.58
W182E	1.17±0.31	0.77±0.32	1.11±0.81	0.89±0.50
W513	1.04±0.30	0.94±0.25	0.87±0.58	0.78±0.29
A629	1.36±0.26	0.91±0.21	0.58±0.20	1.13±0.50
Sorghum-Sudan	4.99±0.44	6.26±1.04	13.97±4.84	11.52±4.96
Wheat	1.31±0.28	0.78±0.20	-	-
German millet	3.00±0.19	2.22±0.32	1.74±0.39	2.58±0.57
White proso millet	2.32±0.41	2.20±0.30	1.24±0.50	0.62±0.35
Fescue	0.93±0.27	0.84±0.25	-	-
Field corn	1.75±0.23	1.75±0.28	1.67±0.67	1.95±1.03
Wis. 900	1.75±0.39	1.59±0.42	1.18±0.68	1.30±0.68
Golden cross	1.15±0.31	1.20±0.38	0.45±0.27	0.75±0.28
Digitaria sanguinalis	4.15±0.51	3.27±0.72	-	-

*Popcorn white cloud; sweet corn Golden cross, Wis. 900, Gold rush; inbred maize W182E, W513, A629; hybrid maize fieldcorn.

S. lipoferum, it seems important that the search continue for proper associations between higher plants and N_2-fixing organisms that may proliferate in the roots or in the rhizosphere around roots. The system remains attractive, becasue it is simple enough to invite exhaustive fixation tests as well as physiological and genetic manipulations. The organism can grow at the reduced pO_2 of the plant root without the elaborate hemoglobin-nodule system of the rhizobia-legume association. The organisms have access to photosynthate, and their fixed nitrogen should be transported readily through the vascular system of the plant. Large numbers of bacterial strains can be tested with a variety of plants, and both bacteria and plants can be selected from mutagenized or natural isolates. It seems particularly important to search for bacterial strains that will establish themselves in large numbers in roots without overwhelming the host plant.

It seems reasonable to say that researchers have only scratched the surface in investigating the possibilities of associative symbioses. Different plants, different microorganisms, different methods of practical husbandry all should be tested. The dependence of the world upon cereal grains as a major food source makes it worthwhile to examine the associative symbioses in some detail and to search for suitable plant-bacterial associations that will fix substantial amount of N_2 from the air. In this way we may decrease the dependence of these crops upon chemically fixed nitrogen.

REFERENCES

1. Albrecht, S.L., Okon, Y., and Burris, R.H. (1977) Plant Physiol. in press.

2. Barber, L.E., Tjepkema, J.D., Russell, S.A., Evans, H.J. (1976) Appl. Environ. Microbiol. 32, 108.

3. Day, J.M., and Döbereiner, J. (1976) Soil Biol. Biochem. 8, 45.

4. Day, J.M., Neves, M.C.P., and Döbereiner, J. (1975) Soil. Biol. Biochem. 7, 107.

5. DePolli, H., Matsui, E., Döbereiner, J., and Salati, E. (1976) Soil. Biol. Biochem. 8, 1.

6. Döbereiner, J., and Day, J.M. (1976) in Proceedings of the First International Symposium on Nitrogen Fixation (Newton, W.E., and Nyman, C.J., eds.) Washington State Univ. Press, Pullman.

7. Döbereiner, J., Marriel, L.E. and Nevy M. (1976) Can. J. Microbiol. 22, 1464.

8. Emerich, D.W., and Burris, R.H. (1976) Proc. Nat. Acad. Sci. U.S.A. 73, 4369.

9. Ludden, P.W., and Burris, R.H. (1976) Science 194, 424.

10. Okon, Y., Albrecht, S.L. and Burris, R.H. (1976) J. Bacteriol. 127, 1248.

11. Okon, Y., Albrecht, S.L., and Burris, R.H. (1976) J. Bacteriol. 128, 592.

12. Okon, Y., Albrecht, S.L., and Burris, R.H. (1977) Appl. Environ. Microbiol. 33, 85.

13. Okon, Y., Houchins, J.P., Albrecht, S.L., and Burris, R.H. (1977) J. Gen. Microbiol. 98, 87.

14. Phillips, D.H., and Johnson, M.J. (1961) J. Biochem. Microbiol. Tech. Engineering 3, 277.

15. Smith, R.L., Bouton, J.H., Schank, S.C., Quesenberry, K.H., Tyler, M.E., Milam, J.R., Gaskins, M.H., and Littell, R.C. (1976) Science 193, 1003.

16. Stadtman, E.R., Ginsburg, A., Ciardi, J.E., Yeh, J., Hennig, S.B., and Shapiro, B.M. (1970) Adv. Enzyme Regulation 8, 99.

17. Tjepkema, J., and Van Berkum, P. (1977) Appl. Environ. Microbiol. 33, 626.

18. Vargas, M.T., and Harris, R.F. (1977) in Meetings of the Am. Soc. Microbiol., New Orleans.

19. Von Bülow, J.F.W., and Döbereiner, J. (1975) Proc. Natl. Acad. Sci. U.S.A. 72, 2389.

20. Wolk, C.P., Thomas, J., and Shaffer, P.W. (1976) J. Biol. Chem. 251, 5027.

TAXONOMY OF THE ROOT-ASSOCIATED NITROGEN-FIXING BACTERIUM *SPIRILLUM LIPOFERUM*

Noel R. Krieg and Jeffrey J. Tarrand

Biology Department, Virginia Polytechnic Institute and State University, Blacksburg, Virginia U.S.A. 24061

For the past two years, Jeffrey Tarrand and I have been concerned with the classification of *Spirillum lipoferum*, a nitrogen-fixing bacterium first described by Beijerinck (2) in 1925. Except for a few scattered reports in the literature (1, 6, 12) this organism was largely forgotten until 1974 when Döbereiner and Day (4) showed that it was associated with the roots of tropical grasses. Because a large number of strains have now been isolated, and because precise communication between investigators is one of the goals of bacterial taxonomy, we began a comprehensive taxonomic study of the *S. lipoferum* group.

The specific objectives of our work were (a) to determine by the use of DNA base composition, and especially by the use of DNA homology experiments for direct comparison of bacterial genomes, how many species are represented by the strains in the *S. lipoferum* group, (b) to learn what phenotypic characteristics can, by correlation with the genetic data, serve to distinguish such species, (c) to provide a general characterization of the organisms, and (d) to investigate the problem of which genus would be the most appropriate for the organisms.

In our initial studies, we wished to determine the overall DNA base composition of a number of strains isolated from different regions and different plants. If significant differences occurred in DNA base composition, this alone would be strong evidence for the existence of more than one species. The determination of the base composition was done by thermal denaturation techniques, with the DNA of *Escherichia coli* B serving as a reference. The melting points (Tm) and mol% guanine + cytosine (G + C) values are presented in Table 1. As indicated, the G + C values

TABLE 1. Melting points (Tm values) and mol% G + C values for the DNA of 11 strains of the *S. lipoferum* group*

Strain	Source	Average Tm, °C	Mol% G + C (± 1%)
Sp 7	*Digitaria*, Km 47, Brazil	98.1	70
Sp 13	" " "	98.0	70
Sp 4	" " "	97.8	70
Sp 35	" " "	98.2	70
Sp M82	Maize, " "	98.4	71
Sp M75	" " "	98.2	70
Sp 51e	Wheat, " "	98.2	70
Sp T60	" " "	98.4	71
Sp 59b	" " "	97.9	70
Sp RG6xx	Wheat, Rio Grande do Sul, Brazil	97.8	70
Sp USA5b	Soil with natural grass cover, Pullman, Washington, U.S.A.	97.6, 97.8	69, 70**

*The DNA was extracted by lysing the cells in 1X saline-EDTA (NaCl, 0.15 M; EDTA, 0.01 M; pH 8.0) with 1% sodium lauryl sulfate (SLS) at 60 °C. After 2 phenol extraction to remove protein, the DNA was purified by the method of Marmur and Doty (11). Bovine pancreatic ribonuclease was used to remove most of the RNA. A concentration of 50 µg/ml was prepared in 0.025% phosphate buffer (pH 6.8) + 0.01% SLS, and the preparations were dialyzed in this buffer together with *E. coli* B reference DNA. Tm values were determined with a Gilford thermospectrophotometer, and the mol% G + C was calculated by the formula

$$\text{mol\% G + C} = \frac{\text{Tm} - 69.3}{0.41}$$

The Tm value of *E. coli* B (90.5 °C) was used to normalize the Tm values of the test strains.

**In the case of the second value, the DNA was prepared by the hydroxylapatite method as described in Table 2.

were all ca. 70%, and this similarity indicated that the strains could all belong to a single species. However, subsequent DNA homology experiments indicated this was not the case.

A rigorous comparison between the DNA obtained from various bacterial strains can be made on the basis of the sequence of bases in the DNA, as it is this sequence which comprises the genetic specificity of an organism. In our DNA homology experiments, we employed essentially the membrane-filter competition method described by Johnson and Cummins (9) in 1972. Crude preparations of

DNA were sonicated and then purified by a method involving specific adsorption to hydroxylapatite (HA). The HA was then washed repeatedly and the DNA finally eluted. Competitor DNA preparations were subjected to an alkaline treatment to destroy residual RNA and also to denature the DNA. Radioactive homologous DNA was labeled by growing reference strains Sp 7 and Sp 59b with 32-P or 33-P and was thermally denatured. In early experiments, DNA for binding to membrane filters was prepared by the method of Marmur and Doty (11); however, when it was found that DNA isolated by the HA method gave similar results, the HA method was subsequently used. All competition experiments were incubated at 25°C below the Tm of the DNA -- i.e., at 73°C -- to minimize non-specific duplex formation.

The results of the DNA homology experiments are presented in Table 2. It can be seen that when strain Sp 7 was used as the reference strains, the other strains fell into two distinct but related groups --- those strains having ca. 70% or higher homology values with Sp 7, and those having homology values in the 30-50% range. In the reciprocal situation where strain Sp 59b was used as the reference, those strains which had exhibited low homology with Sp 7 now exhibited high homology with Sp 59b. Conversely, those strains which had exhibited high homology with Sp 7 now had low homology with Sp 59b.

These results provide strong evidence that two distinct but related species existed in the S. lipoferum group, and also that the two species belonged together in the same genus.

The techniques of DNA homology are painstaking, time-consuming, and require special equipment such as accurate circulating water baths, liquid scintillation counters, and radioisotope facilities. However, once such studies have indicated the relationships among a number of strains, then those phenotypic characteristics which are highly conserved -- i.e., those characteristics which are found to correlate well with the genetic data -- can then be used by other investigators to determine easily the species to which a new isolate would belong.

In the case of our S. lipoferum strains, we have found certain characteristics that were correlated with the two homology groups. These are indicated in Table 3, as well as details of the methods. Other differential characteristics may well exist. In contrast to strains of Group I, strains of Group II appeared to possess some fermentative ability, as judged from the acidification of glucose or fructose media anaerobically, ability to exhibit slight growth in sugar broth and even to form minute colonies on sugar plates anaerobically, and formation of a small amount of gas. Yet Group II grows far better aerobically than anaerobically, and, like Group I, appears to have mainly a respiratory type of metabolism.

TABLE 2. DNA homology values for various strains of *Spirillum lipoferum**

Strain	Source	DNA homology values (%)	
		Reference = Sp 7	Reference = Sp 59b
Sp 7	*Digitaria*, Km 47, Brazil	100	34
Sp 4	" " "	102	38
Sp 13	" " "	94	32
Sp 13v (Pink)	" " "	103	47
Sp 34	" " "	81	43
Sp 35	" " "	74	42
Sp 51e	Wheat, " "	71	41
Sp 52	Sorghum, " "	85	44
Sp 67	Maize, " "	80	30
Sp M75	" " "	71	49
Sp 80	" " "	84	52
Sp 81	" " "	73	41
Sp M82	" " "	73	41
Sp P1	Soil, Peru	72	46
Sp P2	" "	82	52
Sp F4	Millet, Florida	80	35
Sp F6	" "	74	40
Sp Br 8	Soil, Brasilia, Brazil	84	37
Sp Br 11	Maize, " "	67	38
Sp Br 13	Soil, " "	72	41
Sp Br 14	Wheat, " "	81	38
Sp Br 21	" " "	80	Not done
Sp A2	Maize, Ibadan, Nigeria	92	50
Sp A7	Rice, " "	96	33
Sp A8	*Panicum*, " "	81	42
Sp MT 20	Soil, Mato Grosso, Brazil	82	39
Sp T60	Wheat, Km 47, Brazil	85	30

TABLE 2, continued

Strain	Source	DNA homology values (%)	
		Reference = Sp 7	Reference = Sp 59b
Sp MT 21	Soil, Mato Grosso, Brazil	81	35
Sp L 69	Wheat, Londrina, Brazil	83	49
Sp Col 1c	Soil, Colombia	80	34
Sp RG 8a (Pink)	Wheat, Rio Grande do Sul, Brazil	104	37
Sp RG 16a	_Lolium_, " " " " "	77	35
Sp RG 20b	Wheat, " " " " "	73	32
JM 6A2	_Zea mays_, Ecuador	78	49
JM 6B2	" " "	87	48
JM 24B4	_Musa_, "	83	48
JM 28A2	" "	91	46
JM 52B1	_Panicum_, Venezuela	86	40
JM 73B3	_Zea mays_, "	81	45
JM 73C3	" " "	79	40
JM 73C28	" " "	83	49
JM 75A1	_Panicum_, "	77	45
JM 82A1	_Zea mays_, "	90	45
JM 119A4	_Pennisetum_, Florida	91	47
JM 125A2	" "	88	52
Cd	_Cynodon dactylon_, California	100	49
Sp 59b	Wheat, Km 47, Brazil	31	100
Sp USA 5b	Soil, Pullman, Washington U.S.A.	36	70
Sp RG6xx	Wheat, Rio Grande do Sul, Brazil	36	73
Sp RG 8c	" " " " " "	34	72
Sp RG 9c	" " " " " "	35	73
Sp RG 18b	" " " " " "	39	75

TABLE 2, continued

Strain	Source	DNA homology values (%)	
		Reference = Sp 7	Reference = Sp 59b
Sp RG 19a	*Digitaria*, Rio Grande do Sul, Brazil	28	72
Sp RG 20a	Wheat, " " " " "	37	76
Sp Col 2b	Maize, Papayan, Colombia	43	72
Sp Col 3	*Brachiaria*, Colombia	34	76
Sp Col 5	*Hyparrhenia rufa*, Colombia	46	73
Sp Br 10	Soil, Brasilia, Brazil	38	74
Sp Br 17	Maize, " "	30	73
Sp A3a	Grass, Dakar, Senegal	29	70
Sp Sla	Lawn north of Uppsala, Sweden	34	73

*Harvested cells were lysed at 60°C in 0.1 M NaCl + 0.01 M EDTA + 0.1 M Na borate buffer (pH 8) by addition of 1% SLS. The lysate was sonicated briefly and protein was removed by phenol extraction. The DNA was purified by a method involving adsorption to HA, repeated washing of the HA, and subsequent elution in phosphate buffer. For competitors, DNA was sonicated briefly again and the phosphate was removed by dialysis in 0.1X SSC (1X SSC = NaCl, 0.15 M; Na citrate, 0.015 M; pH 7). The DNA was precipitated in 1X SSC by alcohol and was redissolved in 0.02 M NaCl + 0.001 M HEPES buffer (pH 7). To denature the DNA and also to destroy residual RNA, 0.05 ml of 5N NaOH was added per ml, the mixture was incubated at 45 C for 15 min and was then quickly chilled and neutralized. It was then dialyzed with 2.2X SSC, diluted to 1.5 mg/ml, and frozen. Labeled homologous DNA was prepared by growing cultures with 32-P or 33-P and isolating the DNA by the HA method. The DNA was denatured thermally. For preparation of DNA filters, in initial experiments DNA prepared by the method of Marmur and Doty was used, but subsequently DNA prepared by the HA method was used. Membrane filters were prepared as described by Johnson and Cummins (9). Competition reaction were conducted as described by Johnson and Cummins (9), and incubations were carried out at 25°C below the Tm of the DNA -- i.e., at 73 C.

Another useful difference between the two groups has been found by Döbereiner (personal communication) using semi-solid N-free medium containing vitamins but with glucose substituted for malate. Here, Group II strains grew heavily in this medium, whereas Group I strains did not.

Concerning other characteristics of the S. lipoferum strains, many of these are summarized in Table 4. When cultured for 24 h in PSS broth, both Group I and II strains exhibited a similar morphology (see Fig. 1), namely, short, plump, slightly-curved motile rods, ca. 1.0 µm in diameter. Many cells had pointed ends. Poly-β-hydroxybutyrate granules were present. In semi-solid malate medium + 0.005% yeast, however, Group II strains tended to become larger and longer, and many S-shaped or spirillum-shaped cells appeared (see Fig. 2). The long cells seemed to undergo fragmentation eventually into pleomorphic misshapen cells filled with large granules, probably poly-β-hydroxybutyrate. Group II strains also tended to lose their motility in the semi-solid medium. In contrast, Group I strains retained motility for up to a month, and the cells tended to remain a normal size. In old cultures, however, some S-shaped cells could be found (10).

One of the most interesting aspects of the strains was their type of flagellation. Until a short time ago, the cells were considered to be monotrichous, and this type of flagellation indeed does occur when the cells are cultured in liquid media such as PSS broth. However, when cultured on PSS agar at 30°C for 3 days, numerous lateral flagella of shorter wavelength occur in addition to the single polar flagellum. This situation is similar to that described for Pseudomonas stutzeri (5), Vibrio parahaemolyticus (13), and Vibrio alginolyticus (3). Electron micrographs of two representative strains of S. lipoferum are presented in Figs. 3 and 4, showing the two types of flagella.

With regard to the appropriate genus for the S. lipoferum strains, Beijerinck (2) placed the organisms in the genus Spirillum. However, this genus is presently reserved for organisms resembling Spirillum volutans (7) and having a mol% G + C of ca. 38.

If significant DNA homology were to occur between S. lipoferum and a recognized member of an established genus, this would constitute strong evidence for assigning S. lipoferum to that genus. We have tested a number of strains from several genera, such as Aquaspirillum, Derxia, Comamonas, Azomonas, and Pseudomonas, against S. lipoferum, but have failed to obtain any repeatable DNA homology values greater than ca. 20%. We consider such values to have little significance (8). In fact, DNA homology studies are most useful at the species level of classification, not the genus level, and

TABLE 3. Distinction between homology groups on the basis of phenotypic characters

Test	% of strains positive	
	Group I	Group II
Acid from glucose[1,2]	0	100
Acid from mannitol, ribose, and sorbitol[2]	0	80
Acid from glucose and fructose in anaerobic hydrogen jar (slight growth also occurs) [3]	0	100
Biotin required for growth[4]	0	100
Sole C sources (by disc test)[5]		
Ketoglutarate	0	100
Mannitol	0	80, + 7% variable
Sorbitol	0	67, + 13% variable
Ribose	0	40, + 27% variable
Glucose[6]	9	100
In semisolid N-free malate medium + 0.005% yeast extract, cells tend to become wider, longer (often S or spirillum-shaped), and non-motile.	0	100

[1]The medium used had the following composition (g/l): Peptone (Difco), 2.0; $(NH_4)_2SO_4$, 1.0; $MgSO_4.7H_2O$, 1.0; $FeCl_3.6H_2O$, 0.002; $MnSO_4.H_2O$, 0.002; brom thymol blue, 0.025. The pH was adjusted to 7.0 - 7.1 with KOH. Filter-sterilized sugar was added to a final concentration of 1.0%. Cultures were incubated for 4 days at 37°C. A yellow color indicated acid production.

[2]The following medium was used (g/l): Yeast extract (Difco), 0.05; K_2HPO_4, 0.25; $FeSO_4.7H_2O$, 0.01; $Na_2MoO_4.2H_2O$, 0.001; $MnSO_4.H_2O$, 0.002; $MgSO_4.7H_2O$, 0.2; NaCl, 0.1; $CaCl_2.2H_2O$, 0.026; $(NH_4)_2SO_4$, 1.0; biotin, 0.0001; brom thymol blue, 0.0375; agar (Difco), 15.0; pH adjusted to 7.1 with KOH. Filter-sterilized sugar was added aseptically to a final concentration of 1.0%. The medium was added to sterile micro-titer plates (NUNC), ca. 0.16 ml per well. The method of Wilkins et al. (14) was used for inoculation of the wells, by use of a replicator constructed of wire brads projecting from a plastic block, sterilized by alcohol and UV light. The replicator was inoculated from a master micro-titer plate, each well containing a different strain. The replicator was then placed over a micro-titer plate containing sugar medium and pressed onto this. Thus, each brad would inoculate a particular strain into a particular well of sugar medium. After incubation for 72 h at 37°C, the pH of each well was determined with a small pH electrode (Sargent No. S-30070-10) as described by Wilkins et al. (14). The pH of uninoculated wells ranged from 5.7-6.2 (probably from absorption of carbon di-

TABLE 3, continued

oxide). Acid production by a strain was considered positive if the pH of the well was 3.9-5.0. Essentially the same method could be used without a replicator by inoculating wells with a straight inoculating needle.

[3]See footnote 1. Streaked plates of sugar medium were also used.

[4]The medium used had the following composition (g/l): K_2HPO_4, 0.5; succinic acid, 5.0; $FeSO_4.7H_2O$, 0.01; $Na_2MoO_4.2H_2O$, 0.002; $MgSO_4.7H_2O$. 0.2; NaCl, 0.1; $CaCl_2.2H_2O$, 0.026; $(NH_4)_2SO_4$, 1.0. The pH was adjusted to 7.0 with KOH pellets. Media with and without biotin (0.0001 g/l) were prepared. Cultures grown in peptone-succinate-salts (PSS) broth (7) were inoculated into 25 ml of 1/4X nutrient broth and incubated at 37°C for 48 h. The cells were harvested by centrifugation, washed twice in 10 ml of sterile distilled water, and suspended in water to a density of 20 Klett units (blue filter, 16 mm cuvettes). One-tenth ml of this suspension was used to inoculate each 5 ml volume of medium. Incubation was for 48 h at 37°C. In cases where slight growth occurred in medium without biotin, a second serial transfer was made to confirm the biotin requirement.

[5]The medium used had the following composition (g/l): K_2HPO_4, 2.0; $(NH_4)_2SO_4$, 1.0; $MgSO_4.7H_2O$, 0.1; $FeCl_3.6H_2O$, 0.0047; $MnSO_4.H_2O$, 0.0025; $CaCO_3$, 0.001; $ZnSO_4.7H_2O$, 0.00072; $CuSO_4.5H_2O$, 0.000125; $CoSO_4.7H_2O$, 0.00014; H_3BO_3, 0.000031; $Na_2MoO_4.2H_2O$, 0.000245; and agar (Difco, purified), 15.0. Before adding the $CaCO_3$ and biotin, the medium was acidified to pH 2.5 with HCl to dissolve precipitates and was then brought to pH 7.0 with KOH. Cells were prepared as described in footnote 4 above, except that the final suspension was 30 Klett units. Two ml of suspension were used to seed 20 ml of molten medium (45°C) in a petri dish. After the medium had solidified, sterile discs (7.0 mm) punched from Beckman electrophoresis filter paper (Cat. No. 319328) were dipped into 5% solutions of filter-sterilized carbon sources. (Organic acids such as ketoglutarate had been adjusted to pH 7 with KOH before use.) The paper discs were then placed near the periphery of the agar plates (3 discs per plate). The plates were then incubated at 37°C for 72 h. Any visible zone of turbidity around a disc, when the plates were held near a light source, as judged with the naked eye, constituted a positive growth response.

[6]The Group I strains positive for glucose were Sp T60, Sp 67, Sp F6, and Sp 34. These strains were unable to produce an acid reaction from glucose. The growth of Group II strains appeared to be heavier than for the four Group I strains.

TABLE 4. Physiological characteristics of S. lipoferum strains

Test	% of strains positive	
	Group I	Group II
Oxidase, phosphatase, urease, esculin hydrolysis, anaerobic growth with nitrate[1]	100	100
Starch and gelatin hydrolysis, fluorescent pigment[1]	0	0
Indol[2]	0	0
Acid from fructose[3]	100	100
Acid from lactose, maltose, sucrose, rhamnose, cellobiose, erythritol, dulcitol, and melibiose[3]	0	0
Acid from the following sugars:[3]		
arabinose	87	100
galactose	70	93
i-inositol	0	13
Catalase[1]		
strong	87	27
weak	13	40
negative	0	33
Sole carbon sources (disc method)[4]		
succinate, L-malate, oxalacetate, β-hydroxybutyrate, lactate, fumarate, gluconate, pyruvate, glycerol, fructose	100	100
propionate	98	93, + 7% v[5]
galactose	65, + 15% v	80, + 13% v
arabinose	74, + 9% v	67, + 27% v
citrate	96, + 4% v	100
malonate	0	0

[1]Methods of Hylemon et al. (7) were used.

[2]1% tryptone broth was used, with incubation for 48 h.

[3]See footnote 2, Table 3.

[4]See footnote 5, Table 3.

[5]Variable, differs in same strain.

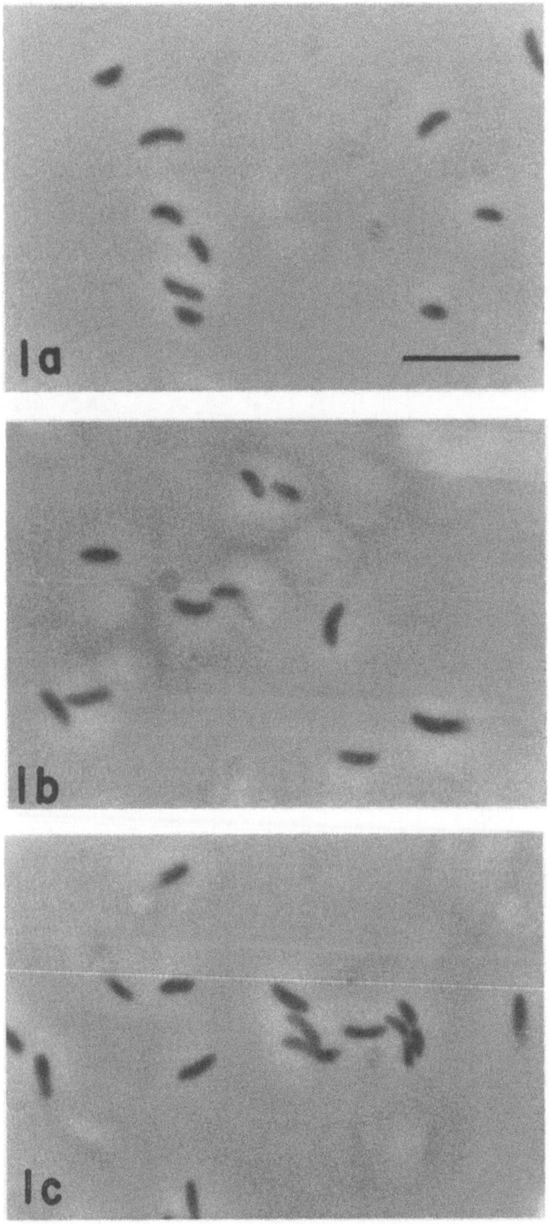

FIG. 1. Appearance of S. lipoferum strains grown in peptone-succinate-salts broth (7) for 24 h at 37°C. A, strain Sp 7. B, strain Sp 59b. C, strain Sp RG 20a. The bar represents 10 μm. Phase-contrast microscopy of wet mounts.

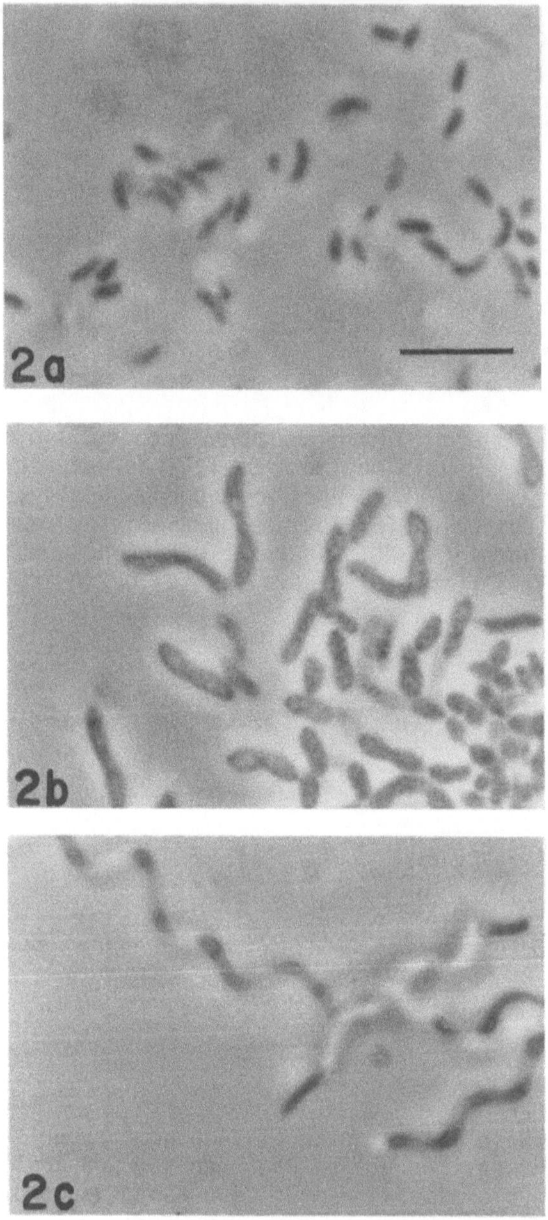

FIG. 2. Appearance of S. lipoferum strains grown in semi-solid N-free malate medium + 0.005% yeast extract at 37°C. A, strain Sp 7 at 48 h. B, strain Sp 59b at 48 h. C, strain Sp RG 20a at 24 h. The bar represents 10 μm. Phase-contrast microscopy of wet mounts.

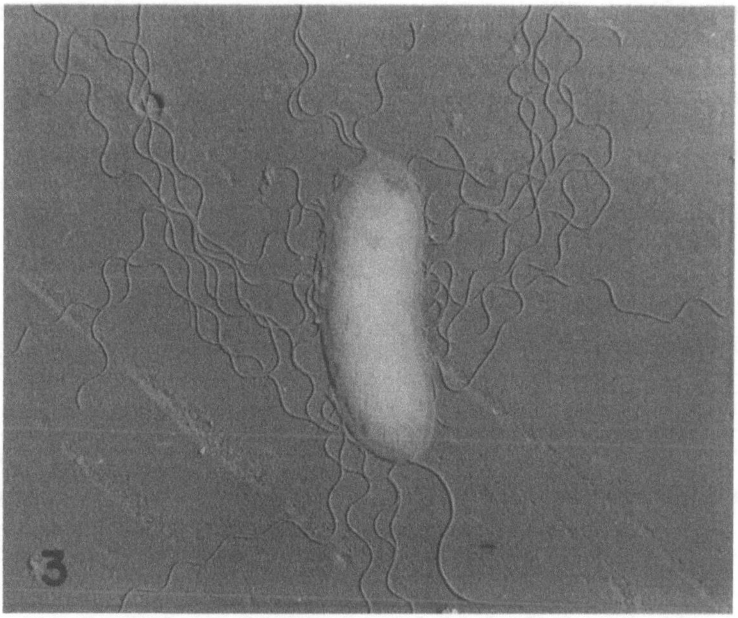

FIG. 3. Electron micrograph of strain Sp 7 grown on agar at 30°C, showing lateral flagella in addition to the single polar flagellum. 17,000 X.

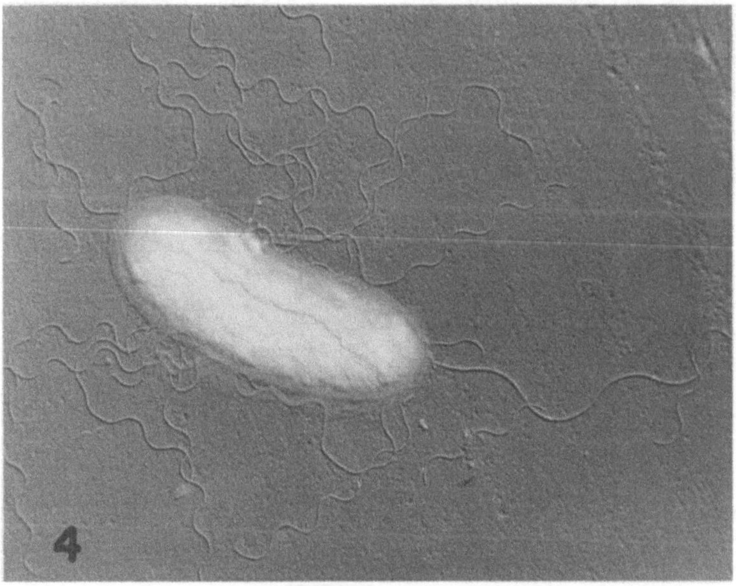

FIG. 4. Electron micrograph of strain Br 17 grown on agar at 30°C, showing lateral flagella in addition to the single polar flagellum. 17,000 X.

lack of significant DNA homology would not preclude assignment of an organism to a particular genus.

Freshwater spirilla other than S. volutans are presently classified in the genus Aquaspirillum (7). Table 5 presents a comparison of this genus with S. lipoferum. In view of the number of differences and atypical characteristics, we suggest that S. lipoferum not be assigned to this genus.

TABLE 5. Comparison of S. lipoferum with the genus Aquaspirillum

Aquaspirillum	S. lipoferum
Mol% G + C = 50 - 65 (Tm)	Mol% G + C = 69 - 71 (Tm)
Typically 1 or more helical turns, but some species are vibrioid.	Mainly vibrioid, but helical cells can occur under certain conditions.
Typically bipolar tufts of flagella, but one species has a single polar flagellum, and another has bipolar single flagella.	Single polar flagellum, but on agar at 30°C also form numerous lateral flagella.
Sugars typically not catabolized, but 3 species attack a very limited variety. These species form acid anaerobically from the sugars but do not grow.	A number of sugars are catabolized. Group II strains form acid anaerobically from sugars and also exhibit slight growth.
Typically do not grow anaerobically with nitrate or denitrify, except for 3 species.	All strains grow anaerobically with nitrate, and many are able to denitrify.
Typically do not require biotin, except for 1 species.	Group II requires biotin.
Typically do not have nitrogenase activity, except for 2 species.	Nitrogenase present.

The nitrogen-fixing genus Derxia, like S. lipoferum, has a mol% G + C of 70 (Tm). However, Derxia is a straight rod, with a single polar flagellum. Moreover, Derxia does not denitrify or even reduce nitrate to nitrite. Unlike S. lipoferum, Derxia produces a copious slime. Unlike the genus Azotobacter, S. lipoferum does not form cysts. With regard to the genera Azomonas and

Beijerinckia, the mol% G + C of these is much lower than that of S. lipoferum. Similarly, the mol% G + C for the genus Vibrio would preclude inclusion in this genus as well.

The flagellation of Pseudomonas stutzeri is similar to that of S. lipoferum; moreover, the mol% G + C of this species is high (61 - 66 by buoyant density), and some Pseudomonas species have a mol% G + C as high as 70. Moreover, the present definition of the genus Pseudomonas indicates that curved rods (but not helical) may be included. Like S. lipoferum, many pseudomonads can carry out denitrification. However, nitrogen-fixers are not presently included in the principal list of Pseudomonas species (5). Also, unlike Group II of S. lipoferum, pseudomonads do not form acid anaerobically from sugars and have no fermentative ability. P. stutzeri seems to be the most similar of the Pseudomonas species to S. lipoferum, but the cells have a smaller diameter, lack poly-β-hydroxybutyrate, and catabolize maltose and sucrose.

It appears to us that the best course would be to assign the S. lipoferum strains to a new genus. We believe that the generic name Azospirillum would be suitable, for the following reasons:

1. The name Spirillum lipoferum, assigned by Beijerinck, has become a familiar one since 1974, and retention of the term "spirillum" in the new name would minimize confusion.

2. S. lipoferum strains, although mainly vibrioid in shape, can exhibit helical forms under certain conditions.

3. Certain spirilla, such as Aquaspirillum aquaticum, Aquaspirillum delicatum, Aquaspirillum metamorphum, and Oceanospirillum japonicum, generally have less than one helical turn.

4. Like spirilla, S. lipoferum strains have poly-β-hydroxybutyrate granules, have mainly a respiratory type of metabolism, and grow well on the salts of organic acids.

In the genus Azospirillum, two species would presently occur. Although Beijerinck's cultures are not longer in existence, it seems likely that his organisms belonged to Group II. Beijerinck referred to the development of the spirilla in solutions of glucose or mannitol inoculated with soil, although the spirilla were later displaced by the growth of Azotobacter and Clostridium. When malate was used as the carbon source, no such displacement occurred. Moreover, Beijerinck also provided a drawing of cells cultured in sugar medium. All of this strongly suggests that Group II organisms were involved. Consequently, we believe that the name Azospirillum lipoferum should be applied to Group II strains.

With regard to the second species in the genus, because Group I strains were first isolated in Brazil, we propose that the species name Azospirillum brasilense would be an appropriate one.

We further propose that A. lipoferum become the type species of the genus Azospirillum, and that strain Sp 59b be designated as the type strain of this species. For A. brasilense, strain Sp 7 should be designated as the type strain.

REFERENCES

1. Becking, J. H. (1963) Antonie van Leeuwenhoek J. Microbiol. Serol. 29, 326.

2. Beijerinck, M. W. (1925) Zentralbl. Bakteriol. Parasitenkd. Infektionskr. Hyg. Abt. 2., 63, 353.

3. De Boer, W. E., Golten, C., and Scheffers, W. A. (1975) Antonie van Leeuwenhoek J. Microbiol. Serol. 41, 385.

4. Döbereiner, J., and Day, J. M. (1976) In W. E. Newton and C. J. Nyman (ed.), Proceedings of the 1st international symposium on nitrogen fixation, p. 518. Washington State Univ. Press, Pullman.

5. Doudoroff, M., and Palleroni, N. (1974) In R. E. Buchanan and N. E. Gibbons (ed.), Bergey's manual of determinative bacteriology, 8th ed., p. 217. Williams and Wilkins, Baltimore.

6. Giesberger, G. (1936) Ph. D. Thesis, Utrecht Univ., Netherlands.

7. Hylemon, P. B., Wells, J. S., Jr., Krieg, N. R., and Jannasch, H. W. (1973) Int. J. Syst. Bacteriol. 23, 340.

8. Johnson, J. L. (1973) Int. J. Syst. Bacteriol. 23, 308.

9. Johnson, J. L. and Cummins, C. S. (1972) J. Bacteriol. 109, 1047.

10. Krieg, N. R. (1976) Annu. Rev. Microbiol. 30, 303.

11. Marmur, J. and Doty, P. (1962) J. Mol. Biol. 5, 109.

12. Schröder, M. (1932) Zentralbl. Bakteriol. Parasitenkd. Infektionskr. Hyg. Abt. 2., 85, 178.

13. Shinoda, S., and Okamotom K. (1977) J. Bacteriol. 129, 1266.

14. Wilkins, T. D., Walker, C. B., and Moore, W. E. C. (1975) Appl. Microbiol. 30, 831.

ABSTRACTS OF ORIGINAL PAPERS

NITROGEN UTILIZATION IN *Phaseolus vulgaris* L.

Avilio A. Franco, João C. Pereira and Carlos A. Neyra

Programa de Fixacão Biológica de Nitrogênio, Convênio CNPq-EMBRAPA-UFRRJ, EMBRAPA, Km 47, Seropédica, Rio de Janeiro, 23460, Brasil.

Nitrate uptake, nitrate reductase activity (NRA) in the leaves and nodule nitrogenase (N_2-ase) activity were determined throughout the growing season in bean plants grown in the field under different regimes of fertilizer-N.

High levels of N (20 Kg N/ha, weekly) reduced nodule mass and N_2-ase activity but resulted in higher levels of NRA. N_2-ase activity began 2-3 weeks after sowing, was maximal at flowering and then declined rapidly. On the other hand, NRA was maximal after flowering. Tissue analysis indicated that appreciable NO_3^- uptake occurs after flowering and that the addition of ammonium nitrate to the soil at the beginning of flowering resulted in even higher levels of NR through most of the pod filling period. The practical implications of this work will be discussed.

BEAN INOCULATION IN THE VALLEY OF MEXICO UNDER UNIRRIGATED CONDITIONS

Roberto Núñez E. and Maria Valdés

Colegio de Postgraduados, Escuela Nacional de Agricultura, Chapingo, Estado de México and Escuela Nacional de Ciencias Biológicas, Instituto Politécnico Nacional, México 17, D.F., México.

Although spontaneous nodulation is observed in beans growing at the Central Valley of Mexico, N_2 fixation is not significant and this crop usually requires nitrogen fertilizer in amounts from 40 to 100 kg/ha.

We report here a two year study in an attempt to find rhizobial strains that may replace the nitrogen fertilizer. For this purpose two varieties of *Phaseolus vulgaris* were inoculated with three different *Rhizobium phaseoli* strains in combination with N and P amendment.

Data from the first year showed the variety Puebla 338 to have better nodulation than variety N-150. The importance of phosphate fertilization was marked, both in nodulation and in the seed yield. Lowest yields in both varieties were observed when plants did not receive chemical fertilizer. The treatment 80-60-0 produced the highest seed yield.

During the second year, there was a better rainfall distribution and the yield response to P was decreased. Again the treatment 80-60-0 produced the highest dry matter and seed yield. Statistical differences were not observed between fertilized treatments and those with 0-60-0 inoculated or uninoculated.

The above results show that efficient *Rhizobium phaseoli* strains are not yet available to be recommended for use as bean inoculants.

THE EFFECT OF LIGHT, DARK, AND TEMPERATURE ON THE ROOT NODULE ACTIVITY (C_2H_2 REDUCTION) OF SOYBEANS (*Glycine max* L. Merr.)

Lee E. Schweitzer and James E. Harper

Department of Agronomy, University of Illinois and U.S. Department of Agriculture, Agricultural Research Service, Urbana, Illinois 61801, U.S.A.

The effects of light, dark, and temperature on the root nodule activity (*in situ* C_2H_2 reduction) of hydroponically grown soybeans were investigated.

Plants were sealed into the covers of 7-liter polyethylene containers equipped with ion exchange columns to insure a stable pH and proper aeration of the nutrient solution. During *in situ* measurement of root nodule activity, a known volume of nutrient solution was removed to expose only the well nodulated upper portions of the roots and acetylene was injected to a partial pressure of 0.1 atm. Samples of the incubation atmosphere were withdrawn at intervals and ethylene production was determined by hydrogen flame gas chromatography. In the current study, soybeans (cv. Calland) were grown for 35 days in controlled environment chambers. A 14-hr photoperiod (750 μE m^{-2} sec^{-1} at plant top) and day/night temperatures of 27/18 were used. At 35 days, light, dark, and temperature treatments were initiated and root nodule activity (C_2H_2 reduction) was monitored.

Root nodules of plants grown at a constant 27 C were found to exhibit no variation in activity in response to the light and dark treatments. In contrast, at diurnal temperatures of 27/18 C, changes in nodule activity paralleled changes in temperature. This effect was independent of light and dark treatment. Nodules on the roots of plants grown in continuous dark at a constant 27 C retained their activity undiminished for a period of 72 hours followed by a gradual decline to zero activity at 8 days. In extended dark conditions, plants lost nodular activity more rapidly when exposed to a constant 27 C than to a 27/18 C diurnal temperature. This response appears to be due to a more rapid utilization of available photosynthate at the higher temperature.

Evidence presented suggests that nodule activity may be supported in part by a stored supply of photosynthate in addition to that made available by recent photosynthetic activity.

INFLUENCE OF HERBICIDES ON THE NODULATION OF SOYBEANS

Antonio Roberto Giardini, Eli Sidney Lopes and Roberto Deuber

Instituto Agronômico de Campinas, Avenida Barão de Itapura, no 1.481, Campinas, São Paulo, Brazil.

A field experiment was conducted over two years in the same place with a soil (dark-red latosol) originally free from *Rhizobium japonicum*. The experimental design was a complete factorial (4×2^2) with four herbicide, two inoculation and two N level treatments. Common dosages of the herbicides trifluralin, vernolate and alachlor were used. Soybean (cv Santa Rosa) seeds were surface sterilized with $HgCl_2$ (0.1%) and inoculated with "turfa" and a mixture of strains (SMS-65 and SMS-313). The nitrogen was applied as urea at a rate of 60 kg N/ha the first year and 90 kg N/ha the second year.

The uninoculated treatments did not nodulate the first year. The statistical analysis showed that nodulation was not influenced by the herbicides when evaluated at 35, 58 and 98 days after sowing. The application of N resulted in a significant decrease in nodule dry weight and number of nodules. Grain yield in the first year was the same for all treatments. In the second year all treatments were nodulated when observed at 46, 70 and 104 days after sowing. The N application did not affect nodulation of the inoculated plants but reduced significantly nodule dry weight of the uninoculated plants. The inoculated plants showed significantly higher grain yields. There was a significant interaction between inoculation and fertilizer-N. Yield increase due to the N application was only verified in the uninoculated treatment. Grain yield was also higher in those plots receiving the herbicide treatment but it was not possible to attribute this effect to weed control. The results also indicate that although nodulation was abundant in the uninoculated plants it was lower than the nodulation caused by the strains used as inoculum.

HOST SPECIFICITY AND STRAIN COMPETITION IN THE *Rhizobium - Vigna sinensis* (L) ENDL. SYMBIOSIS

Newton Pereira Stamford and André Martin Louis Neptune

Universidade Federal Rural de Pernambuco, Recife, and Universidade de São Paulo, Escola Superior de Agricultura, Centro de Energia Nuclear na Agricultura, São Paulo, Brasil.

Vigna sinensis (L) Endl. may nodulate with *Rhizobium* strains isolated from other tropical plant species of the so-called "cowpea cross inoculation group." With respect to the host specificity in *Rhizobium-Vigna sinensis* there are no definite conclusions, especially related with strain competition.

A greenhouse experiment was conducted in order to test the specificity of Rhizobium strains from different origins in cross inoculation with four *Vigna sinensis* cultivars. The experimental design was a factorial 4x5 with 3 replicates, carried out with Leonard jars.

The results indicate that nodule numbers as well as nodule growth was not influenced by the cultivars tested. The nodule numbers and the total nodule weight varies with the origin of the strains and also with individual strains. Strain 5.000 (Rothamsted) was more effective on the cultivars "Seridó," "Sempre verde" and "Garôto" and highest nitrogen fixation was obtained with this strain.

Certain host specificity was observed and only the cultivar "Cariri" showed a different behavior, with strain C-101 fixing more nitrogen. The cultivars "Seridó," "Sempre verde" and "Garôto" showed a close relationship in cross inoculation. The black nodule forming strain C-102 produced black nodules in all cultivars but the other black nodule forming strain C-101 did not form black nodules with the cultivars "Seridó" and "Garôto."

By the linear regression of total plant N versus nodule weight two different lines were obtained for cultivar "Garôto" and another for cultivars "Seridó," "Sempre Verde" and "Cariri," with a large difference between the two slopes. The cultivar "Garôto" showed a low amount of N fixed per unit of nodule tissue.

When all strains used were applied in a mixture there was no effect on both nodule formation and nitrogen fixation. It seems that the homologous strain 5.000 was more competitive in nodule formation.

NITROGEN TRANSFORMATIONS IN SOILS UNDER *Digitaria decumbens* VEGETATION

Paulo A. da Eira, Programa de Fixacão de Nitrogênio, Convênio CNPq-EMBRAPA-UFRRJ-UEPAE, Itaguaí, Seropédica, 23460, Rio de Janeiro, Brazil.

Two greenhouse experiments were carried out with *Digitaria decumbens* to study nitrogen transformations in soil cores with intact plants or with roots only.

In the first experiment there were two nitrogen treatments (with and without NH_4^+) and five weekly samplings. In the second experiment a NO_3^- treatment and two early samplings during the first week were added. At each harvest, pH, NH_4^+ and NO_3^- N and total N in soil, and dry weight and N content in leaves and roots were determined.

Nitrate contents in the soil varied between 0.66 and 12.95 ppm in cores without plants and between 0.32 and 1.28 in cores with intact plants when no nitrate was added. Rapid nitrification of NH_4^+ was observed which was intensified in the presence of roots, in experiment I. In experiment II this could not be observed because NO_3^- was immediately denitrified. Denitrification was faster in cores with roots without leaves but there were also losses in the cores without roots. Digitaria assimilated NH_4^+ and NO_3^-. NH_4^+ required 7 days to be absorbed while NO_3^- required two weeks. There was an increase of nitrate reductase activity in leaves four hours after nitrate application to the soil. Maximal nitrate reductase was reached at four and seven days after application of NO_3^- and NH_4^+ respectively.

From the agronomic point of view it was concluded that if N fertilizer is used on Digitaria, NH_4^+ salts should be prefered because nitrate, besides being assimilated slower and being more subject to denitrification, is also leached easier under field conditions.

EFFECT OF MICRONUTRIENTS ON ESTABLISHMENT OF TROPICAL LEGUMES AND PERSISTENCE ON A RED YELLOW PODZOLIC SOIL HILL PASTURE

H. De-Polli, S. R. De Carvalho, P. F. Lemos and A. A. Franco

Programa de Fixacão Biológica de Nitrogênio, Convênio CNPq-EMBRAPA-UFRRJ-UEPAE, Itaguaí, Km 47, Seropédica, Rio de Janeiro, 23460, Brazil.

An experiment was carried out combining three legumes (Centrosema pubescens, Macroptilium atropurpureum cv. siratro and Stylosanthes guyanensis cv. IRI/1022) with several methods of application of FTE containing: Fe, B, Zn, Cu, Mo, Mn and Co. Ditches made in the natural pasture of melasses grass (Mellinis minutiflora) received basic fertilization of P and the seeds of the legumes treated according to treatment. The experiment was sown in October 1972. One sampling to evaluate establishment was carried out 90 days after sowing. The first harvest to evaluate production was done in July 1973 and the second in May 1974. Then the plots were left under grazing until October 1975. A third harvest was carried out in October 1976. In the first sampling, the number and the dry weight of nodules, and dry weight and total N of the plants were determined. Following harvests the fresh weight and the total nitrogen of each plot and of each component (grasses, legumes and weeds) were determined separately.

The stand, nodulation and total plant N were not affected significantly by treatment in the first sampling. However, there were differences in the persistence of the legumes up to the last harvest due to treatments and legume species. Stylo alone or mixed with grass gave highest yield but was not dependent on the micronutrient fertilization. Siratro persisted in the pasture with micronutrient fertilization with good yields up to the last harvest (1400 x over that of the control with P alone). Centrosema was responsive to micronutrient fertilizer at the initial stage but did not persist well under the experimental conditions up to the last harvest.

The addition of micronutrients increased the total yield and total N (grasses, legumes and weeds) of the plots with siratro and centrosema, but showed no effect on the plots with stylo. Pelleting the seeds of siratro and centrosema with FTE alone was more effective than in mixture with lime or applied directly in the soil. Increase of the proportion of legume in the plot did not decrease the grass production; in one case it increased dry weight and total N of grass plus weeds in the third harvest.

INCREASED N_2 FIXATION (C_2H_2) IN FIELD GROWN MAIZE BY HERBICIDE TREATMENTS

Ivanildo E. Marriel and José Carlos Cruz

Centro Nacional de Pesquisa de Milho e Sorgo, EMBRAPA, Caixa Postal 151, 35700, Sete Lagoas, Minas Gerais, Brazil.

Preliminary experiments performed to test inhibitive herbicide effects on N_2-fixing bacteria in soil revealed that on the contrary, there was an increase of numbers of *Spirillum lipoferum* in soils treated with Atrazin.

To confirm these observations and mainly to investigate the possibility of increasing N_2 fixation in maize, a field experiment was conducted with brachitic maize cv. piranão. Two preemergence herbicides were used, Atrazin (2-chloro-4-ethylamine-6-isopropylamine S-triazine) and Alachlor (2-chloro-2',6'-diethyl-N-mexomethyl-acetanilide) which are of general use in maize cultures. The herbicide levels used were 1.6, 3.2 and 4.8 kg/ha of Atrazin and 1.2, 2.4 and 3.6 kg/ha of Alachlor. Blanks without herbicides which also were not cleared by hand and an additional nitrogen treatment (60 kg N/ha) with the intermediate herbicide levels were included in the experiment. Nitrogenase activity on excised preincubated roots was evaluated by C_2H_2 reduction measurements.

In the laboratory, three strains of *S. lipoferum* were grown in N-free semi-solid malate medium to which 0, 1, 2, 4, 8, and 16 p.p.m. of the two herbicides were added. Effects were evaluated by C_2H_2 reduction rate measurements at early log phase.

Results revealed three times higher nitrogenase activity on maize roots with Alachlor and twice the activity with Atrazin (significant at $p = 0.01$). There was furthermore a linear effect of both herbicides on N% and total N in leaves (fourth leaf). The total nitrogen incorporated into leaf 4 was as high with the maximal dose of Alachlor as that with the intermediate level complemented with 60 kg N/ha.

The effects of the two herbicides on N_2 dependent *S. lipoferum* growth in culture medium was still more affected than the root activity. The strains varied in respect to optimal levels.

These results suggest an entirely new approach of increasing nitrogen fixation in the field, in maize-*S. lipoferum* associations.

EFFECT OF MINERAL NITROGEN AND LEGUMES UPON DRY MATTER PRODUCTION OF "COLONIÃO" GRASS, Panicum maximum Jacq

Herbert Barbosa de Mattos and Joaquim Carlos Werner

Instituto de Zootecnia, Nova Odessa, São Paulo, Brazil.

Five tropical legumes in association with "colonião," (*Panicum maximum* Jacq., var. Laras) were grown in experimental fields (red-yellow podzolic soil) of the Estação Experimental Central do Instituto de Zootecnia, Nova Odessa, São Paulo, to verify the amount of N incorporated into the system. The legumes used were: *Centrosema pubescens*, *Galactia striata*, *Glycine wightii*, *Macroptilium atropurpureum* cv. Siratro and *Stylosanthes guyanensis*. The experimental design consisted of randomized blocks with 4 replications. The size of each plot was 3.0 x 6.0 m. The grass was established by transplanting into rows at 0.5 m intervals. The legumes, which had been inoculated previously, were sown between the rows. Each plot received basic fertilization of P, K, Cu, Zn and Mo. The experiment was conducted over 3 years with 5 harvests each year.

The results showed that the association of *Galactia* with "colonião" resulted in a dry matter yield equivalent to the "colonião" fertilized with 225 Kg N/ha. The association with the other 4 legumes resulted in a dry matter yield equivalent to "colonião" fertilized with 75 Kg N/ha. Furthermore, there was an increase in protein content in the associated "colonião" in relation to fertilization with N in the range of 1-2%. There was also an increase in the level of P and Ca.

ENERGY REQUIREMENTS FOR UPTAKE OF AMMONIUM, DINITROGEN, AND NITRATE BY TROPICAL LEGUMES

W. J. Broughton

Department of Genetics & Cellular Biology, University of Malaya, Kuala Lumpur, Malaysia.

Recent evidence suggests that the energy requirements for the uptake of combined nitrogen or gaseous nitrogen are not different in temperate legumes. In an attempt to analyze the energy requirements for the different forms of nitrogen uptake in tropical legumes, *Centrosema pubescens*, *Glycine max* and *Vigna unguiculata* where chosen as the test plants.

Actual experimentation involved raising the plants supplied different nitrogen sources under controlled energy input systems, and using a plant growth analysis approach to interpret the results. The data are discussed in terms of energy requirements of different plants, and the distribution within the plant of photosynthetic energy for different processes, particularly those involved in nodulation.

POPULATION VARIATION OF NITROGENASE ACTIVITY IN NODULATED ALFALFA

F. Favelukes, R. F. de Schuttenberg and O. M. Aguilar

Cátedra de Química Biológica I, Facultad de Ciencias Exactas, Universidad Nacional de la Plata, 1900, La Plata, Argentina.

In a search for factors that determine the effectiveness of symbiotic N_2 fixation in alfalfa, we have explored the levels of nitrogenase activity in populations of alfalfa, nodulated with an efficient R. meliloti strain. Seedlings of the variety "Intacic" (resistant to aphids) individually planted on sterile agar, were inoculated with strain R 41 and grown in a controlled chamber. At 40 days each plant was examined for growth and nodulation, and assayed for acetylene reduction activity. Afterwards the plants were transferred to vermiculite and later to soil, for cloning.

Out of 64 nodulated plants, 14% had zero acetylene reduction, and the rest had a broad bimodal distribution of activities ranging from 0.1 to 125 nmoles ethylene/hour/plant. The individual values were compared with the other observations for each plant, whereby some weak correlations could be drawn. These and previous results show that populations of alfalfa can be heterogeneous in their nitrogenase activity, and suggest that the phenomenon may be genetically determined. Our findings will be discussed in the light of the known genetic and biochemical determinants of nodule efficiency in N_2 fixation, described in other legumes.

The collaboration of M. G. Castro, A. M. Cortizo, G. Guillén and L. A. Manchi is gratefully acknowledged. Supported by grants from CONICET and CICPBA, Argentina.

NITROGEN FIXATION BY SOME FORAGE LEGUMES UNDER WEST INDIAN CONDITIONS

Belal Ahmed and John Marshall Keoghan

University of the West Indies, Department of Soil Science, St. Augustine, Trinidad, West Indies.

Surveys by various workers have shown that a wide range of potentially useful forage legumes occur naturally in the West Indies. Although the importance of these legumes for West Indian pastures was long ago emphasized, very few studies have been conducted on their nodulation and nitrogen fixing capability.

In the present study, growth, nodulation and nitrogenase activity of some indigenous and introduced forage legumes were examined in Antiguan soils. In a pot experiment, uninoculated scarified seeds of several Stylosanthes spp. were grown in soil cores collected from 12 soil types of Antigua.

Although all legumes, except S. hamata cv. CIAT 122 and S. guyanensis cv. Endeavour, nodulated and exhibited nitrogenase activities in all soils, there was variability in growth and nodule activity due to soil type. In addition, all plants except *S. sympodialis* cv. CIAT 1043, responded to nutrient application. The negative response in *S. sympodialis* is possibly due to its adaptability in low fertility level. Results indicated the presence of a number of effective and relevant rhizobia in the soils studied. In a field trial at two soil sites, uninoculated *S. hamata* cv. local and *T. labialis* showed high nitrogenase activity accompanied by good growth and nodulation. While there was little effect of soil type or irrigation on nodule activity, legume yield was clearly influenced by soil type. *S. hamata* produced higher dry matter yield in Fitches clay (pH 8.2) while *T. labialis* grew better in Ottos clay (pH 6.9). While levels of soil mineral nitrogen apparently did not affect yield in Ottos clay, higher yields of *S. hamata* and *T. labialis* in Fitches clay were always associated with high amounts of soil mineral nitrogen.

SYMBIOTIC NITROGEN FIXATION BY THREE PEANUT TYPES INOCULATED WITH TWO *Rhizobium* spp. STRAINS

Luis B. Ayala B.

Instituto de Investigaciones Agrícolas Generales, Centro Nacional de Investigaciones Agropecuarias, Apartado 2653, Maracay, Venezuela.

The effect of inoculation of three peanut types with two *Rhizobium* strains was studied.

Significant differences in specific nitrogenase activity were detected for peanut types; these effects were not detected when the activity of the enzyme was expressed as total per plant. Differences in specific nitrogenase activity as caused by *Rhizobium* strains were not reflected by total nitrogenase activity, percent total nitrogen in plant or mg of nitrogen per plant. The fact that plants showing larger specific nitrogenase activity had a smaller total amount of nodules and smaller nodules suggests that nodulation acted as a compensatory factor absorbing differences in specific nitrogenase activity from *Rhizobium* strains. The above results make it an important question whether the specific or the total nitrogenase activity is the most meaningful for the evaluation of *Rhizobium* strains and indicate the importance of nodulation in symbiotic nitrogen fixation. Therefore, despite the lack of interaction between *Rhizobium* strain and peanut type, further research would be required to determine whether the evaluation of symbiotic nitrogen fixation should be based on the *Rhizobium* strain or on the *Rhizobium*-legume system. Furthermore, a suggested control of nodulation from the host legume points out the importance of such matters.

Although not statistically significant, plant percent total nitrogen, mg of nitrogen per plant, and dry weight of plant tended to be higher for plants inoculated with strain VIII than those for plants inoculated with strain 10. These results suggest possible differences in effectiveness between the two Rhizobium strains. Significant differences were detected in percent total nitrogen of plants and dry matter from peanut types. Peanut type and Rhizobium strain affected total mg of nitrogen per plant. The effect of peanut types and Rhizobium strains on nodule nitrogenase activity and plant nitrogen and dry matter content were independent. These results did not show the effect of host genotypes on effectiveness of the Rhizobium strains as reported in the literature.

MAXIMIZING N_2 FIXATION IN TROPICAL FORAGES

J. Halliday

Beef Production Program, CIAT, AA 67-13, Cali, Columbia

National and International agencies have taken up the challenge to convert the potential for increased food production into reality.

Latin America has 850 million hectares of acid, infertile soils (oxisols and ultisols). In Brazil they comprise 68% of the land surface; in Colombia 57%;Peru 44%; Venezuela 58%; Guyana 62%; Surinam 62%; French Guyana 94%; Panama 63%; Trinidad 84%. The majority of these areas are jungle, but some 300 million hectares are grassland savannahs such as the Llanos Orientales of Colombia. Beef production is the logical cropping system for such extensive areas but the quantity and quality of native forage is inadequate. Animals take four to five years to reach market weight.

CIAT, the international centre for tropical agriculture, with its headquarters at Cali in Colombia, is charged with increasing beef production in Latin America's acid infertile soils, primarily through year-round forage production.

Rather than attempt to modify soil pH and nutrient status to levels adequate for growth of currently available commercial tropical forages, an interdisciplinary team is collecting, evaluating and aspiring to introduce adapted species tolerant of extreme soil acidity, high exchangeable aluminium, and low available phosphorus.

The objective of the Soil Microbiology section is to maximize N_2-fixation in these adapted forages and, although this implies both legumes and grasses, easily justifiable priority is given to the legume-Rhizobium symbiosis. This project is unique in that it will furnish an inoculation recommendation in respect of each individual germplasm accession nominated as promising for specific sites.

For each plant accession there may be 50-100 strains in the CIAT Rhizobium bank isolated from nodules of that genus. Sequential screening allows selective reduction to a manageable number for field evaluation. In the first stage, strains are tested in tube culture for compatability with the host. In stage two, strains are ranked in order of their nitrogen fixation efficiency on the basis of their performance in Leonard jars in optimal media. Next they are tested under the physical and chemical stress of sterilized site soils in pot culture. The fourth stage is evaluation under the full biological and climatological stresses of the field.

Detailed results are presented which consistently demonstrate the non-applicability of strain recommendations from selection programs in Australia and in the USA to the specialized plant types emerging from CIAT's forage program.

THE EFFECT OF SHADING WITH "ERITRINA" (*Erythrina fusca*) ON THE NITROGEN LEVELS OF SOILS PLANTED WITH COCOA

Maria Bernadeth M. Santana and Francisco Ilton Morais

Setor de Fertilidade, Centro de Pesquisas do Cacau (CEPEC), Km 22, Rodovia Ilhéus/Itabuna, Bahia, Brasil.

This work was conducted in order to determine the importance of N_2-fixation by "eritrina" (*Erythrina fusca*, ex *E. glauca*) in the enrichment of soils planted with cocoa where the leguminous plant has been used for permanent shading.

Composite samples were collected at two soil depths (0-10 and 10-20 cm) and in 9 replications of soils planted with cocoa, with or without "eritrina" shading. The sampling was performed in fertile soil and in chemically poor soil.

The level of N in wet samples was used to interpret the results. Dried soil was used for determination of other chemical characteristics.

In both soils the level of N was higher in areas shaded with "eritrina" ($P < 0.01$) as compared to those not shaded.

The results indicate a possible transfer of the nitrogen fixed by the legume ("eritrina") to the areas occupied by the cocoa plants.

NITROGEN FIXING BLUE-GREEN ALGAE IN RICE SOILS OF SRI LANKA AND THEIR POTENTIAL AS A FERTILIZER IN RICE CULTIVATION

S. A. Kulasooriya and R. S. Y. de Silva

Department of Botany, Peradeniya Campus, University of Sri Lanka, Sri Lanka.

A survey of nitrogen fixing blue-green algae in rice fields of Sri Lanka was undertaken as a prelude to a wider research programme which envisages the exploitation of these organisms as biological fertilizers in rice cultivation.

Soil samples collected from different Districts of the island were first examined microscopically and then inoculated into different algal culture media without combined sources of nitrogen. The algae that grew in such media were observed and wherever possible, were isolated into unialgal cultures. In this study, the ability to grow in isolation in N-free media and the formation of heterocysts during such growth have been used as indicators of their ability to fix nitrogen. Some of these isolates were tested for their ability to reduce acetylene, at the University of Dundee, Scotland, and have given positive results (Prof. W. D. P. Stewart, personal communication).

This survey has been conducted more extensively in the District of Kandy, in Central Sri Lanka, than in the other areas. The results indicate that rice fields in this District possess a very wide flora of heterocystous blue-green algae. We have isolated in unialgal culture 40 different species of heterocystous blue-green algae belonging to 12 different genera. However, blue-green algae seldom predominate the algal flora of these paddy fields. On the other hand, microscopic examination of rice soils in the District of Jaffna, in Northern Sri Lanka, has revealed the presence of many heterocystous blue-green algae as the dominant algal types. These algae form dense blue-green mats, and *Aulosira* spp. is noteworthy in this respect.

The inability of the indigenous blue-green algae to predominate in the rice soils in the wet zone districts like Kandy may be due to the low pH of these soils as compared to those of Jaffna where the soil reaction is neutral to alkaline. Soil inoculation experiments with isolated blue-green algae have confirmed this supposition. Therefore attempts to make blue-green algae colonize rice soils in the wet zone districts may not yield promising results. On the other hand *Azolla*, a water fern containing a nitrogen fixing blue-green alga as an endosymbiont, appears to be a more suitable organism for these soils. Preliminary experiments with *Azolla* have shown that it grows very well in the wet zone and when grown together with a crop of rice has registered a percentage increase in yield above that of a treatment which received 80 lbs. per acre of urea.

VERTICAL DISTRIBUTION OF ALGAE AND ACETYLENE REDUCING ACTIVITY IN AN ALGAL MAT ON A SANDY WATERLOGGED TROPICAL SOIL

P. Reynaud and P. Roger

O.R.S.T.O.M., Microbiologie des Sols, B.P. 1 386, Dakar, Sénégal.

Vertical distribution of algae in a permanent algal mat on a waterlogged sandy soil was studied. Cores including the algal mat and the first centimeter of soil were cut into sections 2 mm high, after fixation with silicagel.

Qualitative and quantitative composition of algal biomass and acetylene reduction activity (A.R.A.) were measured on each section.

Analysis of vertical distribution of the algal species indicated three types of light level dependance: high, neutral or low. The first two types occurred in eucaryotic algae whereas all three types occurred among the procaryotic algae. Our previous studies demonstrated blue-green algal (B.G.A.) sensitivity to high light intensities: the vertical distribution of species can function as a protective mechanism.

Acetylene reduction activity and nitrogen-fixing algal biomass are strongly correlated, the relative value of the latter being higher in the soil than in the algal mat. Maintenance of an algal A.R.A. in some bare soils, after exposure to high light intensities may be due to B.G.A. present below the soil surface.

PLANT PRODUCTIVITY AND NITROGEN FIXATION IN VARIOUS AMAZONIAN SOILS

Luiz Antonio de Oliveira and Rosemary Silvester-Bradley

Instituto Nacional de Pesquisas da Amazonia, C. Postal, 478, 69000-Amazonas-Manaus.

In order to determine areas for the study of the ecological and agricultural aspects of nitrogen fixation in the Amazon, field observations were made of the occurrence of legume root nodulation, Azospirillum spp. and nitrogenase (C_2H_2) activity of a range of samples collected near Marano and along the Rio Negro. In samples collected from latosol, these N_2-fixing agents were absent, even in regenerating forest. However, in samples from terra preta dos Índios, alluvial soil (várzea) and also floating aquatic macrophytes, they were present. However, the vegetation on each soil type was different. Therefore, in order to distinguish soil effects from vegetation effects, surface sterilized seeds of winged bean (Psophocarpus tetragonolobus (L) DC), soybean (variety "Jupiter") and maize were planted in eight 1.6 kg samples of each of várzea, terra preta, sandy latosol, heavy latosol from primary forest and heavy latosol collected three months after the primary forest had been burned. After 42 days the highest nodulation had occurred in terra preta and várzea soils. It was higher in winged bean than in soybean. Nitrogenase (C_2H_2) activity of the intact system was detected only in winged bean. Nitrogenase (C_2H_2) activity of 42 hour enrichment cultures from Azospirillum spp. of maize roots and soil was higher in várzea and burned latosol. Plant productivity was lower in heavy and sandy latosol than in the other three soils. Soil analyses showed large differences between soils, and it was concluded that soil fertility caused large differences in plant productivity and nitrogen fixation. The effect of different Amazonian soils on nitrogen fixation in natural and agricultural systems should be assessed in order that they can be most effectively utilized.

NITROGEN FIXATION (C_2H_2) IN WHEAT

M. Nery, G. T. V. Abrantes, O. dos Santos and J. Döbereiner

Instituto Agronómico do Paraná, C.P. 133, 86100, Londrina, Paraná, Brasil.

Nitrogen fixation in wheat (*Triticum sativum* Lam.) was estimated by the acetylene reduction method using measurements in intact soil-plant systems and excised roots.

There was a highly significant correlation between both methods (r = 0.83). Among 30 wheat cultivars significant differences in nitrogenase activity, assayed in intact soil-plant cores, were observed with activities ranging from 3523 to 111 nmols of C_2H_4 (per g of dry roots per hr). If the theoretical 3:1 (C_2H_2/N_2 reduced) conversion factor is used, a potential daily N_2 fixation of 800 g of N/ha can be estimated.

Ninety percent of the enrichment cultures inoculated with root pieces or soil particles yielded the N_2-fixing bacterium, *Spirillum lipoferum*.

Nitrogenase fertilizer applications of 50 kg/ha did not affect the N_2-ase activity assayed on excised roots of three cultivars; however, 75 kg/ha inhibited the N_2-ase activity. In 20 wheat cultivars planted in a soil with an initial pH of 4.3, liming to reach pH 6.5 increased N_2 fixation. N_2-ase activity was related to the stages of development of the plant, with maximum activity occurring at the stage of grain filling.

VERIENTAL DIFFERENCES AFFECTING NITROGENASE ACTIVITY IN RHIZOSPHERE OF SUGARCANE

Alaides Puppin Ruschel and Renato Ruschel

Centro de Energía Nuclear na Agricultura, Universidade de São Paulo/Comissão de Energía Nuclear and PLANALSUCAR/IAA, Piracicaba, São Paulo, Brasil.

Nitrogenase (C_2H_2) activity was estimated in roots, germinated cuttings, whole plants of sugarcane (2 months old) in intact and disturbed systems, under low oxygen and normal (air) atmosphere.

Strong evidence of a genetic effect on nitrogenase activity in the sugarcane rhizosphere was observed. This activity was high for NA56-62 and CB46-47 varieties and almost nil for the CB41-76 variety. Correlation of rhizosphere nitrogenase activity and N-content of leaves was postulated. The rate of nitrogenase activity increased with time on NA56-62 and CB46-47. Nitrogenase activity was observed in roots and germinated cuttings without roots and shoots.

BACTERIA IN THE ENDORHIZOSPHERE OF MAIZE IN BRAZIL

David G. Patriquin and Johanna Döbereiner

Biology Department Dalhousie, University of Halifax, N.S. B3H 4J1, Canada and Programa de Fixação Biológica de Nitrogênio, Convênio CNPq-EMBRAPA, UFRRJ, Itaguaí, Seropédica, 23460, Rio de Janeiro, Brazil.

Establishment of bacteria in the endorhizosphere of grasses has been described as a stage in the progressive decomposition of roots, or as in the case of associations of tropical grasses with *Spirillum lipoferum* as a possible primitive N_2 fixing symbiosis. We examined roots of field-grown tropical maize for endorhizosphere bacteria following incubation in pH 7, 0.05 molar phosphate buffer containing 0.625 g/l malate and 1.5 g/l 2,3,5 triphenyl tetrazolium chloride. Production of prominent formazan crystals by the bacteria occurred in both the presence and absence of added NH_4^+, and thus was not specific for N_2-fixing organisms. But this technique was useful because reduction of tetrazolium facilitated selection of regions of viable bacteria for sectioning, and enhanced bacterial outlines as observed in sections. Bacteria were most common in the inner cortex and in xylem and pith tissues in the stele, and occurred there without necessary collapse of the outerlying tissues. Bacteria in the stele remained viable after 6 h treatment of roots with chloramine-T, indicating that the endodermis was intact. *S. lipoferum* was isolated from surface-sterilized roots, and *S. lipoferum*-like organisms were observed in the roots, but other bacterial morphologies were also seen. Infection appears to occur initially in secondary roots, spreading longitudinally into the main roots. Highest nitrogenase activities were always observed in mature roots with many laterals, in *Digitaria*, maize and sorghum.

Sterile-grown sorghum and wheat seedlings inoculated with *S. lipoferum* showed cortex infection and tetrazolium reduction similar to the field plants.

TAXONOMY OF THE DIAZOTROPH OF TROPICAL GRASSES CALLED *Spirillum lipoferum*

Max E. Tyler, James R. Milam and David A. Zuberer

Department of Microbiology and Cell Science, University of Florida, Gainesville, Florida, 32611, U.S.A.

Thirty strains of *Spirillum lipoferum* isolated by J. Döbereiner of Brazil and by us have been examined by an array of morphological, physiological and biochemical techniques. All isolates were recovered from roots or rhizospheres of *Panicum*, *Zea*, *Pennisetum*, *Musa* or other plants in Brazil, Ecuador, Florida or Venezuela, and all reduce acetylene to ethylene in pure culture.

Methods for determinination of phenotypic characteristics where those of Stanier et al., 1966, J. Gen. Microbiol. 43:159. Ratios of guanine + cytosine were determined by a modification of the method of Preston, J. F., and Boone, D. R., 1973, FEBS Letters, Vol. 37.

All isolates are straight or slightly curved, Gram negative rods with poly-β-hydroxybutyrate inclusions and a dominant monopolar flagellum, with numerous smaller lateral flagella, which develop only at 30° C on agar medium. A few exhibit cellular shapes slightly different from the majority. There are two physiological groups, based on mode of respiration and several other reactions. Phenotypically, most of the strains are intermediate between described species of Spirillum and Pseudomonas. DNA base ratios substantiate this assessment.

NITRATE-DEPENDENT NITROGENASE ACTIVITY IN Azospirillum spp. UNDER LOW OXYGEN TENSIONS

D. Barry Scott and Christine A. Scott

Programa de Fixação Biológica de Nitrogênio, Convênio CNPq-EMBRAPA UFRRJ-UEPAE, ITAGUI, Seropédica, 23460, Rio de Janeiro, Brasil.

Neyra and Van Berkum (Can. J. Microbiol. 23, 1977) have recently demonstrated NO_3^- dependent nitrogenase activity in Azospirillum brasilense, strain Sp 7 (ATCC no 29145) under O_2 limited conditions. The reduction of NO_3^- to NO_2^- under these conditions coincided with an increase in N_2 (C_2H_2) fixation. We have confirmed this result, using known O_2 tensions, with strains of A. brasilense that denitrify (Sp 7, Sp 13) and strains that do not denitrify (Sp 82, Br 14), and with A. lipoferum (strain USA 5b). We have also observed that the latter two strains of A. brasilense are able to dissimilate N_2O, suggesting that this subgroup may have the capability to dissimilate NO_2^- to N_2O under different growth conditions or may be deficient in the dissimilatory nitrate reductase.

All strains of Azospirillum that were examined for NO_3^- dependent nitrogenase activity exhibited a lag phase of 20-30 min for the reduction of NO_3^- to NO_2^- and a lag of about 1 hour for C_2H_2 reduction. Addition of chloramphenicol (50 ug/ml) to cultures of Sp 82, that had been preincubated for 3 h under microaerophilic conditions (pO_2, 0.015-0.020 atm), resulted in an inhibition of the NO_3^- dependent nitrogenase activity (pO_2<0.002 atm). This confirms that the lag phase corresponds to the period of induction of the dissimilatory nitrate reductase.

In order to clarify the role of NO_3^- in this process similar experiments were carried out using nitrate reductase negative mutants of A. brasilense (Sp 13 and Sp 34) and A. lipoferum (USA 5b). These mutants exhibited low rates of N_2 fixation (C_2H_2) under limiting O_2 tensions (pO_2, 0.001-0.0015 atm) and NO_3^- had no effect on this activity, confirming that NO_3^- reduction is necessary for the NO_3^- enhanced nitrogenase activity. These results suggest that under anaerobic conditions NO_3^- can act as an alternative e^- acceptor, allowing the synthesis of sufficient ATP for N_2 (C_2H_2) reduction.

Preliminary experiments suggest that under anaerobic conditions NO_3^- is dissimilated to ammonia. Cultures of A. lipoferum (USA 5a) released NH_4^+ into the medium following complete reduction of NO_3^- to NO_2^- under both N_2 and helium.

NITRATE REDUCTASE NEGATIVE MUTANTS OF Azospirillum spp. Krieg

Luiz Mauro S. Magalhães, Carlos A. Neyra and Johanna Döbereiner

Programa de Fixacão Biológica de Nitrogênio, Convênio CNPq-EMBRAPA-UFRRJ, EMBRAPA, Km 47, Seropédica, Rio de Janeiro, 23460, Brazil.

All strains of A. lipoferum and half of A. brasilense examined so far are able to dissimilate NO_3^- to NO_2^- and N_2. The strains of A. brasilense which do not denitrify (under oxygen-limited conditions) do accumulate NO_2^-. Under microaerophilic conditions, all strains from both species are able to grow in the presence of NO_3^- but do not fix N_2. Thus, the isolation of nitrate and nitrite reductase negative mutants of Azospirillum spp. is of great scientific interest particularly with regard to N_2-fixation and denitrification. It might also become of great practical interest if strains are obtained which can fix nitrogen in the presence of high nitrate levels in soils and plant roots.

Large test tubes with deep layers of nutrient agar containing 0.1% $KClO_3$ were inoculated with 10^8 cells of A. brasilense or A. lipoferum. After two weeks, single colonies were picked, replicated and stored at 10° C in semisolid potato medium cultures.

Spontaneous ClO_3^- negative mutants occurred at frequenices of $4x10^{-6}$; $16x10^{-6}$ and $13x10^{-6}$ in Azospirillum brasilense (nir^+) (denitrifying strain), A. brasilense nir^- (non-denitrifying strain) and A. lipoferum, respectively. All these mutants reduced C_2H_2 equal to that of the wild type. Seventy-five percent, 50%, and 85% respectively of the ClO_3^- mutants were nitrate reductase negative (nr^-). These mutants fixed N_2 in the presence of 10 mM NO_3^- while the wild types did not (semisolid malate medium). Most of the nr^- mutants were also nir^- but some retained the nitrate reductase. Furthermore two mutants, only, lost their nitrate reductase and became similar to A. brasilense (nir^-) supporting the existence of a non-denitrifying subgroup.

FORCED ASSOCIATION OF Spirillum lipoferum WITH TISSUE CULTURES OF SOME TROPICAL GRASSES

Vimla Vasil, Indra K. Vasil, David A. Zuberer and David H. Hubbell

University of Florida, Gainesville, Florida, 32611, U.S.A.

Attempts are being made to establish forced symbiotic associations between callus and suspension cultures of centipede grass, pearl millet, and sugarcane, with Spirillum lipoferum, and to introduce the bacteria directly into protoplasts isolated from roots, leaves, and suspension cultures of various cereal grass species. The technique involves placing the bacteria directly over callus tissue cultures, or inducing bacterial uptake by the protoplasts with the aid of a solution containing polyethylene glycol.

Continued growth of callus cultures of sugarcane, inoculated with the bacteria, is observed on media devoid of any exogenous source of nitrogen or containing markedly reduced levels of nitrogen supplied as glutamine. The bacteria continue to reduce acetylene at a high level in such inoculated cultures. Electron microscopic evidence shows that the bacteria continue to live and multiply in the intercellular space system of the callus tissue in culture. Cells of Spirillum lipoferum can also be introduced into protoplasts isolated from roots of sorghum or suspension cultures of sugarcane, but we have no evidence so far that the bacteria are biologically viable and functional within the protoplasts. Only continued growth of callus cultures or protoplasts on media devoid of any exogenous nitrogen, or very low in nitrogen, will demonstrate that a true symbiotic association has been achieved. There are indications of this in our pearl millet and sugarcane callus cultures, but it is too early to claim that we have forced a symbiotic association between Spirillum lipoferum and grass species.

EFFECT OF DIFFERENT ENERGY SOURCES ON GROWTH AND NITROGEN FIXING ABILITY OF Rhizobium japonicum STRAINS

Kuzhiparambil Prakasan

Department of Agrobiological Sciences, CCT, Universidade Federal da Paraíba, Campus de Areia, Paraíba, Brasil.

Attempts have been made from time to time to find out a suitable method for determination of the efficiency of Rhizobium species from their easily determinable characteristics. Studies of physiological characters with a view to determine the degree of efficiency are rather limited. Physiological characters like acid production, inability to metabolize nitrite, failure to reduce triphenyl tetrazolium chloride and high capsule formation have been found to be associated with efficient Rhizobium species.

The present investigation reports on the effect of different energy sources on growth and nitrogen fixation ability of 13 strains of R. japonicum. The common mannitol yeast extract agar medium was used for maintaining the bacterial strains in the laboratory. The growth studies were made in liquid medium (the same composition excluding agar). Mannitol was substituted with different carbon sources namely glucose, galactose, glycerol, mannose, arabinose, maltose, raffinose, sorbitol, fructose, xylose, pyruvate, lactose, sucrose, ribose, succinate, citrate, malate, lactate, and acetate as the sole source of energy. The calcium carbonate was substituted with calcium sulphate to avoid the initial turbidity of the medium. The growth was compared turbidometrically using a Klett-Summerson photoelectric colorimeter with a blue filter (400-465 nm). The quantity of glucose consumed was estimated colorimetrically. Experiments to test the efficiency of nitrogen fixation were conducted in sterilized pots. The nitrogen content of inoculated and uninoculated plants was determined by microkjeldahl method. Each treatment was replicated three times. The ratio of the nitrogen content of inoculated to that of uninoculated plants was taken as the efficiency of the inoculated strain.

Out of 20 carbon sources tested, mannitol, glucose, glycerol, mannose, arabinose, galactose, ribose, glucose, and xylose were found to be good energy sources. In most cases turbidity obtained with the various substrates was not related with the efficiency of the strains since either slime or pigment production affected the turbidity. The quantity of glucose consumed was found to have a positive correlation with the efficiency of nitrogen fixation, in pot cultures ($Y=2.063\ x + 1.059$).

SCREENING OF *Rhizobium* spp ISOLATES OF NATIVE AND INTRODUCED LEGUMES IN THE TROPICS

M. Julian Quintero and R. Garza T.

Departmento de Forrajes, Instituto Nacional de Investigaciones Pecuarias, Km 15.5, Carretera México-Toluca, México 10, D.F. México.

Seed of *Glycine wightii* Var. Cooper and Clarence, *Pueraria phaseoloides*, *Centrosema pubescens*, *Stylosanthes guyanensis* Var. Schofield, *Clitoria ternatea* and *Macroptilium atropurpureum* were inoculated with 9 *Rhizobium* spp strains.

At 90 days *Macroptilium*, *Centrosema* and *Clitoria* plants showed best growth, *Pueraria*, *Glycine* and *Stylosanthes* plants showed poor growth. *Macroptilium* and *Clitoria* were blooming and podded.

Rhizobium K_2 strain, isolated from *Centrosema pubescens* was better with *Centrosema* (185 nodules/plant) *Clitoria* (116 nodules/plant) and *Macroptilium* (96 nodules/plant). *Rhizobium* K_4 strain, isolated from *Pueraria phaseoloides* produced 100, 60 and 24 nodules/plant in the same hosts, respectively. *Stylosanthes* and the two *Glycine* species showed no nodulation. There was a signficant increase in dry matter and the number of nodules between inoculated and uninoculated plants.

NITROGEN FIXATION BY FREE-LIVING ORGANISMS IN TROPICAL RICE SOILS

V. Rajaramamohan Rao, P. B. B. N. Charyulu, D. N. Nayak and C. Ramakrishna

Laboratory of Soil Microbiology, Central Rice Research Institute, Cuttack-752 006 (Grissa) India.

Recent studies on biological nitrogen fixation in natural ecosystems have almost exclusively used two techniques, one involving the indirect, and highly sensitive acetylene reduction assay and the other involving the more direct and reliable $^{15}N_2$ tracer. The conclusions on nitrogen gains based on exclusively one technique can often be misleading. For instance, in the indirect assay of nitrogen fixation by acetylene reduction, the amount of nitrogen fixed derived by extrapolation from nmoles of ethylene formed is often exaggerated. For meaningful quantitation of the nitrogen-fixing potential in a system, assay by both techniques is essential. Both $^{15}N_2$ and acetylene reduction techniques were empolyed in these studies for determining the overall contribution by free-living bacteria in the nitrogen economy of tropical rice soil amended with rice straw and mineral fertilizers.

The data demonstrated that the addition of rice straw certainly stimulated nitrogen fixation in rice soils, but the extent of nitrogen fixation varied with the technique employed for the assay. Thus, the average indigenous nitrogen fixation during 30 day period amounted to 55 kg N/ha by acetylene reduction as compared to a value of 17 kg N/ha by $^{15}N_2$ technique. Likewise, during the same period, by acetylene reduction assay, 132 kg N/ha and 77 kg N/ha were fixed in soils amended with 10 and 5 tons/ha rice straw respectively, as compared to the values of 79 kg N/ha and 24 kg N/ha by ^{15}N technique. Evidently, quantitation by $^{15}N_2$ technique would provide a more realistic estimate of N_2-fixation in soils.

Thus, application of rice straw significantly enhanced the nitrogen-fixing potential in a paddy soil. Active N_2 fixation in soil was noticed up to the flowering stage of rice plants with a maximum activity at the tillering stage.

N_2-fixing efficiency was studied in Indian rice soils including two unique acid sulfate saline soils (Pokkali and Kari) following application of rice straw to soils under both flooded and non-flooded conditions. *Bacillus polymyxa* and *Clostridium* sp. occurred in larger numbers in alluvial and laterite soils than in *Pokkali*. Despite high activity and salinity, *Pokkali* soil harboured active nitrogen-fixing associations. Preliminary investigations reveal the ubiquitous presence of *Spirillum* in all the above soil types under both upland and lowland conditions. *Spirillum* was not very active in an extremely acid sulfate saline *Kari* soil; but liming the soil enhanced the population and nitrogen-fixing ability of these isolates. The results suggest that *Spirillum* could be a potential contributor to the nitrogen economy of rice soils, especially under tropical Indian conditions of agriculture.

TUDY OF THE EFFECT OF DIFFERENT DOSES OF NITROGEN FROM TWO SOURCES ON NODULATION, EFFICIENCY OF N_2 FIXATION, AND YIELD OF BEANS (*Phaseolus vulgaris*).

G. Alcántar, S. Alcalde, M. Valdés and I. Cortés

CIMMYT Laboratorios, Londres 40, Apartado Postal 6-641, México 6, DF y Colegio de Postgraduados, Rama de Suelos, E.N.A., Chapingo, México

A greenhouse experiment was conducted with a genotype of bean, a strain of *Rhizobium* and two sources of nitrogen ($(NH_4)_2SO_4$ and $Ca(NO_3)_2 \cdot 4H_2O$). Five doses of each: 0, 15, 30, 45 and 60 ppm. The soil was sterilized with methyl bromide; three plants were repeated per pot.

Ten samples of soil weekly were made to observe the content of NH_4^+ and NO_3^- as well as its effect on the nodulation. Five samplings of plants were also examined for dry weight and nitrogen content besides number, dry weight and hemoglobin content of nodules.

There was a trend of both ions to decrease in soil with time. Growth curves of plants showed differences for plants furnished nitrates versus those given NH_4^+; higher yield of nitrogen was extracted from the plants treated with nitrates.

The level of hemoglobin in the nodules was not lowered with nitrogen fertilizers. However, the weight and number of nodules was repressed during the early phases of their formation by nitrogen fertilizer.

The negative effect on the intensity of infection and nodule growth was present only in treatments with more than 30 ppm of nitrogen. Differential effects were observed with nitrates and ammonia.

INDUCTION OF CARBOHYDRATE MUTANTS IN *Rhizobium*

A. Arias, C. Cerveñanský, A. Gardiol and C. Martinez-Drets

Division de Bioquímica, Instituto de Investigaciones Biológicas Clemente Estable, Montevideo, Uruguay

Carbohydrates have been widely employed in growing rhizobia but the mechanism of its utilization has not yet been completely clarified. Mutations affecting the carbohydrate metabolism would be of biochemical interest and no knowledge has previously been reported.

A mutant isolated from R. meliloti L5-30 was selected after N-methyl-N'-nitro-N-nitrosoguanidine (NSG) mutagenesis, followed by penicillin treatment. This mutant was unable to utilize or grow on: D-mannitol, D-fructose, D-sorbitol, D-mannose, D-xylose, D-ribose, D- and L-arabitol. It grew on: D-glucose, L-arabinose, glycerol, pyruvate and succinate. Biochemical analysis showed that it was deficient in phosphoglucose isomerase (pgi); whereas levels of glucose 6-phosphate dehydrogenase, gluconate 6-phosphate dehydrogenase, glucokinase where similar in parent and mutant.

The mutant reverted to complete restoration of the wild-type phenotype, and the revertant isolated from fructose contained a higher level of pgi than the wild-type.

These results suggest the organism is a single gene mutant, and the pleiotropic effects could be explained by previous metabolic studies of rhizobia. The carbohydrate mutant could be a useful tool in future studies on polysaccharide structure and its possible relationship with nitrogen fixation.

SURVIVAL OF NITROGEN FIXING BACTERIA IN "CERRADO" SOILS

A. R. Ascenção, L. de Vasconcelos, M. F. F. Faria and A. Drozdowicz

Instituto de Microbiologia da Universidade Federal do Rio de Janeiro, Ilha do Fundão, 20000, Rio de Janeiro, Brasil.

Selective culture media were used to study the survival of nitrogen-fixing bacteria in treated and untreated "Cerrado" soil.

Derxia gummosa survived for 40 days, Spirillum lipoferum SP7 survived 6 days and Spirillum lipoferum SP82 40 days.

The survival of these bacteria was diminished in autoclaved soil. D. gummosa was unable to survive more than one day in soil which was autoclaved for 2 hours at 121°C. S. lipoferum SP7 survived five days, and S. lipoferum SP82 seven days in autoclaved soil. Addition of large quantities of calcium to the autoclaved soil counteracted this inhibitory effect and allowed longer survival. Addition of fertilizers (NPK) also resulted in increased survival.

THE INITIATION OF ACETYLENE REDUCTION IN ISOLATED ROOTS OF MAIZE: EFFECT OF CARBON, OXYGEN AND MINERAL NITROGEN SOURCES

J. Ivo Baldani, Pedro A. Pereira, Carlos A. Neyra and Johanna Döbereiner

Programa de Fixação Biológica de Nitrogênio, Convênio CNPq-EMBRAPA-UFRRJ, UEPAE, ITAGUAI, Km 47, Seropédica 23460, Rio de Janeiro, Brasil

Since roots removed from the soil show a variable latent period in the reduction of acetylene, experiments were carried out with the purpose of determining the role of carbohydrates, mineral-N and oxygen in the initiation of acetylene reduction in the association of bacteria (Azospirillum spp.) with maize roots removed from the soil.

In all the experiments roots were collected in the field and were added to bottles containing phosphate buffer. The treatments consisted of: a) application of N_2-fixing bacteria (A. brasilense and A. lipoferum) grown in liquid NFB medium with NH_4Cl, which were centrifuged and washed before application; b) application of solutions containing NO_3^-, NO_2^-, or NH_4^+ as sources of mineral-N; c) application of solutions containing malate and bicarbonate as sources of carbon; d) injection of adequate air to give the levels of oxygen required. All the treatments were initiated after substitution of the gas phase by N_2.

The duration of the "lag" was greater in all the treatments containing mineral-N. The effect of NH_4^+ and, to a lesser extent NO_2^-, was more pronounced than that of NO_3^-; in the duration of the "lag" and in the rate of reduction of acetylene. Under limiting oxygen conditions, the rate of acetylene reduction was also reduced. Although greater rates were obtained with 0.3% oxygen, the "lag" was not altered. There was also an interaction between NO_3^- and O_2.

The rate of acetylene reduction was also stimulated by the application of bicarbonate and malate in roots with and without the application of Azospirillum spp. A shorter "lag" (6 hours) was obtained with one treatment of bicarbonate. The application of Azospirillum spp. resulted in much higher rates of acetylene reduction and a "lag" of 4 hours. N_2-fixing bacteria (Azospirillum spp.) added to inert root substitutes (simulating roots) did not fix N_2.

ACETYLENE REDUCTION ACTIVITY IN THE RHIZOSPHERE OF RICE: METHODS OF ASSAY

Robert M. Boddey, Peter Quilt and Nazeer Ahmad

Department of Soil Science, University of the West Indies, St. Augustine, Trinidad.

Current methods of assessing nitrogen fixing activity in the rhizosphere of rice, based on acetylene reduction activity (A.R.A.) are critically reviewed. Both in situ and excised root assays are discussed.

The in situ assay developed by the authors is described in detail. This 24 hour assay involves sealing the plant/soil system in a plastic bag enclosure and injecting acetylene to a concentration of approximately 10% of the total gas volume. A known volume of propane is injected as an internal standard allowing estimation of the volume of the enclosure and correction for gas leakage. Acetylene reduction due to algae in the soil surface is inhibited by using the herbicide Propanil.

At the end of the assay the plant was uprooted, the roots washed free of soil, and subjected to an excised root A.R.A. assay similar to that described by Day, Neves and Döbereiner (1975, Soil Biol. Biochem. 7 (2):107-112). This overall procedure allowed the comparison of the two assay methods for each plant. A weak, but significant, correlation of the two acetylene reduction assays (in situ and excised root) was obtained.

Results of the optimum partial pressure of oxygen and acetylene for maximum A.R.A. are reported, based on work with the excised root assay.

SURVIVAL OF Rhizobium IN DIFFERENT PEAT CULTURES

C. Cervenansky

Division of Biochemistry, Instituto de Investigaciones Biológicas Clemente Estable, Montevideo, Uruguay.

Peat cultures have been widely employed in the production of legume inoculants in Uruguay. We here report the first experiments in which autoclaved- or γ ray-sterilized peat was used as a carrier.

The survival of commercially used strains of Rhizobium in inoculants prepared with peats maintained under different storage conditions was studied.

Peats obtained from the two larger Uruguayan peat-beds were assayed for their chemical characteristics (organic matter, organic carbon, N, Cl^-, $SO_4^=$, pH) and prepared according to Roughley (1970). The peats were milled, dried and packed in polyethylene bags which were carefully sealed. They were sterilized by autoclaving at 120°C for 3 hours or by γ radiations at a dose of 5.0×10^6 rad. Controls of sterility were performed every several weeks. Each bag was aseptically inoculated with the broth culture and incubated for 10 days at 28°C. Two microorganisms were selected as a representative of the strains employed in Uruguay: R. meliloti U185 and R. japonicum E 45. The inoculants were stored at 4°C or 20°C for more than 6 months, and samples were taken at different periods. Survival of organisms was determined by the plant dilution method which was compared with the viable cell counts. Controls with non-sterilized peat were run in each experiment.

These studies confirmed previous investigations that peat sterilization improved (up to 100 fold) the survival of rhizobia inoculants. Sterilization by γ radiation was superior to autoclaving. No significative differences in the increase of survival was detected with the two types of peat employed as a carrier in Uruguay.

These results suggest that an important improvement in the production of legume inoculants in Uruguay may be obtained in the future using sterilized peats.

THE DISCOVERY OF ROOT NODULES IN NEW SPECIES OF NONLEGUMINOUS ANGIOSPERMS FROM PAKISTAN AND THEIR SIGNIFICANCE

Ashraf H. Chaudhary

Department of Biological Sciences, Quaid-i-Azam University, Islamabad, Pakistan.

The growing importance of biological nitrogen fixation prompted a survey of the presence of root nodules in nonleguminous Angiosperms in Pakistan. The study revealed presence of root nodules on a number of previously unexplored non-leguminous species. The present investigation describes the nature, morphology and physiology of the nodules and the associated endophytes in some of these plants.

Light microscopy, microtomy and scanning electron microscopy were used for morphological and anatomical studies of the nodules and the detection of the endophyte. The nitrogen fixing potential was assessed by growth experiments in nitrogen-deficient cultures and by acetylene-reduction assay.

Presence of root nodules was recorded in 12 new plants, by which Datisca cannabina and Juncus articulatus are of particular significance. The nodules of D. cannabina are typically of the Alnus type forming dichotomously-branched coralloid structures 3-4 cm in diameter. The actinomycetal endophyte with hyphal, vesicular and bacteriod stages is located in the enlarged cortical cells. The detached nodules reduced acetylene to ethylene at 5.45 μmoles/gm fresh nodule wt/hr indicating their nitrogen fixing capacity. The already known 13 genera having Alnus type nodules comprise trees and woody shrubs, while D. cannabina is a herb which can be easily propagated by seeds and thus has importance as useful research material as well as from the ecological point of view.

J. articulatus, a monocotyledonous plant from Swat, Pakistan, was found to bear root nodules which are dichotomously branched and 1-2 cm in diameter. In sections, the roots and nodules were filled with fungus hyphae. A bacterial isolate has been obtained in cultures. In pot experiments J. articulatus seedlings inoculated with crushed nodules are showing a normal and healthy growth in nitrogen deficient sand cultures, whereas the control seedlings died in 3-4 weeks showing nitrogen deficiency symptoms. Sedges as a group are not commonly known to have symbiotic associations and thus the presence of root nodules in J. articulatus is of great significance. Further investigations on the nature of this association are in progress.

(Supported by USDA, PL-480 Grant FG-Pa-231.)

RHIZOBIUM STUDIES IN TANZANIA

M. S. Chowdhury

Department of Soil Science, University of Dar es Salaam, P.O. Box 643, Morogoro, Tanzania

Legumes constitute one of the major sources of protein for subsistence farmers in Tanzania. A survey was made with respect for grain yield and nodulation by indigenous rhizobia of certain legumes.

Most of the legumes were found to be nodulated. In certain localities, no nodulation was observed in some legumes especially when they were not grown previously. Preliminary investigations on the effect of rhizobial inoculation was carried out with cowpea, soybean and alfalfa. Seed inoculation improved the dry weight and nitrogen content of cowpeas in pot experiments, but inoculation failed to improve the grain yield of cowpeas in the field. Soybeans responded to inoculation at certain localities only. Application of phosphorus increased the nitrogen uptake in soybean. Both inoculation and lime pelleting improved forage and crude protein content of alfalfa.

NON-SYMBIOTIC NITROGEN FIXATION IN SOME SOILS OF TRINIDAD AND TOBAGO

P. Collins

Soils Department, University of the West Indies, Trinidad, W.I.

(No summary received)

IMPORTANCE OF COTYLEDONARY PHOTOSYNTHESIS ON SYMBIOSIS AND EARLY GROWTH OF *Stylosanthes guyanensis*

Avilio A. Franco and Sebastião M. Souto

Projeto de Fixação Biológica de Nitrogênio, Convênio CNPq-EMBRAPA -UFRRJ, EMBRAPA, Km 47, Seropédica, Rio de Janeiro 23460, Brazil

Two greenhouse experiments were carried out to investigate the importance of cotyledonary photosynthesis on symbiosis and early growth of two *Stylosanthes guyanensis* cultivars, compared with *Phaseolus vulgaris*, with and without nitrogen fertilization.

Covering the cotyledon showed no effect on *P. vulgaris* growth but decreased drastically the cotyledonary area, area of the leaves, N_2-ase (C_2H_2) activity, weight, and total plant nitrogen of both *Stylosanthes* cultivars tested.

The evaluation of N_2-ase (C_2H_2) activity in plants, with or without the cotyledon covered, of the cv. Schofield, with or without nitrogen fertilization indicated that the limitation imposed by cotyledon covering was through restriction of photosynthetic products and not by any other growth factor; and that the restriction of symbiosis was a consequence of competition for energy (low C/N ratio).

The *Stylosanthes* plants which received nitrogen fertilization at sowing, showed higher N_2-ase (C_2H_2) activity after the early plant growth than the plants without nitrogen fertilization.

EFFECTS OF OXYGEN ON NITROGENASE ACTIVITY IN A STERILIZED INTACT GRASS SYSTEM INOCULATED WITH *Azospirillum spp.*

José L. M. de Freitas and Johanna Döbereiner

Programa de Fixacão Biológica de Nitrogênio, Convênio CNPq-EMBRAPA-UFRRJ, EMBRAPA, Km 47, Seropédica, Rio de Janeiro, 23460, Brazil

Numerous reports on nitrogenase activity under controlled test tube conditions with grasses indicated poor reproducibility of results. Moisture conditions seemed to interfere on one hand and diffusion problems on the other. In the field, nitrogenase activity was never found in young roots. In the present experiment we show that the lack of oxygen protection probably is the problem in these young roots.

Surface sterilized seed systems of *Brachiaria mutica*, a forage grass, were pre-germinated on agar and planted into twice sterilized large test tubes with vermiculite/sand mixture (2:1). Major and minor elements including 20 ppm of NO_3^- were supplied. After two weeks the cotton plugs were replaced by two new cotton plugs which were introduced into the tube leaving the leaves free. At this time the cultures were inoculated with 1 ml of *Azospirillum spp.* cultures grown in N-free malate medium. Blanks were inoculated with heat-killed cultures. After four further weeks the leaves were placed into the tubes which were closed with rubber plugs; the gas phase was replaced with N_2 and the desired oxygen tension adjusted and maintained according to gas chromatographic measurements.

No C_2H_2 reduction occurred in such tubes under air, although *Azospirillum* was observed to proliferate on the root surface and in inter and intra cellular sites. No effect of inoculation on the plants was observed. But when the pO_2 was reduced, after a lag of 3 to 6 h (probably due to nitrogenase synthesis) linear rates of C_2H_2 reduction were obtained in all tubes. Blanks inoculated with killed bacteria or tubes without plants inoculated with live *Azospirillum* never showed any activity.

Strains varied in their tolerance to oxygen. *A. brazilense* strains which denitrify were most sensitive while the strains which do not denitrify were less sensitive. The three *A. lipoferum* strains tested were intermediate. Maximal nitrogenase activities for these three groups were reached at pO_2 0.001, 0.004 to 0.01 and 0.002 to 0.004 atm, respectively.

Temperature requirements of these intact systems were similar to those of pure cultures. Plant grown at 30°C when assayed at 26°C showed less than half of the activity of plants assayed at 30°C.

Although this system does not represent a true replicate of field or pot grown plants, it will be useful for many studies on plant/bacteria interactions.

A METHOD FOR DETERMINING THE AMOUNT OF NITROGEN FIXED IN THE FIELD USING ^{15}N

Maurice Fried and Luis Mellado

Kärntnerring 11-13, A-1011, Vienna, Austria

There is a real need for the direct measurement of the amount of nitrogen fixed by microorganisms in association with crops grown in the field both for legumes and non-legumes. Acetylene reduction is both short term and non-specific. The proposed method gives an integrated value over the life time of the crop, is specific for dinitrogen fixation, and is a direct measure.

The method uses the concept that when a plant is confronted with two or more sources of a nutrient it will take nutrient from each source in direct proportion to the amounts available ffrom each source. In a given experimental situation this can be confirmed by using two rates of nitrogen addition and testing for the constancy of soil supply using ^{15}N-labelled fertilizer. In these situations the nitrogen fixed from the atmosphere can be considered one of the sources and by suitable experimental design and the use of ^{15}N-labelled fertilizer nitrogen the proportion of the nitrogen in the plant derived from the atmosphere can be determined. This proportion multiplied by the total amount of nitrogen taken up by the plant gives the amount of nitrogen fixed.

TAXONOMICAL POSITION OF Cicer Rhizobia

Y. D. Gaur and A. N. Sen

Division of Microbiology, Indian Agricultural Research Institute, New Delhi 110012, India

There had been contradictory reports regarding the cross-inoculation group to which *Cicer arietinum* (the most important pulse crop of India occupying 8 million hectares of land) should belong and the "species" to which it should be classified. To settle this issue extensive cross-inoculation tests were performed and detailed examination of the biochemical characteristics of *Cicer* rhizobia were carried out.

Cross-inoculation tests were conducted under aspetic condition using 71 isolates of *Cicer* rhizobia of different origin and 287 isolates of *Rhizobium* obtained from 52 species of leguminous plants belonging to all the cross-inoculation groups. Host legumes comprised 88 different species of leguminous plants (including *Cicer arietinum* L.) representing all the eight existing cross-inoculation groups and belonging to various tribes and sub-tribes of order Fabaceae (Papilionaceae) and Minosaceae.

Out of 71 isolates of Cicer rhizobia tested, 18 were found to nodulate only two species (out of 87 species) of legumes, viz. Sesbania bispinosa and Sesbania sesban. On the other hand, none of the Rhizobium isolates obtained from different legume hosts could nodulate Cicer arietinum except one isolate of Sesbania bispinosa (out of 8 isolates) which nodulated Cicer arietinum.

Thus conclusive evidence has been obtained that Cicer arietinum or its associated Rhizobium cannot be grouped with any of the existing cross-inoculation groups - not even with pea or clover groups which was the belief held until now. A loose non-reciprocal type of relationship with Sesbania (cowpea group) precludes its separation into an independent cross-inoculation group.

The special position of Cicer arietinum and its specific Rhizobium is also shown by a detailed study of its (rhizobial isolates) biochemical characteristics. It was seen that about 50% of the isolates of Cicer rhizobia showed characteristics of slow growing rhizobia and less than 50% of the isolates behaved like fast growing rhizobia.

Cicer arietinum and its specific Rhizobium can, thus, be considered as a link between the ancestral, primitive form of slow-growing cowpea and miscellany group and those of the evolved forms of fast-growing rhizobia belonging to clover, pea, and bean groups.

DYNAMICS OF NITROGEN MINERALIZATION IN SOILS OF THE ARGENTINA REPUBLIC

Nélida Giambiagi

Sínstesis Química, Viamonte 1465, Buenos Aires, Argentina

In order to obtain the best use of nitrogen in soil, it is necessary to rationalize the different stages of its cycle.

Studies were made in Argentina to obtain data on optimum conditions for mineralization in the course of the year in different ecological regions. Their dynamic was not the same over the country, showing particular characteristics in different regions.

Temperature permits nitrification all year in the places of the country studied up to now, with greater intensity in summer than in winter, according to the studies made in the laboratory. Some soils, show minimal values in the summer months.

Moisture stress seems less important than temperature because there was nitrification even at the wilting point. After dry periods nitrification was especially high. Mineral elements are supplied either by the soil or by fertilizers. Ammonium added to some soils as fertilizer is readily mineralized. Each of these factors may be in equilibrium with the others or may be predominant.

NITROGENASE ACTIVITY OF Spirillum lipoferum IN AXENIC CULTURE AND IN MIXTURE WITH NON-N_2-FIXING BACTERIA AND Rhodotorula

Luiz Antonio Graciolli and Alaides Puppin Ruschel

Centro de Energia Nuclear na Agricultura, Universidade de São Paulo, Comissão de Energia Nuclear, Piracicaba, São Paulo, Brazil

Nitrogenase activity of a Spirillum lipoferum strain in pure culture and in a mixture with non-N_2-fixing bacteria and/or Rhodotorula was studied using N-free media containing Na malate. For the same period of incubation and the same inoculum level, mixtures of S. lipoferum with non-N_2-fixing bacteria or with Rhodotorula showed nitrogenase activity higher than S. lipoferum in pure culture.

EXPERIMENTAL EVIDENCE ON THE PRESENCE OF SYMBIOTIC BACTERIA IN Coffea, WITH THE CAPACITY OF TRIPHENYL TETRAZOLIUM CHLORIDE REDUCTION

Rubén Hernandes-Gil

Departamento Botánico, Facultad de Ciencias Florestales, Universidad de los Andes, Mérida, Venezuela

Symbioses occurring in the root nodules of leguminous plants have been thoroughly studied, but very few reports have been published on symbioses in the roots of grasses and other non-leguminous crop plants.

It was observed in free-hand sections, that Coffea arabica L., showed an abundant symbiotic bacterium in the parenchyma cells of the stem, petiol, leaf and root. The endophyte was isolated in semi-solid Na-malate medium and it was a Gram-negative curved cell, containing minute lipid body inclusions, and measuring between 0.5-0.6 µm x 2.5 - 4 µm (average). The bacterium was characterized by its motility. The culture always showed a white pellicle below the surface of the Na-malate agar. The bacterium was catalase positive and showed ability to reduce triphenyl tetrazolium chloride. Coffea arabica L. roots showed a very active triphenyl tetrazolium chloride reducing ability that could be associated with the capacity of the tissue for fixing nitrogen. More work is being done to assay acetylene reduction by Coffea roots.

EFFECT OF CALCIUM AND pH ON THE RESPONSE OF N_2-FIXATION TO Mo FERTILIZER IN Phaseolus vulgaris

N. T. Laera, J. M. Day and A. A. Franco

Programa de Fixação Biológica de Nitrogênio, Convênio CNPq-EMBRAPA-UFRRJ, EMBRAPA, Km 47, Seropédica, 23460, Rio de Janeiro, Brazil

Early experiments have shown no response of several Phaseolus vulgaris L. cultivars growing in a Mo deficient acid soil (pH 5.0), to Mo fertilization. Responses were obtained after the pH was raised above 5.4. In the present work several soils were tested to study the extent of this problem. The separate effects of pH and calcium were also studied in one soil.

In early experiments, with red yellow podzolic soil, alluvial soil, and two red latosol soils from the "cerrado," good results were observed when both lime and Mo fertilizer were added together. The fourth soil, with an initial pH = 5.9, showed a good response to Mo without liming. No additional response to the Mo fertilizer was observed with liming. One latosol soil from "cerrado" with poor plant growth showed less response to the Mo fertilizer.

The addition of calcium ($CaSO_4$) at low pH benefited plant growth more than N_2 fixation, with or without Mo addition. Without calcium addition, both plant growth and N_2 fixation increased with increasing pH, reaching highest values above pH 5.8.

FIELD SELECTION OF STRAINS OF *Rhizobium japonicum*

Margarita Sicardi de Mallorca and Mirta Barate de Bertalmio

Laboratorio de Microbiología de Suelos y Control de Inoculantes, Plan Agropecuário, Bulevar Artigas, 3802, Montevideo, Uruguay.

Every year this laboratory has to provide industry with strains of *Rhizobium japonicum* to be used as commerical inoculants of *Glycine max*, which so far have been polyvalent.

A field test of selection in *Glycine max* (vars. CTS-18 and Hill) included 6 strains of *R. japonicum* of different origins, a nitrogen treatment and a control without inoculation and nitrogen application. The determinations included: number, weight and size of the nodules, percentage of N at different stages of crop development and grain yield. The results obtained stress the need for periodic field evaluations of the potentially useful strains, as well as greenhouse studies.

POTENTIAL OF FIXATION OF N_2 AND INCORPORATION OF MINERAL-N IN SOYBEANS

Ivanildo E. Marriel and Johanna Döbereiner

EMBRAPA, Km 47, 23460, Seropédica, Rio de Janeiro, Brazil

The capacity of soybean (*Glycine max* L. Merril) to incorporate N_2 through symbiosis with *Rhizobium* in comparison to the incorporation of mineral-N was studied with greenhouse grown plants. Plants were grown in large pots (6 kg soil) on mobile benches to expose the plants to free air during the day. Assuming that the integrated nitrogenase activity gives and estimate of the amount of N_2 fixed the following conclusions can be made.

In well-nodulated plants without mineral-N, the total-N incorporated into the plants and seeds was equivalent to the value estimated by acetylene reduction (i.e. 1.44 g N/pot), but in the poorly-nodulated cultivar, 70% of the N was obtained from the soil. When 75 ppm of N was applied (in 3 applications), the N_2 fixed was equivalent to two-thirds of the total N incorporated (1.5 g N/pot) and was similar to the amount incorporated by the well-nodulated cultivar. The poorly-nodulated cultivar with 75 ppm of N also incorporated 1.5 N/pot, half of it being obtained from the soil and half from the symbiosis. With 150 ppm of mineral-N the well-nodulated cultivar behaved similarly to the poorly-nodulated cultivar with 75 ppm but seems to have incorporated a little more than the mineral-N application.

Nitrogen application reduced nodulation and specific activity but had no significant effect on plant weight or N content. Mineral-N in high doses (75 and 150 ppm) resulted in a higher incorporation into the seeds and therefore increased yields. It is suggested that these results be confirmed with N^{15} and a study of the physiology of the transference of the assimilated N, from different sources, to the seeds be made.

STUDIES ON BIOLOGICAL FIXATION OF NITROGEN BY THE ASSOCIATIVE SYMBIOSIS AZOTOBACTERIACEAE-GRAMINEAE IN ARGENTINA

Aníbal H. Merzari

Centro de Radiobiologia, Facultad de Agronomia, Universidad de Buenos Aires, Argentina

(Summary not received)

INTERACTION OF NITROGEN FERTILIZER WITH NITROGEN FIXATION (C_2H_2) IN SORGHUM

Miriam Nery

Instituto Agronômico do Paraná, C.P. 133, 86100, Londrina, Paraná, Brazil

The potential for nitrogen fixation in *Sorghum vulgare* was estimated by the acetylene reduction method in intact soil plant systems and in excised roots. Effects of increasing levels of mineral nitrogen fertilizer (0, 30, 60 and 90 kg N/ha) were studied on excised roots of eight cultivars planted in the field with 60 kg P_2O_5 and 20 kg K_2O as basic fertilizer. Nitrogen was applied as ammonium sulphate, one-third at planting and two-thirds 60 days later. Four assays were performed during the growth cycle of the plants. Significant differences between cultivars ($p = 0.01$) were observed and the interaction of N levels with cultivars and of growth cycle with cultivars was also significant ($p = 0.05$). Plant breeding for N_2 fixation and simultaneous use of nitrogen fertilizer seems therefore possible.

Nitrogenase activity in intact systems was assayed in pots with latosol mixed with stable manure (2:1). The pots were placed into polyethylene bags which were closed around the stem of the plants with string and the opening filled with a 1.5% agar solution. The volume of the bags was estimated by the C_2H_2 peak obtained in vessels with known volume containing a pot with soil. Nitrogenase activities of up to 1000 nmoles C_2H_4/h per pot (2 kg) were observed during grain filling stage. Activities of intact systems and excised roots showed the same growth cycle pattern with maximal activities during grain filling. Individual activities of excised roots and intact systems did not correlate well.

INFLUENCE OF MO AND N ON BEAN CULTIVATION

I. P. de Oliveira, F. J. P. Zimmermann, J. G. C. da Costa, N. K. Fageria, G. E. Wilcox, I. M. J. de O. Zimmermann

CNPAF-EMBRAPA, BR-153 Km 4 Goiânia-Goiás, Brazil

In a field experiment, the influence of molybdenum and nitrogen on yield and yield components on beans (*Phaseolus vulgaris*) was studied. Tested in all combinations were two levels of molybdenum (0 and 14.5 g Mo_2O_3/ha), and three levels of nitrogen (0, 30 and 60 kg N/ha). Yield, was significantly affected by application of molybdenum and nitrogen. Similarly, number of nodules and 100 seeds weight were significantly affected by application of nitrogen and molybdenum. However, number of pods/plant was not affected by application of nitrogen and molybdenum.

EFFECT OF MO, ZN, AND B ON *Phaseolus vulgaris* L. CULTIVATION

I. P. de Oliveira, F. J. P. Zimmermann, J. G. C. da Costa, N. K. Fageria, G. E. Wilcox and I. M. J. de O. Zimmermann

A field experiment was carried out to study the influence of micronutrients on bean (*Phaseolus vulgaris* L. cvs. Tayhu and Tambó) yield and yield attributing characteris. Tested in all combinations were two levels of zinc (0 and 30 kg $ZnSO_4$/ha), two levels of molybdenum (0 and 14.5 g Mo_2O_3/ha), and two levels of boron (0 and 7,5 kg Boric Acid/ha). Yield was significantly affected with combination of zinc, molybdenum and boron but there was no effect of single element application. Yield attributing characters were not affected with the application of micronutrients in the study.

NITROGENASE ACTIVITY, ASSIMILATION OF NO_3^- AND DENITRIFICATION IN *Brachiaria* CORES

Pedro A. A. Pereira, Carlos A. Neyra and Johanna Döbereiner

Programa de Fixacão Biológica de Nitrogênio, Convênio CNPq-EMBRAPA-URFJ-UEPAE, Itaguaí, 23460 Rio de Janeiro, Brazil

Intact soil-plant cores (1.7 liter) were taken from the field with five genotypes of *Brachiaria* and embedded within saran bags in large clay pots with sand. The grass was cut and the cores supplied with a basic fertilization of: 50-61-28 ppm of P-K-Ca and minor elements. 100 ppm of N were applied to the nitrogen treatments.

Regrowth was allowed for seven weeks and then 100 ppm nitrate-N were applied to half of the cores, 24 hours before assays.

The saran bags were closed around the stems and open spaces filled with 2.5% agar solution. 15% C_2H_2 was injected, and C_2H_4 production followed during 24 hours. N_2O formation was assayed after 18 hours in the nitrogen fertilized pots only. Nitrate reductase activity was determined in the leaves of the plants in cores and also from the field, in treatments with NO_3^- and without NO_3^-.

Significant differences ($p = 0.01$) between *Brachiaria* genotypes were observed for nitrogenase activity, denitrification and nitrate reductase activity. 100 ppm NO_3^- eliminated almost all nitrogenase activity. In the cores without NO_3^-, daily mean core activities between 142 and 725 nmoles C_2H_4/h were observed. A pronounced day-night cycle with maximal N_2-ase activities between 9 a.m. and 1 p.m. was observed in the three most active species suggesting the involvement of photosynthesis in N_2 fixation.

Denitrification varied between 0.63-3.80 μmoles N_2O/h. This means that 0.13 to 0.79% of the NO_3^- applied was denitrified in 24 hours. There was a highly significant correlation ($r = 0.98$) between N_2-ase activity in cores without N and denitrification in cores with N, with different genotypes. Tanner grass and a hybrid of Tanner grass fixed most N_2 and denitrified least while *Brachiaria ruziziensis* showed highest denitrification and lowest N_2ase activity. These differences seem to reflect variation in efficiency in NO_3 assimilation between genotypes.

The results of this experiment show that the three main factors determining nitrogen metabolism in grasses should be taken into consideration in forage grass breeding.

BIOLOGICAL FIXATION OF ATMOSPHERIC NITROGEN IN SOME BRAZILIAN "CERRADO SOILS"; PRELIMINARY OBSERVATIONS

José Roberto Rodrigues Peres, Allert Rosa Suhet and Djalma M. Gomes de Souza

EMBRAPA/CPAC, Km 18 BR 020, Rodovia Brasília-Fortaleza, Caixa Postal 70.0023, CEP 70.600, Planaltina, Distrito Federal, Brazil

Soils of the "cerrados" of Brazil occupy an approximate area of 180 million hectares. Despite the limitations to food production, agriculture is becoming established successfully in these soils.

Among the research topics studied in these soils, nitrogen fertilization has been considered for several years. It has already been evidenced that the "cerrado" soils vary considerably in relation to their natural supplying power for nitrogen. While the plots without nitrogen in a Dark Red Latosol produced more than 3 tons of corn/ha in four consecutive crops, the zero nitrogen plots in the Red Yellow Latosol produced only 0.3 tons/ha. The corn production obtained with the application of an economical level (approximately 50 kg of N/ha) were well below the yielding potential of the varieties used.

Biological nitrogen fixation was measured by the acetylene reduction technique in two experiments with corn. In the experiment conducted on the Dark Red Latosol some of the varieties from a group of 36 presented a much higher level of acetylene reduction than others. In the other experiment, (Red-Yellow Latosol) where nitrogen levels were studied, acetylene reduction was only observed on plants from the plots receiving zero and 50 kg of N/ha.

Nitrogen fixation was also evaluated for soybeans. When this crop is grown on "cerrado" soils, during the first year of cultivation poor nodulation occurs due to problems not yet identified. In one experiment the number of nodules formed was much larger in the plots where the seeds were inoculated with a larger quantity of inoculant (2000 g of inoculant/50 kg of seed versus 250, 500 and 1000 g), although nodulation was not satisfactory. In another experiment 4 varieties were inoculated with 10 *Rhizobium* strains and 2 commercial inoculants. Nodulation was low but the commercial inoculant treatments were superior. In one experiment with 4 varieties and one line, only the line presented a satisfactory number of nodules. Among the varieties, 3 did not nodulate and one presented a low occurrence of nodules.

ISOLATION OF *Spirillum lipoferum* BEIJERINCK FROM TROPICAL GRASSES IN MEXICO

M. Julian Quintero and R. Garza T.

Departamento Forrajes, Instituto Nacional de Investigaciones Pecuarias, SARH, Km 15.5, Carretera México-Toluca, México 10, D.F., México.

Digitaria decumbens Var. Transvala, *Panicum maximum*, *Pennisetum purpureum* and *Saccharum officinarum* were sampled in order to isolate native *Spirillum* strains. Fifty 0.5-1.0 cm rootlets were investigated from five plants of each species. These were surface-sterilized with 1% $HgCl_2$ acidified with concentrated HCl and were seeded in serum flasks with Na-malate nitrogen-free semi-solid medium.

Bacterial growth was subcultured on petri dishes with the same medium to which 1.5% agar was added. Characteristic *Spirillum* colonies were selected and their identification attempted. The Kjeldahl method was used to determine fixed N after culturing the isolates on nitrogen-free medium at 37°C without shaking for 72 h.

Three strains were selected according to their higher N_2-fixing capacity: One isolate from Panicum maximum with 4500 μg N_2 fixed, one from Digitaria decumbens with 3800 μg N_2 fixed and one from Saccharum officinarum with 3400 μg N_2 fixed.

AEROBIC AND ANAEROBIC N-FIXING MICROORGANISMS IN SOIL-GROWN SUGAR-CANE AS AFFECTED BY CANE PLANT, N-FIXERS INOCULATION AND SOIL-ADDITION OF CANE LEAVES

Alaides Puppin Ruschel

Centro de Energia Nuclear na Agricultura, Universidade de São Paulo, Comissão de Energia Nuclear, Piracicaba, São Paulo, Brazil

Effects of the sugar cane plant, addition of cane leaves to the soil, and of inoculation of free-living N_2-fixing bacteria on aerobic and anaerobic N_2-ase activity in soil were studied in a greenhouse experiment.

Plant effects on nitrogenase activity were observed only 3 months after planting, whereas the effect of inoculation appeared 30 days after the initial treatment. Cane leaves did not affect N_2-fixers. Yield and total nitrogen increased by addition of cane leaves and by inoculation.

CLOSED-SYSTEM NITROGEN BALANCE STUDIES IN SUGAR CANE UTILIZING ^{15}N AMMONIUM SULPHATE

Alaides Puppin Ruschel, Eichii Matsui, José Orlando Fo and Valdomiro Brittencourt

Centro de Energia Nuclear na Agricultura, Universidade de São Paulo/ Comissão Nacional de Energia Nuclear; PLANALSUCAR/Instituto de Açúcae e do Álcool e Escola Superior de Agricultura Luiz de Queiroz, Universidade de São Paulo, Brasil, Piracicaba, São Paulo, Brasil

Utilization of nitrogen by sugarcane was studied with ^{15}N-ammonium sulphate. The evaluation of possible biological nitrogen fixation through ^{15}N-balance and by use of a "working estimate" is discussed. The percentage of N utilization varies from 21.0 to 24.8 for 100 and 200 kg N/ha applied, respectively.

EFFICIENCY OF Rhizobium INOCULATION ON Phaseolus vulgaris L. a. EFFECT OF NITROGEN SOURCES AND VARIETIES

Alaides Puppin Ruschel and Siu Mui Tsai Saito

Centro de Energia Nuclear na Agricultura, Universidade de São Paulo Comissão de Energia Nuclear, Piracicaba, São Paulo, Brasil

The effect of Phaseolus vulgaris varieties and sources of nitrogen (ammonium sulphate, sodium nitrate, urea - 50 kg N/ha and urea - 100 kg N/ha) on symbiotic nitrogen fixation and on bean yield have been studied in two experiments under field conditions, with the following treatments: control, Rhizobium inoculation, nitrogen fertilizer added to the soil and inoculation plus nitrogen added to soil. Nodulation (weight and nitrogenase-ethylene-activity of nodules), weight and nitrogen content of plants were determined in three different times during plant development in one experiment and at one time in the second one. Seed yield and N-content of seed were analysed.

Rhizobium inoculation increased weight and total nitrogen of plant and seed when compared to control; however, this effect was not higher than in treatments with added N. Differences on weight and total N of tops among the varieties, observed 50 days after sowing, were related to yield, but not to nitrogen in the seed. Yield was not affected by different sources of nitrogen, but Rhizobium inoculation decreased seed nitrogen when $(NH_4)_2SO_4$ and urea were used.

NUMBERS OF Azospirillum spp. ASSOCIATED WITH THE ROOTS OF FIELD GROWN MAIZE

C. A. Scott, F. M. M. Magalhães, V. L. dos Santos Divan and D. B. Scott

Programa de Fixacão Biolõgica de Nitrogênio, Convênio CNPq-EMBRAPA-UFRRJ, EMBRAPA, Km 47, Seropédica, 23460, Rio de Janeiro, Brazil

High nitrogenase activity has been found in maize roots at silk emergence and during flowering but not at other stages of growth by the isolated root method (Neyra, Pereira, von Bülow and Döbereiner, 1976, XI Reun. Brasil.Milho e Sorgo, Piracicaba-SP). In order to see whether the numbers of Azospirillum correspond to changes in nitrogenase activity during the growth cycle, counts of Azospirillum were made in the roots of field grown maize during the life cycle of the plant by inoculating serial dilutions of the crushed roots into semisolid malate media (Okon, Albrecht and Burris, 1977, Appl. Environ. Microbiol. 33). Formation of a pellicle and reduction of C_2H_2 were the criteria used for the presence of Azospirillum. Numbers of these N_2-fixing bacteria were then determined using most probable number tables. To determine relative numbers of Azospirillum tightly associated with the maize roots, different sterilization agents were used for different times.

During the summer Azospirillum numbers from field grown maize showed no significant change throughout the life cycle of the plant. At 3, 6, 8 and 12 weeks (flowering) after planting Azospirillum numbers were 3.5×10^5, 2.5×10^5, 6.0×10^4 and 1.5×10^5/g wet weight of root while total bacterial numbers were: 2×10^6, 2.5×10^8, 2.0×10^6 and 4.5×10^5 respectively. Sterilization of the roots by 1% hypochlorite ("água sanitária," 2.5 min) at 12 weeks decreased both Azospirillum numbers and total bacterial numbers by 100 fold. Azospirillum numbers were 10 to 100 fold lower in maize grown over the winter months. 1% Chloramine T (2.5 min, 30 min, 1 hour) sharply decreased the numbers of Azospirillum associated with the maize roots throughout the growth cycle. Enrichment cultures of root pieces showed a similar effect from sterilization from Chloramine T.

These results suggest that the total number of Azospirillum associated with the roots of maize does not change throughout the growth cycle of the plant. Sterilization of roots suggests that many of the bacteria are associated with the outer root tissue.

OCCURRENCE OF Azospirillum spp. IN SOILS AND ROOTS

Maria Fátima Sarro da Silva and Johanna Döbereiner

Programa de Fixação Biológica de Nitrogênio, Convênio CNPq-EMBRAPA-UFRRJ, EMBRAPA, Km 47, Seropédica, 23460, Rio de Janeiro, Brazil

Soil samples from various European Countries, Hawaii and from Brazil, and roots of various Gramineae, legumes, and tuber plants were examined for Azospirillum incidence. Nitrogenase activity (C_2H_2) of early log phase enrichment cultures in N-free, semisolid malate medium was taken as semiquantitative estimate. Four enrichment cultures were prepared from each soil or root sample. Preliminary observations on the frequency of A. brasilense cultures and A. lipoferum were based on microscopic examination in older cultures where A. lipoferum presents large involution forms.

In comparison with the tropics Azospirillum occurrence was very poor in European soils. Only 4 out of 50 samples contained the organism, 3 of which where A. lipoferum and one A. brasilense. The Brazilian soils in 68% of the samples contained A. brasilense and in 40% A. lipoferum. Soils in Hawaii seemed still more favorable to A. brasilense. Soils under grasses contained more Azospirillum than others.

Washed roots of maize, rice and forage grasses gave the most active enrichment cultures followed by wheat, sugar cane and tuber plants. Legume roots yielded less active cultures with washed roots and no activity with surface sterilized roots. Roots from tuber plants, forage grasses, sugar cane and wheat maintained some viable Azospirillum after 30 sec. sterilization (cortex infection?) while rice and specially maize produced active cultures even after 1 h sterilization of the roots (stele infection?). No clear cut difference between Azospirillum species was found in relation to host plant or site.

Rhizobium INOCULATION COMPARED WITH NITROGEN FERTILIZATION ON YIELD AND PROTEIN CONTENT OF TEN SOYBEAN VARIETIES

Newton Pereira Stamford and Silvia Torres de Barros

Universidade Federal Rural de Pernambuco e EMBRAPA, Recife, Pernambuco, Brazil

A field experiment was carried out on a hydromorphic soil (Serie Curado), testing ten soybean varieties. The work was developed in order to attempt the introduction of the soybean legume in the "zona de mata" of Pernambuco, and also for a comparative study of the nitrogen supplied by symbiotic fixation and fertilization with mineral N.

The experimental design was a split plot with 3 replicates. The treatments were the following: 1) inoculation with strain R-54a; 2) inoculation with strain SM-1b; 3) inoculation with strain CM-1795; 4) application of commercial inoculant; 5) nitrogen fertilization (50 kg of N/ha as ammonium sulfate); 6) uninoculated plot as a control.

There was an apparent difference between treatments, and it was evident that inoculation or mineral fertilization of soybean was necessary to increase yield and protein content. The application of commercial inoculant and nitrogen fertilizer showed the greatest yields, compared with isolated strains and the control treatments. The results suggested that the isolated strains were not efficient for all varieties used. Host specificity, was observed with strains CB-1795 being the most promising.

The black skinned varieties (Otootan, Santa Maria and Alianca Preta) and the light skinned variety (Pelicano) were the best ones, showing an average yield of over 2,000 kg/ha. This suggests the possibilities for establishment of soybeans in the physiographic region of Pernambuco known as the "Zona da Mata."

EFFECT OF OXYGEN ON THE ENERGY EFFICIENCY OF N_2 FIXATION BY *Spirillum lipoferum* IN BATCH AND CONTINUOUS CULTURE

M. A. T. Vargas and R. F. Harris

EMBRAPA, Cerrado Center, Brasília, D. F., and University of Wisconsin, Madison, U.S.A.

The effect of dissolved oxygen tension (DOT) on the growth rate, growth yield, energy efficiency and cell composition of *S. lipoferum* was evaluated in stirred batch and continuous cultures at pH 6.8 and 30°C using a growth limiting level of 0.1 electron equivalent (e^- eq) succinate/l (0.83 g/l) as the C and e^- donor source, and N_2 or NH_4^+ as the N source. DOT was controlled using an O_2-stat with an autoclavable O_2 electrode immersed in the culture, and premixed 1% and 4% O_2 in N_2 as the O_2 supply.

At pO_2 0.003 ± 0.001 atm, N_2-fixing batch cultures showed a growth rate of 6 hr doubling time (td), a stored organic carbon level of 4-7 e^- eq/mole cell N (19-34 g poly-β-hydroxybutyrate/mole cell N), and a growth yield of 27 mg N_2 fixed/g succinate. Under limiting O_2, e.g. pO_2 << 0.0005 atm. and td 7 hr., stored C increased to 13-16 e^- eq/mole cell N and growth yield declined to 21 mg N_2 fixed/g succinate. Similarly, under excess O_2, e.g. pO_2 0.01 ± 0.003 atm., td increased to 23 hr (presumably because of nitrogenase inhibition), stored C increased to 24-27 e^- eq/mole cell N and growth yield declined to 13 mg N_2 fixed/g succinate. Stationary phase incubation of stored C-loaded cells at pO_2 0.0005 atm. in the absence of succinate, resulted in subsequent endogenous use of stored C for additional N_2 fixation. N_2-fixing continuous cultures (td = 10 hr) gave a maximum growth yield of 28 mg N_2 fixed/g succinate with a stored C level 3-6 e^- eq/mole cell N at pO_2 0.003 $\pm$ 0.003 atm.; with decreasing O_2 there was a progressive decline in growth yield and an associated increase in stored C, which was accompanied by complete succinate use until extreme O_2 deficiency occurred.

Comparative continuous culture growth yields on NH_4^+ and N_2 at td 10 hr and DOT 0.005 $\pm$ 0.005 atm., corrected for assimilatory electron requirements for N_2 reduction, indicated an apparent energy efficiency of 35 e^- eq succinate dissimilated/mole N_2 fixed (33 moles ATP/mole N_2, for ATP/e^- = 0.93). Endogenous N_2 fixation showed an energy efficiency of about 50 e^- eq stored material dissimilated/mole N_2 fixed.

NITROGEN CONTENT AND FLUCTUATIONS IN "CERRADO" SAVANNA VEGETATION IN THE WESTERN LLANOS

Maurício Vera

Departamento de Biología, Facultad de Ciencia, Universidad de los Andes, Mérida, Columbia

The nitrogen requirements for the maintenance of savanna vegetation have to be covered by natural processes like biological nitrogen fixation, incorporation by rain, decomposition of organic matter or the translocation of this element to other organs of the plant to avoid losses during shedding of leaves.

The present paper proposes to identify in a "cerrado" savanna composed of trees of the genera Byrosonima, Curatella, and Bowdichia, the mechanism which allows these trees to maintain growth during many years in soils of very low fertility and frequent burning that destroys all organic matter deposited on the soil.

Only a small fraction of the organic matter, that which is deposited after the burning, during the rainy season, is decomposed. Fluctuations in nitrogen content in fresh leaves of the trees indicate that a translocation mechanism exists which takes the nitrogen to other parts of the plant before the older leaves fall. This mechanism results in a greater nitrogen economy.

SYMBIOSIS OF TROPICAL LEGUMES IN PERU

José Gómez C.

Museo de Historia Natural, Universidad Nacional Mayor de San Marcos, Lima, Peru

Studies on symbioses of 4 legumes were conducted in IVITA - Pucallpa and in the Museo de Historia Natural both of the Universidad Nacional Mayor de San Marcos, Peru. The work was conducted in both the field and the laboratory. It was determined that Rhizobium japonicum does not occur in soils of Peru but there were abundant populations of cowpea strains for Centrocema pubescens, Stylosanthes guyanensis and Pueraria phaseoloides. The soybeans, Improved Pelikan, Pelicano and Jupiter interact better, symbiotically, with the strain CIAT-51. The pressure of cattle on Stylosanthes guyanensis seems to have no effect on the number of nodules. The practice of rotation between soybean and rice, and soybean and cotton are satisfactory.

NITRATE ABSORPTION, N_2 FIXATION AND N RE-DISTRIBUTION IN COWPEA

E. L. Pulver and H. C. Wien

International Institute of Tropical Agriculture, P.M.B. 5320, Ibadan, Nigeria

Cowpeas are capable of utilizing both soil and atmospheric N_2 to support growth. TVu 3629 (indeterminate, intermediate growth habit) accumulated 43% of its total reduced nitrogen (TRN) in the pre-flowering stage and 4552 (determinate, erect) 50%. Nitrate absorption from the soil accounted for a high portion of TRN accumulated during vegetative growth. The relative amount of nitrate uptake was estimated by assaying (in vivo) for nitrate reductase (N.R.) activity in leaf tissue. The specific activity of N.R. was at a maximum during early vegetative growth and declined with plant age. TVu 3629 and 4552 exhibited similar N.R. activities and rates of decline with plant development. Both cowpea types were accumulating N at a rate of 50 mg/plant;day during vegetative growth.

During the early reproductive phase, both cultivars started depending upon N_2 fixation to satisfy their N requirements. An alternative procedure to acetylene reduction was developed in which the quantity of N_2 fixed was estimated by determining the amount of amino-N present in the stem exudate. In both cowpea and soybean, results demonstrated that the amount of amino-N in the exudate was closely correlated with the actual rate of TRN accumulation. Maximum rates of TRN accumulation and N_2 fixation were observed at flowering and decreased rapidly during pod development. TVu 3629 produced 40% of its TRN during the two weeks from flowering to rapid pod fill, whereas the same period lasted one week for 4552 and 20% of its TRN accumulation. TVu 3629 not only reached a higher rate of fixation than 4552 but also maintained a moderate rate for a longer duration.

The rates of TRN accumulation and the amino-N in the exudate greatly decreased during rapid pod development. The N requirement for pod growth exceeded the supply from N_2 fixation between 5-6 weeks after emergence for TVu 4552 and one week later for TVu 3629. The point at which the N demand by the pods exceeded the N supply coincided with leaf senescence, indicating that the N demand was being fulfilled by re-distribution of N from other plant organs. In both cowpeas, approximately 40% of the TRN in the seeds was re-translocationed from leaf tissue and 45% synthesized during the pod-filling period. At maturity, 40% of the TRN accumulated by cowpea remained in non-seed tissue.

Even though the rate of TRN accumulation, quantity of amino-N in the exudate, and nitrate reductase activity were all very low during the pod filling period most, if not all, the TRN produced during this phase of growth was translocated directly to the seed.

SYMPOSIUM PARTICIPANTS

Walquíria de Bem Gomes Alcântara (Instituto de Zootecnia)
R. Heitor Penteado, 56, Nova Odessa, São Paulo-Brasil

G. Alcantar (CIMMYT)
Londres 40, Colonia Juarez, México, Apartado Postal 6-642-México

Maria Fatima Alves (Faculdade de Ciências Agrárias do Pará)
Av. Perimetral S/No., Cx. Postal 917, CEP 66000, Belém-Pará-Brasil

C. S. Andrew (C.S.I.R.O.)
Mill. RD, St. Lucia, QHO-Austrália

Alicia Arias (Instituto de Investigaciones Biológicas Clemente Estable), Av. Italia 3318, Montevideo-Uruguay

L. B. B. Ayala (PONAIAP-MAC. IIAG-CENIAP)
CENIAP, El Limon, Maracay, Aragua - Venezuela

Jacques Balandreau (Centro de Pédologie du CNRS)
BP5, F54500, Vandoeuvre, Nancy - France

José Ivo Baldani (EMBRAPA)
Km 47, Antiga Rio-São Paulo, Rio de Janeiro-Brasil

F. J. Bergersen (C.S.I.R.O.)
PO Box 1600, Canberra 2601-Austrália

C. C. Black
University of Georgia, Athens, Georgia 30602-USA

W. Brill, Dept. Bacteriology, Univ. Wis.
Madison, Wisconsin 53706-USA

Joachim F. W. von Bülow
Universidade de Brasília, Departamento de Engenharia Agronômica, 70000 - Brasília, DF-Brasil

R. H. Burris
University of Wisconsin, Department of Biochemistry, Madison, Wisconsin 53706-USA

Linda Styer Caldas (Universidade de Brasília)
SHIN QI 2/10, case 9, 70000, Brasília-DF-Brasil

José Monteiro Carriel, Instituto de Zootecnica, Heitor Penteado no. 56, Nova Odessa, São Paulo-Brasil

J. Gómez Carrion (Museo de Historia Natural, Universidad Nacional Mayor de San Marcos) - Av. Arenales, 1256, Apartado 1109
Lima-Peru

Osvaldo Héctor Caso (Centro de Ecofisiologia Vegetal)
Serrano 661 - 1414, Capital Federal-Argentina

F. Bermudez de Castro
Facultad de Ciencias, Universidad de Salamanca, Salamanca-Spain

Delnida Martinez Cataldo (Fundação IBGE, Assessoria da Presidência)
Ed. Venancio II, 2º andar, SDS, 70000, Brasília, DF-Brasil

Admar Cervellini (Centro de Energia Nuclear na Agricultura)
Av. Centenario S/No., Cx. 96, Piracicaba, São Paulo-Brasil

Mauro da Conceição (EMBRAPA)
Km 47, Antiga, Rio-São Paulo, Rio de Janeiro-Brasil

Francisco Pereira Cupertino
Departamento de Biologia Vegetal, Universidade de Brasília
70000, Brasília, DF-Brasil

John M. Day
Rothamsted Expt. Station, Harpenden-England

Helvécio DE-Polli (EMBRAPA)
Km 47, 23460, Seropédica, Rio de Janeiro-Brasil

Johanna Döbereiner (EMBRAPA)
Km 47, Antiga, Rio-São Paulo, Rio de Janeiro-Brasil

M. G. C. McDonald Dow
National Academy of Sciences, 2101 Constitution Avenue, N.W., Washington, D.C. 20418-USA

Paulo Augusto da Eira (EMBRAPA-UEPAE de Itaguaí)
Km 47, Itaguaí, Rio de Janeiro-Brasil

Ana María Quadrelli de Escuder (EPAMIG)
Cx. Postal 295, 35.700, Sete Lagoas, Minas Gerais-Brasil

H. J. Evans
Department of Botany, Oregon State University, Corvallis, Oregon 97331-USA

Gabriel Favelukes
Facultad de Ciencias Exactas, Universidad Nacional de La Plata
Calles 47 y 115, 1900, La Plata-Argentina

Oswaldo Augusto Curado Fleury Filho
SQS 306, Bloco F, Apto. 201, Brasília, DF-Brasil

Ozório O. M. da Fonseca (EMBRAPA)
Km 47, Antiga, Rio-São Paulo, Rio de Janeiro-Brasil

Balduino França Filho (PLANTAAGRO)
Av. Goiás 671, Jataí, Goiás-Brasil

Avilio A. Franco (EMBRAPA)
Km 47, 23460, Rio de Janeiro-Brasil

Maurice Fried (FAO/IAEA)
Kaerntnerring 11, Vienna A, 1010-Austria

Nélida Giambiagi (Síntesis Química)
Viamonte 1465, Buenos Aires-Argentina

A. R. Giardini (Instituto Agronômico de Campinas)
Av. Barão de Itapura 1481, CP 28, Campinas 13100, São Paulo-Brasil

H. Daniel Ginzo (CEVEG)
Serrano 661, 1414, Capital Federal-Argentina

N. Mora de Gonzalez (Universidad Nacional de Colombia, Departamento de Química) - Transu 44 No. 22A68, Bogotá-Colombia

P. Graham (Equipo de Frijol)
Centro Internacional de Agricultura Tropical, AA 67.13, Cali-Colombia

H. Douglass Gross
North Carolina State University, Crop Science Department, Raleigh, North Carolina 27607-USA

Walter Vierira Guimarães
Universidade Federal de Viçosa - Dept. de Microbiologia, Viçosa, Minas Gerais-Brasil

Daniel Javier Belalcazar Gutierrez (Universidad Nacional de Colombia, Facultad de Ciencias Agropecuarias, Palmira), Cra 12, 1-69 Cali-Colombia

J. Halliday
CIAT, AA 67-13, Cali-Colombia

James E. Harper (USDA-ARS)
University of Illinois, 160 Davenport Hall, Urbana, Illinois, 61801-USA

R. F. Harris, Dept. of Soils
University of Wisconsin, Madison, Wisconsin 53706-USA

Ezechias Paulo Heringer (Fundação IBGE)
Av. W-3 Sul, Quadra 706, Bloco 11, Casa 61, Brasília,DF-Brasil

A. Hollander
Associated Universities, Inc., 1717 Massachusetts Ave., N.W., Washington, D.C. 20031-USA

Eiyti Kato (PLANAGRO-Planej. Agropecuarios Ltda)
CLS 203, Bl. B, loja 5, Brasília, DF-Brasil

Roger Knowles (MacDonald Campus of McGill University)
Ste. Anne de Bellevue, Québec, HoA ICO-Canada

João Kolling (Instituto de Pesquisas Agronômicas - IPAGRO)
Rua Gonçalves Dias, 570, Campinas, São Paulo-Brasil

N. R. Krieg (Virginia Polytechnic Institute & State University)
Blacksburg, Virginia 24060-USA

J. M. Krul (Production Primaire au Sahel, Projet hollandais)
B. P. 1704, Bamako-Mali

S. A. Kulasooriya (University of Sri Lanka)
University Quarters, Upper Hantane, Peradeniya-Sri Lanka

D. Lawrence Chatel (Department of Agriculture Western Australia)
Department of Agriculture, Jarrah Rd., South Perth, West Australia, Australia

M. V. Lemos (Faculdade de Ciências Agrárias e Veterinárias)
Rod. Carlos Tonanni s/n 14870, Jaboticabal, São Paulo-Brasil

Vilneyde Mabel Q. G. Lima
Laboratórie de Microbiologia, Universidade de Brasília, Brasilia, DF-Brasil

L. Longeri
Departamento de Microbiología, Universidad de Concepción, Casilla 272, Concepción - Chile

Eli Sidney Lopes (Instituto Agronômico de Campinas)
Av. Barão de Tapura 1481, C.P. 28, Campinas, São Paulo-Brasil

Kara Baranga Luzindana
C.R.E.N., Kinshasa, B.P. 868, Kinshasa XI-Zaire

José Octávio Machado (Faculdade de Ciências Agrárias e Veterinárias), Jaboticabal, São Paulo-Brasil

Antonia Celso Magalhães (Instituto de Biologia-UNICAMP)
Cidade Universitária, Barão Geraldo, Campinas, São Paulo-Brasil

Ivanildo Exódio Marriel (Centro Nacional de Pesquisas de Milho e Sorgo), Sete Lagoas, CP 151, CEP 35700, Minas Gerais-Brasil

F. Munevar Martinez (Instituto Colombiano Agropecuario)
Apartado Aereo 151123, Bogota-Colombia

G. Martinez-Drets (Instituto de Investigaciones Biológicas Clemente Estable), Avenida Italia 3318, Montevideo-Uruguay

Herbert Barbosa de Mattos (Instituto de Zootecnia)
Rua Heitor Penteado 56, Nova Odessa, São Paulo-Brasil

Pedro P. Meaurio (Ministério de Agricultura y Ganedería)
15 de Agosto 1395, Asunción-Paraguay

Valmira Vieira Mecenas (GDF/SSS - Granja das Oliveiras)
Anexo do Buriti, 4º andar, Secretaria de Serviços Sociais, Brasília, DF-Brasil

Cecília Gonçalves de Medeiros (INPA)
Rua Antonio de Farias 548, Apto. 03, Piedade, Jaboatão, Pernambuco-Brasil

Milton Thiago de Mello (Universidade de Brasília)
Instituto de Ciencias Biológicas, Universidade de Brasília, DF-Brasil

Anibal Humberto Merzari
Facultad de Agronomia, Universidad de Buenos Aires
Av. San Martin 4453 - 1417, Buenos Aires-Argentina

James R. Milam (University of Florida)
1053 McCarty Hall, Gainesville, Florida 32605-USA

Roberto Meirelles de Miranda (Universidade de Brasilia)
Departamento de Engenharia Agronomica, UnB, Brasília, DF-Brasil

Francisco Antonio Monteiro (Instituto de Zootecnia)
Rua Heitor Penteado 56, Nova Odessa, São Paulo-Brasil

Lauro Morhy
Departamento de Biologia Celular, Universidade de Brasilia, 70000 Brasília, DF-Brasil

Luis Mroginski (Facultad de Ciencias Agrarias, Departamento de Botanica), Casilla Correos 209, Corrientes-Argentina

Wilson Nakamura (Fundação Zoobotânica)
QNA 18, lote 01, Taguatinga, Distrito Federal-Brasil

André Martin Louis Neptune (Escola Superior de Agricultura "Luiz de Queiroz"), USP, CENA, LSG, 13400 Piracicaba, São Paulo-Brasil

Miriam Nery
Rua S. Lucia 175, Larajeiras, Rio de Janeiro, 20,000-Brasil

Carlos Neyra (EMBRAPA)
Km 47, Seropédica, Rio de Janeiro-Brasil

Itamar Pereira de Oliveira (CNPAF-EMBRAPA)
Br. 153, Km 4, Goiânia, Goiás-Brasil

José Eduardo de Oliveira (Instituto de Biociências, UNESP)
Rua 10, no. 2527, CEP 13500, Rio Claro, São Paulo-Brasil

Luiz Antonio de Oliveira (INPA)
Caixa Postal 478, Manaus, Amazonas-Brasil

Rosa da Glória Brito de Oliveira (Instituto de Microbiologia)
UFRJ - Ilha do Fundão, Rio de Janeiro-Brasil

Helio Yassuride Ono (Secretaria Especial do Meio Ambiente-SEMA)
Coordenação de Estações Ecológicas, Ministério do Interio, 5º andar, Brasília, DF-Brasil

Herbert Zurita Ouzudo (Estación Experimental de Saavedra-CIAT)
Santa Cruz-Bolivia

Tereza Vaz Parente (Departamento de Engenharia Agronômica, Universidade de Brasília), SQS 310, Bl. G, Apto. 310, Brasília, DF-Brasil

David G. Patriquin (Dalhousie University)
Halifax, Nova Scotia-Canada

F. de Pedrosa (Universidade Federal do Paraná)
C.P. 939, Curitiba, CEP 80.000, Paraná-Brasil

Benedito Alisio da Silva Pereira (Fundação Nacional do Indio, DGPC)
Ed. Alvorada, SCS 6º andar, Brasília, DF-Brasil

E. Coelho Pereira
SQS 108, Bl. B., apto. 205, Brasília, DF-Brasil

José Roberto Rodriques Peres (EMBRAPA/CPAC)
BR-20, Km 18, Brasilia, Planaltina 70100, DF-Brasil

Geraldino Peruzzo (CNPTrigo-EMBRAPA)
Br. 285, Km 174, Passo Fundo, Rio Grande do Sul, CP 569, CEP 99100-Brasil

Francisco das Chagas Dias Pinto (Comissão Estadual de Planejamen to Agricola), Anexo do Palácio do Buriti, 14º andar, sala 13, Brasília, DF-Brasil

Ignacio Porzecanski
CNPGC-EMBRAPA, Campo Grande, Mato Grosso-Brasil

Edward L. Pulver (International Instit. of Tropical Agriculture)
P. M. B. 5320, Ibadan-Nigeria

Manuel Julian Quintero (INIP-SARH)
México-Teluca, Km 15 1/2, México D.F.-México

V. R. Rao (Laboratory of Soil Microbiology)
Central Rice Research Institute, Cuttack, 753 006-India

P. A. Reynaud (O.R.S.T.O.M.)
BP. 1386, Dakar-Senegal

Antonio Carlos Felix Ribeiro (CEPA-D.F.)
Anexo do Palácio do Buriti, 14º andar, sala 13, Brasília, DF-Brasil

K. Dale Ritchey (CPAC)
C. Postal 70.0023, Planaltina, Distrito Federal-70.600-Brasil

Maria das Graças Ribeiro Rodrigues (Departamento de Parques e Jardins), NOVACAP, Setor Bancário Norte, Brasília, DF-Brasil

C. Rodriguez-Barrueco
Centro de Edafologia y Biologia Aplicada, C.S.I.C.
Salamanca-Spain

Alaides Puppin Ruschel (Centro de Energia Nuclear na Agricultura)
Av. Martins Francisco 119, Piracicaba, São Paulo - Brasil

Sin Mui Tsai Saito (CENA)
Av. Centenário s/n, Piracicaba, São Paulo-Brasil

Anajulia E. H. Salles (FIBGE)
SDS, ed. Venâncio II, 2º andar, Brasília, DF-Brasil

Maria do Csrmo T. Sampaio (Faculdade de Ciências Agrárias do Pará), Av. Perimetral s/n, C.P. 917, CEP 66.000, Belém-Brasil

Lillian Frioni de Santiago (Facultad de Agronomía y Veterinária)
Universidade Nacional de Rio Cuarto, 5800, Rio Cuarto-Argentina

Hélio Lopes dos Santos (EMBRAPA)
CNP Gado Leite, Coronel Pacheco, Minas Gerais-Brasil

Stanley C. Schank
University of Florida, 2189 McCarty Hall, Gainesville, Florida-USA

Roze M. Schunke (EMBRAPA-CNPGC)
Br. 263, Km 4, Campo Grande, Mato Grosso-Brasil

Christine Scott
Km 47, Seropédica, 23460, Rio de Janeiro-Brasil

David Barry Scott (EMBRAPA)
Km 47, Seropédica, 23460, Rio de Janeiro-Brasil

Clinton C. Shock (Instituto de Pesquisa IRI)
CP 91, CEP 15990, Matão, São Paulo-Brasil

Ady Raul da Silva (EMBRAPA-CPAC)
SQS 111, Bl. F., Apto. 502, Brasília, DF-Brasil

Fernando Carvalho da Silva (Universidade de Brasília)
SQS 406, Bl. K., Apto. 101, Brasília, DF-Brasil

Wanlou Coelho e Silva (UEPAE)
Km 09, Rodovia Brasília-Anápolis, Distrito Federal-Brasil

Cleverson Siqueira (EMBRAPA-CNP Gado de Leite)
Cel. Pacheco, Minas Gerais-Brasil

Edna Riemke de Souza
Rua São Francisco Xavier, 524, Vila Izabel, Rio de Janeiro-Brasil

Newton Pereira Stamford (Universidade Federal de Pernambuco)
Rua Dr. Paulo Pinto, 1072, Piracicaba, São Paulo-Brasil

W. D. P. Stewart, Dept. Biology
University Dundee, Scotland-U.K.

Allert Rosa Suhet (EMBRAPA-CPAC)
Km 18, BR 020, Rodovia Brasília-Fortaleza, Brasília, DF-Brasil

Rosemary Sylvester-Bradley
INPA, CP 478, 69000, Manaus, Amazonas-Brasil

Tiago Gomes Teixeira Neto (DPJ-NOVACAP)
SQS 403, Bloco Q, apto. 203, Brasília, DF-Brasil

Enrique Escudero Torres
Ministério da Agricultura y Ganadería, Quito-Equador

Maria Valdes
Escuela Nacional de Ciencias Biológicas, Plan de Ayala y Carpio, México 17, DF-México

Laura da E.A.R.T. de Vasconcelos (UFRJ-Institute de Microbiologia)
Rua das Laranjeiras 109, apto. 701, Laranjeiras, Rio de Janeiro-Brasil

Mauricio Vera M. (Universidad de Los Andes)
U.L.A., Facultad de Ciencias, Departamento de Biologia, Merida, Edo. Merida, Venezuela

Reynaldo Luiz Victoria (CENA)
C.P. 96, 13400, Piracicaba, São Paulo-Brasil

Leonidas Galo Tobar Villacis
Dept. Agronomia, Universidade de Brasília, Brasília, DF-Brasil

P. B. Vose (CENA)
C.P. 96, Piracicaba, São Paulo-Brasil

Shin R. Wang (EMBRAPA-CNP Soja)
Cx. Postal 1061, 86100, Londrina, Paraná-Brasil

S. H. West
University of Florida, 1022 McCarty Hall, Gainesville, Florida 32611-USA

D. Zuberer
University of Florida, McCarty Hall, Gainesville, Florida 32611-USA

STUDENT PARTICIPANTS

Margarida R. de Almeida (Soc. Ens. Sup. de N.I.)
Av. Abilio Augusto Tavora 2134, Nova Iguaçú, Rio de Janeiro-Brasil
Res.: Rua Aladir de Mello, 113, Jardim Tropical, Nova Iguaçú, 20000-Rio de Janeiro, RJ-Brasil

Luis Fernando Alves (Universidade de Brasilia)
SQS 106, Bloco D, apto. 605, 70000, Brasília, DF-Brasil

Luiz A. Ambrosio (CENA)
Centro de Energia Nuclear na Agricultura, ESALQ, USP, Piracicaba, São Paulo-Brasil

Mariluza A. G. E. Barros (Universidade de Brasília)
Colina, Bloco D, apto. 23, 70000, Brasília, DF-Brasil

Roberto Antonio Goié Blavia (UFRRJ)
Km 47, antiga Rio-SP, Seropédica 23.420, Rio de Janeiro-Brasil

R. M. Boddey (Department of Soil Science)
University of West Indies, St. Augustine-Trinidad

Klinger Gomes do Carmo (Universidade de Brasília)
SQS 408, Bloco C, apto. 305, 70000, Brasília, DF-Brasil

Manoel B. Castro (Fundação Técnica Educacional Souza Marques)
Av. Ernani Cardoso 345, Cascadura, 20000, Rio de Janeiro-Brasil

Joana Regina S. Cavalcante (Universidade de Brasília)
SQN 412, Bloco A, apto. 101, 70000, Brasília, DF-Brasil

Lia D. C. Chagas (Universidade de Brasília)
SQS 111, Bloco A, apto. 201, 70000, Brasília, DF-Brasil

Nadia Maria Conceição (Faculdade le Filosofia, Ciências e Letras de Nova Iguaçú), Av. Abilio Augusto Tavora, 2134, Nova Iguaçú, Rio de Janeiro-Brasil
Res: Av. Estácio de Sá, 118/102, Icaraí, Niterói, Rio de Janeiro-Brasil

José Maurício Teixeira Ferro Costa (Instituto de Biologia-UER)
Rua Humberto de Campos 974/903, Leblon, 20000, Rio de Janeiro, RJ-Brasil

Roberto de Castro Cunha (Universidade de Brasília)
H.C.G.N. 713, Bloco R, casa 7, 70000, Brasília, DF-Brasil

Vera Lúcia Divane (EMBRAPA)
Km 47, antiga Rio-São Paulo, Seropédica, Rio de Janeiro-Brasil

Carlos Augusto Fernandes (Universidade de Brasília)
SQS 205, Bloco B, apto. 106, 70000, Brasília, DF-Brasil

Félix H. Franca (Universidade de Brasília)
SQN 403, Bloco 15, apto. 307, 70000, Brasília, DF-Brasil

Valdecir B. de França (Universidade de Brasília)
Quadra 8, Bloco E, casa 14, Brasília, DF-Brasil

José Luiz Morgado Freitas (EMBRAPA)
Km 47, antiga Rio-São Paulo, Seropédica, Rio de Janeiro-Brasil

Maria Teresa Von Gal (UNESP, Campus de Jaboticabal)
R. Décio, 66, São Paulo, Capital, Brasil

Luiz A. Graciolli (CENA)
C. P. 96, Piracicaba, São Paulo-Brasil

Marília de Q. Dias Jacome (Universidade de Brasília)
SQS 307, Bloco J, apto. 405, 70000, Brasília, DF-Brasil

Neves Terriani Laéra (UFRRJ)
Km 47, antiga Rodovia Rio-São Paulo, Seropédica, 23.460, Rio de Janeiro-Brasil

José Aluízio R. Lara (Universidade de Brasília)
SQS 107, Bloco E, apto. 405, 70000, Brasília, DF-Brasil

Fátima da S. Leal (UFRRJ)
Km 47, antiga Rio-São Paulo, Seropédica, Rio de Janeiro-Brasil

Irlanda T. Lima (UFRRJ)
Km 47, antiga Rio-São Paulo, Seropédica, Rio de Janeiro-Brasil

José Císino M. Lopes (Universidade de Brasília)
Quadra 16, Conj. R., casa 9, Sobradinho, DF-Brasil

Bonifácio P. Magalhães (Universidade de Brasília)
Quadra 101, Bloco C, apto. 403, S. HCES, 70000, Brasília, DF-Brasil

Luís Mauro Magalhães (EMBRAPA)
Km 47, antiga Rio-São Paulo, Seropédica, Rio de Janeiro-Brasil

Alfredo Henrique Mager (EMBRAPA)
Km 47, antiga Rio-São Paulo, Seropédica, Rio de Janeiro-Brasil

Mario Barbosa Marques (UFG)
Rua C, 83, Quadra 169, lote 17, Setor sudoeste, Goiânia, Goiás-Brasil

Marília de Freitas Marreco (Universidade de Brasília)
SQS 114, Bloco D, apto. 502, 70000, Brasília, DF-Brasil

Fernando José Rios de Melo (UFRRJ)
Km 47, antiga Rio-São Paulo, Seropédica, 23460-Rio de Janeiro-Brasil

João Carlos Moraes (EMBRAPA)
Km 47, Seropédica, 23460-Rio de Janeiro-Brasil

Fátima Maria Moreira (EMBRAPA)
Km 47, antiga Rio-São Paulo, Seropédica, 23460-Rio de Janeiro-Brasil

Scheila Oberstern (UERJ)
Rua São Francisco Xavier, 524, Vila Izabel, 20000-Rio de Janeiro-Brasil

Jairo A. de Oliveira (CENA-ESA Luiz de Queiroz)
Centro de Energia Nuclear na Agricultura, Piraciacaba, São Paulo-Brasil

Jorge A. F. C. de Oliveira (UFRRJ)
Km 47, antiga Rio-São Paulo, Seropédica, 23460-Rio de Janeiro-Brasil

Maria Mota Passos (UFRRJ)
Praia Congonhas do Campo, 181, Rio de Janeiro-Brasil

João Carlos Pereira (EMBRAPA)
Km 47, antiga Rio-São Paulo, Seropédica, Rio de Janeiro-Brasil

Pedro Antonio A. Pereira (EMBRAPA)
Km 47, antiga Rio-São Paulo, Seropédica, Rio de Janeiro-Brasil

Violeta de F. Pereira (Universidade de Brasília)
SQI 3/12, Bloco F, apto. 102, Brasília, DF-Brasil

Mauro Vierira de Queriroz (ESAL)
Rua Afonso Pena 167, Lavras, Minas Gerais, Brasil

Pedro Paulo Queiroz (UERJ)
Rua São Francisco Xavier, 524, Vila Izabel, Rio de Janeiro-Brasil

Maria Cristina M. C. Ribeiro (Universidade de Brasília)
SQN 410, Bloco C, apto. 302, 70000, Brasília, DF-Brasil

Marcia F. M. Sá (UFRRJ)
Km 47, antiga Rio-São Paulo, Seropédica, Rio de Janeiro-Brasil

Elvira Maria Breier Saraiva (UERJ)
Rua São Francisco Xavier, 524-Vila Isabel, Rio de Janeiro-Brasil

Leonardo Melco Sfeir (Escola Superior de Agricultura de Lavras)
Av. Getúlio Vargas, 1076, Jacarezinho, Paraná-Brasil

Reinaldo A. da Silva (Soc. de Ensino Superior de Nova Iguaçú)
Av. Abilio Tavora, 2134, Nova Iguaçú, Rio de Janeiro-Brasil
Res.: R. Rabelo da Silva, 181, Anchieta, Rio de Janeiro-Brasil

José Eustáquio da Silva (Universidade de Brasília)
SQN 408, Bloco N, apto. 105, 70000, Brasília, DF-Brasil

Maria Fatima R. da Silva (EMBRAPA)
Km 47, antiga Rio-São Paulo, Seropédica, Rio de Janeiro-Brasil

Paulo C. Tarchetti (UFViçosa)
QND 55, Casa 13, Taguatinga, DF-Brasil

Angelo Testa (UFRRJ)
Km 47, antiga Rio-São Paulo, Seropédica, Rio de Janeiro-Brasil

Olga Maria M. Villela (Universidade de Brasília)
SQS 306, Bloco J., apto. 203, 70000, Brasília, DF-Brasil

Antonia G. T. Volpon (UFRRJ)
Km 47, antiga Rio-São Paulo, Seropédica, Rio de Janeiro-Brasil

INDEX

Acetylene
- lag in reduction, 18, 357
- method for measuring reduction, 280-281, 357-358
- reduction, 18-19, 77, 310-311, 354
- reduction in coffee, 364
- reduction in maize roots, 356-357
- reduction in rice, 357

Activating factor
- for Fe protein of *S. lipoferum*, 307

Agents fixing N_2
- algae, 41-43

Alfalfa
- N_2 fixation and alfalfa strain, 342

Algae
- (see blue green algae)
- ammonia excretion, 230
- assimilation of ammonia, 231
- distribution, 42-43
- in soil, 43-44, 49-50
- light, 54
- moisture, 45, 48, 56-59
- molybdenum, 59-60
- nitrogen excretion, 231-232
- nitrogenase activity, 48
- nitrogen fixers, 41
- occurrence, 41
- pH, 51-52
- rates of N fixation, 48-50, 60-61
- review, 41-63
- significance, 60-61
- soluble N, 58-59
- species, 43
- structure, 46-47
- temperature, 51, 53-54

Alnus-type symbiosis
- compatibility plant and endophyte, 123-128
- genera, 121-122
- growth substances, 128-132
- inoculation, 123-128
- nodule induction, 128
- nodules, 359
- review, 121-133

Aluminum
- calcium uptake, 146
- dry matter, 144
- phosphorus, 147-148
- soil concentration, 142
- symbiotic system, 142-149

Ammonia
- assimilation in leguminous nodules, 226-229
- assimilation in N_2 fixing systems, 223-235
- excretion, 239
- export from bacteroids, 227

Anaerobes
- classification, 27

Animals
- N_2 fixation in gastro-intestinal tracts, 32-34
- table of fixation, 33

Aquaspirillum
- comparison with *S. lipoferum*, 330

Asparagine
asparaginase, 227-228
export from nodules, 227, 230
synthetase, 227
Aspartate
aminotransferase, 228
Assay methods
acetylene reduction, 357-358
Assimilation
ammonia, 220, 223-235
regulation, 224-226
ATP
binding, 195-196
production, 256
requirement for nitrogenase, 213
role in nitrogenase activity, 195
Azolla
with rice, 21, 41, 346
Azospirillum
(see Taxonomy of Spirillum lipoferum; Spirillum lipoferum)
classification, 26
grass associations, 17-18, 20-21, 77, 361-362
nitrate-dependent nitrogenase, 350-351
nitrate reductase negative mutants, 351
numbers on roots of maize, 371
occurrence in soils and roots, 372
on maize roots, 347
oxygen, 361-362
sorghum, 113
suggested new genus, 331-332
Azotobacter
applications, 240-241
classification, 26
genetics, 240
regulation, 240
Bacillus
N_2 fixer, 26
with wheat, 21
Bacterial strain variation
peanuts, 343-344
Bacteroids
carbon substrates, 259
respiration, 255-256, 258-259
stimulation by oxyleghemoglobin, 253, 255-256
Beans
inoculation, 182-184, 335
molybdenum, 367
N and nodulation, 355
N utilization, 335, 367
Blue-green algae
(see algae)
furnish N in rice soils, 345-346
light, 347
N_2-fixing mat on soil, 346-347
Boron
deficiency, 162
Brachiaria
nitrogenase activity, assimilation NO_3^- and denitrification, 367-368
plant genotype effects, 20
Calcium
pH, 140-142
CAM metabolism, 95, 97
Carbon dioxide
concentration and N_2 fixation, 112
Carbon monoxide
effect on respiration, 257-258
inhibitor, 257
Cellulose
to support N_2 fixation, 33

Cerrado
agricultural potential, 5
area, 8
deficiencies, 6
N content of cerrado vegetation, 374
N_2 fixation, 368-369
pasture, 10
potential frontier, 2, 13
soils, 7, 10, 356
survival of bacteria, 356
vegetation, 6
water, 11
Chlorine
role, 164
Cicer arientinum
taxonomy of rhizobia, 362-363
Cobalt
deficiency, 164
Cocoa
N and shade from Erythrina fusca, 345
Coffee
symbiotic bacteria, 364
Combined nitrogen
suppression of nodulation, 177
Control
(see regulation)
Copper
deficiency, 163-164
role, 163-164
Corynebacterium, 26
Cowpea
nitrate absorption and N_2 fixation, 375-376
C_4 Photosynthesis
carbon dioxide, 101-102
efficiency of N use, 102-104
light, 97, 100-101
nitrogen assimilation, 104-109
review, 96-109
temperature, 97, 99, 101
Crop rotation
management, 88
traditional practice, 16
Cross reactions
nitrogenase components, 201-203
Cytochrome P-450
ATP production, 254-255
Denitrification
in Brachiaria, 367-368
Digitaria
association with Azospirillum, 17
nitrogen transformations, 338-339, 349
DNA
base composition, S. lipoferum, 317-319
homology, S. lipoferum, 319-323
Ecology of legume-Rhizobium symbiosis
review, 173-190
symbiotic N_2 fixation, 176-177
Efficiency of N_2 fixation
relative, 214-217
Energy
requirement for N_2 fixation and use of ammonia and nitrate, 213, 341
sources for R. japonicum, 352-353
sources in heterotrophs, 27
Enterobacteriaceae
classification, 26
Environmental conditions
unfavorable, 85-89
Erythrina fusca
transfer of N to cocoa, 345
Excretion
ammonia, 232
fixed N, 231-232
Facilitated diffusion
leghemoglobin, 253
oxygen, 253
Forage legumes
cultivated species, 69-70
grown in West Indies, 342-343
maximizing N_2 fixation, 344-345
production, 69
Stylosanthes, 342-343
Frankia
nodulation of non-legumes, 123, 128

Free-living bacteria
occurrence, 25
taxonomy, 25-27
Fritted trace elements (FTE)
use as nutrient source, 161, 167-168
Genetics
Azotobacter, 240
in grass associations, 292-293
Klebsiella pneumoniae, 238-239
nif genes, 239
regulation of N_2 fixation, 237-245
Rhizobium, 241-242
variability and efficiency of N_2 fixation, 76
Gene transfer
nif genes, 239
Glutamate
dehydrogenase, 105, 223-225, 232, 307
synthase, 105, 116, 223-227, 307
Glutamine
oxoglutarate amino transferase (GOGAT), 224-229, 231-232
synthetase, 105, 116, 223-228, 232, 238, 270-272, 307
synthetase adenylylation, 224, 226, 238
Grass associations
(see Nitrogen Fixation in grasses; *Spirillum lipoferum*)
Azospirillum, 361-362
limiting factors, table, 298
N_2 fixation, 76-77, 83, 275-302
N_2 fixing bacteria, 17, 21
nitrogenase activity, 294-297
Green manure
agricultural practice, 16-17
Growth substances
nodulation of non-legumes, 128-132
Herbicides
increased N_2 fixation in maize, 340
on maize, 21
nodulation of soybeans, 337
Hill reaction
nitrite as oxidant, 106-107
Host specificity
alfalfa, 342
Vigna sinensis, 337-338
Hydrogen
evolution, 209-210, 214-216
inhibitor of N_2 fixation, 209
metabolism and N_2 fixation, 209-222
production, 30
recycling, 216-219
to support N_2 fixation, 29
uptake, 210, 216-219
Hydrogenase
ATP synthesis, 210
reactions, 213
recycling H_2, 217-220
Inhibitors
of CH_4 metabolism, 31
tight binding, 200-201
Inoculation
beans, 182-184
cultivated legumes, 179-185
peanuts, 184-185
soybeans, 180-182
specificity, 76
with *Rhizobium*, 360
Iron
deficiency, 163
role, 163
Iron (Fe) protein
nitrogenase, 192, 194-195
Klebsiella pneumoniae
genetics, 238-239
regulation, 237-239

Leghemoglobin
absorption spectrum, 250
bacteroids, 253, 255
carbon monoxide, 252-253
concentration, 247, 256
discovery, 247
enhances bacteroid activity, 253
extraction and purification, 248-249
maintaining O_2 concentration, 20, 220, 247
oxygenation, 247, 251-253
properties, 248
respiration of bacteroids, 254-256
role, 248, 251-253
synthesis, 249
Legume-Rhizobium symbiosis
amount of N_2 fixed, 65-66
beans, 68-69
distribution, 65
food crops, 67
forage legumes, 69
inoculation, 14
nitrogen losses, 66
nodulation and growth substances, 132
peanuts, 67-68
protein production, 66-67
review, 65-73
soybeans, 66-67
tropical legumes in Peru, 374-375
Legumes and acid soils
review, 135-160
Light
algae, 54, 347
intensity and grasses, 284, 309-312
soybeans, 336
Lime
correcting deficiencies and toxicities, 15-16
growth depression, 88
Limiting factor analysis
for grass-bacteria associations, 202-203
Maize
acetylene reduction, 18, 35-36
Azospirillum numbers on roots, 371-372
increased N_2 fixation with herbicides, 340
roots reduce acetylene, 356-357
Manganese
deficiency, 162
dry matter production, 151
excess, 149-153
nodule tissue weight, 152
phosphate effect, 152-153
role, 162
tolerance, 149, 152
toxicity, 15, 88, 149-153, 162
Methionine sulfoximine
derepression for nitrogenase synthesis, 239, 307-308
Methods
N_2 fixation measured with ^{15}N, 362
Methylobacteria
classification, 26
systems supported by methane, 29-31
Micronutrients
Boron, 162
Chlorine, 164
Cobalt, 164
Copper, 163-164
establishing legumes, 339-340
Iron, 163
legumes, 15
Manganese, 162
Molybdenum, 164-167
requirements of legume-Rhizobium symbiosis, 161-171
review, 161-171
Zinc, 161-162
Moisture
algae, 45, 48, 56-59

Molybdenum
algae, 59-60
bean cultivation, 367
calcium and pH, 364-365
deficiency, 15, 164-167
measurement, 165
nitrogenase, 192-194
pH, 166-167
regulation, 238
requirement, 165, 167
role, 164
supplementation, 16
Molybdenum Iron (MoFe) protein
nitrogenase, 192-194
substrate binding, 197
Mutants
carbohydrate, 355-356
reversion, 356
Nitrate
absorption in cowpea, 375-376
assimilation, 111, 114-116
dissimilation, 20
electron acceptor, 351
supports nitrogenase at low pO_2, 350-351
Nitrate reductase
crabgrass, 105
localization, 105
negative mutants of *Azospirillum spp.*, 351
Nitrite
Hill oxidant, 106-107
reductase localization, 105
Nitrogenase
activity in grass associations, 294-297
algal activity, 48
ATP binding, 195-196
catalytic site, 203-204
complex, 198-201
cross reactions, 201-203
electron flow, 191-192, 195-198, 212
energy requirements, 212
Fe protein, 192, 194-195, 210
MoFe protein, 192-194, 210
reaction, 211
Nitrogenase (continued)
S. lipoferum, 307-308
substrate binding, 192, 197
systems, 191-207
tight binding inhibitors, 200-201
Nitrogenase and H_2 production
inhibition by N_2, 211
nodules, 209-211, 214-216
Nitrogen assimilation
C_4 plant, 104-109
products assimilated, 220
Rhizobium, 269-270
Nitrogen fertilizer
bean cultivation, 367
energy, 13
influence on N_2 fixation, 366-367
inhibiting N_2 fixation, 86-87
requirement, 14
supplement, 15
Nitrogen fixation by *Rhizobium*
applications, 265
continuous cultures, 267
cultural conditions, 264
in laboratory culture media, 263-274
nitrogen assimilation, 269-270
nitrogen-limited cultures, 267-268
nitrogen sources, 264-265, 268-270
oxygen concentration, 265, 267-272
rates of fixation, 268
regulation, 270-272
repression by ammonia, 268
strains, 265-267
Nitrogen fixation in grasses
associated bacteria, 277-279
bacteria, 293-294
factors influencing nitrogenase activity, 276
light intensity, 284
limiting factors, 275-302
measurement of fixation, 280
photosynthetic potential, 285

Nitrogen fixation in grasses (continued)
plant genetic factors, 292-293
soils and plants, 283, 290, 292
temperature, 285-286, 288-289
water, 286-287, 290-291
Nitrogen fixation rates
algae, 48-50, 60-61
grasses, 77
legumes, 65-66, 179
^{15}N as tracer, 354
Rhizobium, 268
soybeans, 365-366
tropics, 178
Nitrogen fixers
algae, 41
table, 28
Nitrogen transformations
in cerrado vegetation, 374
mineralization in soils of Argentina, 363
under Digitaria decumbens, 338-339
^{15}N methods
in field use, 362
Nodulation
suppression by combined nitrogen, 177, 355
tropical legumes, 178-179
Nodules
H_2 evolution, 214-216
Non-leguminous symbiosis
cross-infectivity, 123-128
genera, 121-122
growth substances, 128-132
inoculation, 123-128
new species, 359
review, 121-133
Non-symbiotic N_2 fixation
in soils, 360-361
Nutrient elements
calcium, 140-142
deficiencies, 136
nodulation, 136
Nutrient uptake
efficiency, 85
Oxidative phosphorylation
ATP supply, 212
Oxygen
and energy efficiency of S. lipoferum, 373-374
favorable concentration, 86, 309
grass plus Azospirillum, 361-362
inactivation of nitrogenase, 20
supply and N_2 fixation, 247-261
Oxyhydrogen reaction
in bacteroids, 217
nodules, 210, 217
Panicum maximum ("Colonião grass")
associated legumes increase yield, 341
Paspalum notatum
associated N_2 fixation, 18
Peanuts
bacterial strain variation, 343-344
inoculation, 184-185, 343
productivity, 67-68
pH
calcium, 140-142
fixed N, 138
nodulation, 137-140
phosphorus use, 155
Rhizobium, 137
S. lipoferum, 303
soil, 137-140
Phaseolus vulgaris
inoculation, 335, 370-371
N and nodulation, 355
N utilization, 335
response to Mo, 364-365
response to Zn, B, 367
Phosphate
deficiency, 15, 153-156
requirement for S. lipoferum, 303
uptake, 155
Phosphoenol pyruvate
carboxylase, 229-230

Photosynthesis
C_4 pathway, 96-109
CAM metabolism, 95, 97
diversity, 95-98
efficiency, 98, 101
growth of Stylosanthes, 360-361
interaction with N_2 fixation, 111-113
interaction with nitrate assimilation, 111, 114-116
light, 97, 100-101
nitrate assimilation, 114-116
Panicum, 97
review, 95-110
temperature, 97, 99, 101
Photosynthetic bacteria
classification, 27
Photosynthetic efficiency
C_4 plants, 98, 101
grasses, 83
legumes, 82
N_2 fixation, 81-83
Plant associations
with N_2 fixers, 34-37
Plant influence on N_2 fixation
experimental variability, 89-90
genetic variability, 76
grasses, 77-78, 85
legumes, 76-78
maize, 78-81
N_2 fixation rates, 78
nutrient uptake, 85
oxygen, 86
planting density, 88
review, 75-94
roots, 84-85
vegetative and reproductive growth, 83
Planting density
yield, 88
Plant residues
to support N_2 fixation, 31-34
wood, 32
Poly-β-hydroxybutyrate
in S. lipoferum, 305, 323
Protein production
legumes, 66-67
Protoplasts
invasion by S. lipoferum, 352
Regulation
Azotobacter, 240
genetics, 237-243
Klebsiella pneumoniae, 237-239
N_2 fixation, 237-245
Rhizobium, 241, 270-272
Rhizobium
applications, 242
carbohydrate mutants, 355-356
Cicer rhizobia taxonomy, 362-363
genetics, 241-242
inoculation of beans, 371
inoculation of soybeans, 372-373
in Tanzania, 360
N_2 fixation in laboratory culture media, 263-274
(see nitrogen fixation by Rhizobium)
regulation, 241
screening, 242, 353
sources of energy, 352-353
strains, 14, 343, 365
survival in peat cultures, 358
taxonomy, 174-176
Rhizosphere
maize, 349
sugar cane, 348
Rice
acetylene reduction, 357-358
assay for N_2 fixation, 354
fixation by free-living organisms, 354
N_2 fixation, 21
Roots
influence on N_2 fixation, 84-85
nutrient uptake, 85
Screening
Rhizobium strains, 242, 253, 365

Shipworms
associated N_2 fixers, 33-34
Soil
aluminum, 142-149
grass associations, 283, 290, 292
N mineralization, 363
of the Amazon and N_2 fixation, 347
pH, 137-140
phosphorus, 153-156
physical condition, 88
Sorghum
fertilizer and N_2 fixation, 366-367
inoculated with S. lipoferum, 349
Soybeans
inoculation, 180-182, 372-373
light, 336
N_2 fixation on cerrado soils, 368-369
potential fixation of N_2, 365-366
productivity, 66-67
temperature, 336
Spartina
association with bacteria, 21
Spirillum lipoferum
(See Taxonomy of Spirillum lipoferum)
acetylene reduction, 310-311
activating factor, 307
activity of nitrogenase, 364
ammonia, 304
association with grasses, 309-313
classification, 26
cytochromes, 306
derepression, 307-308
field trials, 310, 312-313
fixation of $^{15}N_2$, 303-304
glutamate dehydrogenase, 307
glutamate synthase, 307
glutamine synthetase, 307
grass associations, 17-18
Spirillum lipoferum (cont.)
growth rate, 304-305
isolation from grasses, 369-370
in plant tissue, 35
light, 309-312
maize, 349
nitrogenase, 307-308
oxygen, 309, 373-374
pH, 303
phosphate, 304
physiology and biochemistry, 303-315
poly-β-hydroxy butyrate, 305
preincubated roots, 310
rice, 354
substrates, 303, 305-306
temperature, 309-312
wheat, 348
with tissue cultures, 351-352
Strain competition
Vigna sinensis, 337-338
Stylosanthes guyanensis
photosynthesis and growth, 360
Substrates
binding, 192, 197
cellulose, 33
endogenous for bacteroids, 257
for respiration, 259
for S. lipoferum, 303, 305-306
to support N_2 fixation, 30
Sugar cane
N-fixing microorganisms, 370
nitrogenase activity in rhizosphere, 348
Taxonomy
Cicer rhizobia, 362-363
leguminous plants, 174
Rhizobium, 174-176
Taxonomy of Spirillum lipoferum
Azospirillum, suggested new genus, 331-332
comparison with Aquaspirillum, 330
DNA base composition, 317-319, 350

Taxonomy of Spirillum lipoferum (continued)
DNA homology, 319-323
flagellation, 323, 329
morphology, 327-329
phenotypic characters, 324-325
physiological characteristics, 326, 350
review, 317-329
Temperature
algae, 51, 53-54
grass associations, 285-286, 309-312
soybeans, 336
Termites
associated N_2 fixers, 33-34
Tissue cultures
association with S. lipoferum, 351-352
Toxicity
manganese, 149-153
Trees
N_2 fixation, 16
Uptake of N_2, NH_3 and NO_3^-
energy required, 341
Vegetative vs. reproductive growth
grasses, 83-84
legumes, 83
Water
grass associations, 286-287, 290-291
Wheat
N_2 fixation in association, 348
Zinc
cultivation of Phaseolus vulgaris, 367
deficiency, 161-162

MIX
Papier aus verantwortungsvollen Quellen
Paper from responsible sources
FSC® C105338

If you have any concerns about our products,
you can contact us on
ProductSafety@springernature.com

In case Publisher is established outside the EU,
the EU authorized representative is:
Springer Nature Customer Service Center GmbH
Europaplatz 3, 69115 Heidelberg, Germany

Printed by Libri Plureos GmbH
in Hamburg, Germany